Sławomir J. Grabowski (Ed.)

Analysis of Hydrogen Bonds in Crystals

This book is a reprint of the Special Issue that appeared in the online, open access journal, *Crystals* (ISSN 2073-4352) from 2015–2016 (available at: http://www.mdpi.com/journal/crystals/special_issues/analysis_hydrogen_bonds_crystals).

Guest Editor
Sławomir J. Grabowski
Kimika Fakultatea, Euskal Herriko Unibertsitatea UPV/EHU
San Sebastian, Spain

Editorial Office
MDPI AG
St. Alban-Anlage 66
Basel, Switzerland

Publisher
Shu-Kun Lin

Assistant Managing Editor
Huimin Lin

1. Edition 2016

MDPI • Basel • Beijing • Wuhan • Barcelona

ISBN 978-3-03842-245-7 (Hbk)
ISBN 978-3-03842-246-4 (PDF)

Articles in this volume are Open Access and distributed under the Creative Commons Attribution license (CC BY), which allows users to download, copy and build upon published articles even for commercial purposes, as long as the author and publisher are properly credited, which ensures maximum dissemination and a wider impact of our publications. The book taken as a whole is © 2016 MDPI, Basel, Switzerland, distributed under the terms and conditions of the Creative Commons by Attribution (CC BY-NC-ND) license (http://creativecommons.org/licenses/by-nc-nd/4.0/).

Table of Contents

List of Contributors VII

About the Guest Editor XI

Preface to "Analysis of Hydrogen Bonds in Crystals" XIII

Sławomir J. Grabowski
Analysis of Hydrogen Bonds in Crystals
Reprinted from: *Crystals* **2016**, *6*(5), 59
http://www.mdpi.com/2073-4352/6/5/59 1

Anas Tahli, Ümit Köc, Reda F. M. Elshaarawy, Anna Christin Kautz and Christoph Janiak
A Cadmium Anionic 1-D Coordination Polymer {[Cd(H$_2$O)$_6$][Cd$_2$(atr)$_2$(μ_2-btc)$_2$(H$_2$O)$_4$] 2H$_2$O}$_n$ within a 3-D Supramolecular Charge-Assisted Hydrogen-Bonded and π-Stacking Network
Reprinted from: *Crystals* **2016**, *6*(3), 23
http://www.mdpi.com/2073-4352/6/3/23 7

Christian Heering, Bahareh Nateghi and Christoph Janiak
Charge-Assisted Hydrogen-Bonded Networks of NH$_4^+$ and [Co(NH$_3$)$_6$]$^{3+}$ with the New Linker Anion of 4-Phosphono-Biphenyl-4'-Carboxylic Acid
Reprinted from: *Crystals* **2016**, *6*(3), 22
http://www.mdpi.com/2073-4352/6/3/22 21

Sofiane Saouane and Francesca P. A. Fabbiani
Structural Elucidation of α-Cyclodextrin-Succinic Acid *Pseudo* Dodecahydrate: Expanding the Packing Types of α-Cyclodextrin Inclusion Complexes
Reprinted from: *Crystals* **2016**, *6*(1), 2
http://www.mdpi.com/2073-4352/6/1/2 40

Thomas Gelbrich, Doris E. Braun, Stefan Oberparleiter, Herwig Schottenberger and Ulrich J. Griesser
The Hydrogen Bonded Structures of Two 5-Bromobarbituric Acids and Analysis of Unequal C5–X and C5–X' Bond Lengths (X = X' = F, Cl, Br or Me) in 5,5-Disubstituted Barbituric Acids
Reprinted from: *Crystals* **2016**, *6*(4), 47
http://www.mdpi.com/2073-4352/6/4/47 57

Pierre Baillargeon, Édouard Caron-Duval, Émilie Pellerin, Simon Gagné and Yves L. Dory
Isomorphous Crystals from Diynes and Bromodiynes Involved in Hydrogen and Halogen Bonds
Reprinted from: *Crystals* **2016**, *6*(4), 37
http://www.mdpi.com/2073-4352/6/4/37 76

Guido J. Reiss
Constructor Graphs as Useful Tools for the Classification of Hydrogen Bonded Solids: The Case Study of the Cationic (Dimethylphosphoryl)methanaminium (*dpma*H+) Tecton
Reprinted from: *Crystals* **2016**, *6*(1), 6
http://www.mdpi.com/2073-4352/6/1/6 93

Leonardo H. R. Dos Santos and Piero Macchi
The Role of Hydrogen Bond in Designing Molecular Optical Materials
Reprinted from: *Crystals* **2016**, *6*(4), 43
http://www.mdpi.com/2073-4352/6/4/43 105

Niall J. English
Diffusivity and Mobility of Adsorbed Water Layers at TiO_2 Rutile and Anatase Interfaces
Reprinted from: *Crystals* **2016**, *6*(1), 1
http://www.mdpi.com/2073-4352/6/1/1 124

Steve Scheiner
Dissection of the Factors Affecting Formation of a CH···O H-Bond. A Case Study
Reprinted from: *Crystals* **2015**, *5*(3), 327–345
http://www.mdpi.com/2073-4352/5/3/327 134

Jelena P. Blagojević, Goran V. Janjić and Snežana D. Zarić
Very Strong Parallel Interactions Between Two Saturated Acyclic Groups Closed with Intramolecular Hydrogen Bonds Forming Hydrogen-Bridged Rings
Reprinted from: *Crystals* **2016**, *6*(4), 34
http://www.mdpi.com/2073-4352/6/4/34 156

Boaz G. Oliveira, Edilson B. Alencar Filho and Mário L. A. A. Vasconcellos
Comparisons between Crystallography Data and Theoretical Parameters and the Formation of Intramolecular Hydrogen Bonds: Benznidazole
Reprinted from: *Crystals* **2016**, *6*(5), 56
http://www.mdpi.com/2073-4352/6/5/56 175

Ibon Alkorta, Janet E. Del Bene and Jose Elguero
$H_2XP:OH_2$ Complexes: Hydrogen *vs.* Pnicogen Bonds
Reprinted from: *Crystals* **2016**, *6*(2), 19
http://www.mdpi.com/2073-4352/6/2/19 190

Jing Wang, Jiande Gu, Md. Alamgir Hossain and Jerzy Leszczynski
Theoretical Studies on Hydrogen Bonds in Anions Encapsulated by an Azamacrocyclic Receptor
Reprinted from: *Crystals* **2016**, *6*(3), 31
http://www.mdpi.com/2073-4352/6/3/31 205

Antonio Bauzá and Antonio Frontera
$RCH_3\cdots O$ Interactions in Biological Systems: Are They Trifurcated H-Bonds or Noncovalent Carbon Bonds?
Reprinted from: *Crystals* **2016**, *6*(3), 26
http://www.mdpi.com/2073-4352/6/3/26 219

Halina Szatylowicz, Olga A. Stasyuk, Célia Fonseca Guerra and Tadeusz M. Krygowski
Effect of Intra- and Intermolecular Interactions on the Properties of *para*-Substituted Nitrobenzene Derivatives
Reprinted from: *Crystals* **2016**, *6*(3), 29
http://www.mdpi.com/2073-4352/6/3/29 236

Filip Sagan, Radosław Filas and Mariusz P. Mitoraj
Non-Covalent Interactions in Hydrogen Storage Materials $LiN(CH_3)_2BH_3$ and $KN(CH_3)_2BH_3$
Reprinted from: *Crystals* **2016**, *6*(3), 28
http://www.mdpi.com/2073-4352/6/3/28 259

Sławomir J. Grabowski
$[FHF]^-$—The Strongest Hydrogen Bond under the Influence of External Interactions
Reprinted from: *Crystals* **2016**, *6*(1), 3
http://www.mdpi.com/2073-4352/6/1/3 283

List of Contributors

Ibon Alkorta Instituto de Química Médica (CSIC), Juan de la Cierva, 3. E-28006 Madrid, Spain.

Pierre Baillargeon Département de chimie, Cégep de Sherbrooke, 475 rue du Cégep, Sherbrooke, QC J1E 4K1, Canada.

Antonio Bauzá Department of Chemistry, Universitat de les Illes Balears, Crta de Valldemossa km 7.5, 07122 Palma de Mallorca (Baleares), Spain.

Jelena P. Blagojević Department of Chemistry, University of Belgrade, Studentski trg 12-16, 11000 Belgrade, Serbia.

Doris E. Braun Institute of Pharmacy, University of Innsbruck, Innrain 52c, 6020 Innsbruck, Austria.

Édouard Caron-Duval Département de chimie, Cégep de Sherbrooke, 475 rue du Cégep, Sherbrooke, QC J1E 4K1, Canada.

Janet E. Del Bene Department of Chemistry, Youngstown State University, Youngstown, OH 44555, USA.

Yves L. Dory Laboratoire de synthèse supramoléculaire, Département de chimie, Institut de Pharmacologie, Université de Sherbrooke, 3001 12e avenue nord, Sherbrooke, QC J1H 5N4, Canada.

Leonardo H. R. Dos Santos Department of Chemistry and Biochemistry, University of Bern, Freiestrasse 3, 3027 Bern, Switzerland.

Jose Elguero Instituto de Química Médica (CSIC), Juan de la Cierva, 3. E-28006 Madrid, Spain.

Reda F. M. Elshaarawy Institut für Anorganische Chemie und Strukturchemie, Universitätsstr. 1, 40225 Düsseldorf, Germany; Faculty of Science, Suez University, 43533 Suez, Egypt.

Niall J. English School of Chemical and Bioprocess Engineering, University College Dublin, Belfield, Dublin 4, Ireland.

Francesca P. A. Fabbiani Georg-August-Universität Göttingen, GZG, Abt. Kristallographie, Goldschmidtstr. 1, 37077 Göttingen, Germany.

Radosław Filas Department of Theoretical Chemistry, Faculty of Chemistry, Jagiellonian University, R. Ingardena 3, Cracow 30-060, Poland.

Edilson B. Alencar Filho Colegiado de Farmácia, Universidade Federal do Vale do São Francisco, Petrolina, Pernambuco 56304-917, Brazil.

Célia Fonseca Guerra Department of Theoretical Chemistry and Amsterdam Center for Multiscale Modeling, Vrije Universiteit Amsterdam, De Boelelaan 1083, 1081 HV Amsterdam, The Netherlands.

Antonio Frontera Department of Chemistry, Universitat de les Illes Balears, Crta de Valldemossa km 7.5, 07122 Palma de Mallorca (Baleares), Spain.

Simon Gagné Département de chimie, Cégep de Sherbrooke, 475 rue du Cégep, Sherbrooke, QC J1E 4K1, Canada.

Thomas Gelbrich Institute of Pharmacy, University of Innsbruck, Innrain 52c, 6020 Innsbruck, Austria.

Sławomir J. Grabowski Kimika Fakultatea, Euskal Herriko Unibertsitatea UPV/EHU, and Donostia International Physics Center (DIPC), P.K. 1072, San Sebastian 20080, Spain; IKERBASQUE, Basque Foundation for Science, Bilbao 48011, Spain.

Ulrich J. Griesser Institute of Pharmacy, University of Innsbruck, Innrain 52c, 6020 Innsbruck, Austria.

Jiande Gu Drug Design & Discovery Center, State Key Laboratory of Drug Research, Shanghai Institute of Materia Medica, Chinese Academy of Sciences, Shanghai 201203, China.

Christian Heering Institut für Anorganische Chemie und Strukturchemie, Universitätsstraße 1, 40225 Düsseldorf, Germany.

Md. Alamgir Hossain Department of Chemistry and Biochemistry, Jackson State University, Jackson, MS 39217, USA.

Christoph Janiak Institut für Anorganische Chemie und Strukturchemie, Universitätsstraße 1, 40225 Düsseldorf, Germany.

Goran V. Janjić ICTM, University of Belgrade, Njegoševa 12, 11000 Belgrade, Serbia.

Anna Christin Kautz Institut für Anorganische Chemie und Strukturchemie, Universitätsstr. 1, 40225 Düsseldorf, Germany.

Ümit Köc Institut für Anorganische Chemie und Strukturchemie, Universitätsstr. 1, 40225 Düsseldorf, Germany.

Tadeusz M. Krygowski Department of Chemistry, Warsaw University, Pasteura 1, 02-093 Warsaw, Poland.

Jerzy Leszczynski Interdisciplinary Nanotoxicity Center, Department of Chemistry, Jackson State University, Jackson, MS 39217, USA.

Piero Macchi Department of Chemistry and Biochemistry, University of Bern, Freiestrasse 3, 3027 Bern, Switzerland.

Mariusz P. Mitoraj Department of Theoretical Chemistry, Faculty of Chemistry, Jagiellonian University, R. Ingardena 3, Cracow 30-060, Poland.

Bahareh Nateghi Institut für Anorganische Chemie und Strukturchemie, Universitätsstraße 1, 40225 Düsseldorf, Germany.

Stefan Oberparleiter Institute of General, Inorganic and Theoretical Chemistry, University of Innsbruck, Innrain 80, 6020 Innsbruck, Austria.

Boaz G. Oliveira Instituto de Ciências Ambientais e Desenvolvimento Sustentável, Universidade Federal da Bahia, Barreiras, Bahia 47808-021, Brazil.

Émilie Pellerin Département de chimie, Cégep de Sherbrooke, 475 rue du Cégep, Sherbrooke, QC J1E 4K1, Canada.

Guido J. Reiss Institut für Anorganische Chemie und Strukturchemie, Heinrich-Heine-Universität Düsseldorf, Düsseldorf 40225, Germany.

Filip Sagan Department of Theoretical Chemistry, Faculty of Chemistry, Jagiellonian University, R. Ingardena 3, Cracow 30-060, Poland.

Sofiane Saouane Georg-August-Universität Göttingen, GZG, Abt. Kristallographie, Goldschmidtstr. 1, 37077 Göttingen, Germany.

Steve Scheiner Department of Chemistry and Biochemistry, Utah State University, Logan, UT 84322-0300, USA.

Herwig Schottenberger Institute of General, Inorganic and Theoretical Chemistry, University of Innsbruck, Innrain 80, 6020 Innsbruck, Austria.

Olga A. Stasyuk Faculty of Chemistry, Warsaw University of Technology, Noakowskiego 3, 00-664 Warsaw, Poland.

Halina Szatylowicz Faculty of Chemistry, Warsaw University of Technology, Noakowskiego 3, 00-664 Warsaw, Poland.

Anas Tahli Institut für Anorganische Chemie und Strukturchemie, Universitätsstr. 1, 40225 Düsseldorf, Germany; Faculty of Science, Al-Furat University, Deir Ezzor, Syria.

Mário L. A. A. Vasconcellos Departamento de Química, Universidade Federal da Paraíba, João Pessoa, Paraíba 58036-300, Brazil.

Jing Wang Interdisciplinary Nanotoxicity Center, Department of Chemistry, Jackson State University, Jackson, MS 39217, USA.

Snežana D. Zarić Department of Chemistry, University of Belgrade, Studentski trg 12-16, 11000 Belgrade, Serbia; Department of Chemistry, Texas A&M University at Qatar, P.O. Box 23874, Doha, Qatar.

About the Guest Editor

Sławomir Janusz Grabowski was born in Warsaw, Poland (1956) and received his M.Sc. degree (1981) and his Ph.D. (1986) from the University of Warsaw. He received his D.Sc. (habilitation, 1998) at the Technical University of Łódź, Poland. Since 1986, he has been working at the University of Białystok and since 2002 at the University of Łódź, Poland (as a full professor since 2005). He has been employed in a number of universities as postdoctoral researcher or professor: E.T.H. Zentrum, Zürich, Switzerland (1987), University of Uppsala, Sweden (1988), University of Grenoble, France (1992), Jackson State University, Jackson MS, USA (during summers in 2003–2008), Fukuoka University, Japan (2009, awarded by the Japanese Society for the Promotion of Science). In 2009, he moved to San Sebastian and is currently employed as the Ikerbasque Research Professor at the University of the Basque Country, Spain. Dr. Grabowski has authored or co-authored about 170 papers, 11 book chapters, and has edited a book on hydrogen bonding (Hydrogen Bonding—New Insights Ed. Grabowski, S.J, Springer, Dordrecht 2006) and he has been a guest editor of several journal special issues concerning the hydrogen bond interaction or related topics. His work encompasses the analysis of hydrogen bonding and Lewis acids - Lewis base interactions in the gas phase and crystals. He is a member of the Editorial Boards of the *Journal of Physical Organic Chemistry*, *Computational and Theoretical Chemistry* and *Crystals*.

Preface to "Analysis of Hydrogen Bonds in Crystals"

Crystal structures are a source of information on geometries of species that form networks in solids, in addition to providing information on the nature of interactions that occur between them. Hydrogen bonding is one of the most important interactions responsible for the arrangements of molecules and ions in crystals. For this reason, this special issue collects new, interesting, and important findings and ideas on the role of the hydrogen bond in crystals.

The contributions of this issue may be classified into three groups: the first concerns new crystal structures where interesting arrangements of species are described; the second group considers crystal structures and their observed phenomena; and the last group is strongly related to a discussion on theoretical results, where often hydrogen bond interactions are compared with other interactions (halogen bond, pnicogen bond, carbon bond, etc.).

Despite the fact that the last group of articles concerns the results of calculations, it is strongly associated with experiment since the results are compared with experimental observations. Besides other interesting phenomena, the hydrogen bond interactions are well described in the theoretical work.

I would like to thank all authors whose contributions are included in this issue for their excellent work, and for their inspiring and interesting articles.

Sławomir J. Grabowski
Guest Editor

Analysis of Hydrogen Bonds in Crystals

Sławomir J. Grabowski

Abstract: The determination of crystal structures provides important information on the geometry of species constituting crystals and on the symmetry relations between them. Additionally, the analysis of crystal structures is so conclusive that it allows us to understand the nature of various interactions. The hydrogen bond interaction plays a crucial role in crystal engineering and, in general, its important role in numerous chemical, physical and bio-chemical processes was the subject of various studies. That is why numerous important findings on the nature of hydrogen bonds concern crystal structures. This special issue presents studies on hydrogen bonds in crystals, and specific compounds and specific H-bonded patterns existing in crystals are analyzed. However, the characteristics of the H-bond interactions are not only analyzed theoretically; this interaction is compared with other ones that steer the arrangement of molecules in crystals, for example halogen, tetrel or pnicogen bonds. More general findings concerning the influence of the hydrogen bond on the physicochemical properties of matter are also presented.

Reprinted from *Crystals*. Cite as: Grabowski, S.J. Analysis of Hydrogen Bonds in Crystals. *Crystals* **2016**, *6*, 59.

1. Introduction

Numerous important findings concerning the hydrogen bond are related to crystal structures. For example, one of the first definitions of the hydrogen bond was proposed by Pauling in his monograph on the chemical bond, and it is as follows: *"Under certain conditions an atom of hydrogen is attracted by rather strong forces to two atoms, instead of only one, so that it may be considered to be acting as a bond between them. This is called the hydrogen bond"* [1]. Pauling also has pointed out that the hydrogen atom is situated between the most electronegative atoms and that it usually interacts more strongly with one of them, forming the covalent bond; the other interaction of hydrogen with the next electronegative atom is much weaker and mostly electrostatic in nature. It is important that Pauling, describing the hydrogen bond, refers directly to crystal structures, to fluorine compounds, to clathrate compounds, and to structures of alcohols or carboxylic acids. He describes the effect of the hydrogen bond on the physical properties of substances, and in detail, the cooperativity H-bond effects in crystals, particularly the cooperativity in HF and HCN compounds (the term "cooperativity" that appeared later in the literature does not occur in this monograph; however, exactly this phenomenon was described and discussed [1]).

One can mention other important monographs; however, this case is completely concerned with the hydrogen bond and mainly addresses the crystal structures, including the monographs of Pimentel and McClellan [2], Jeffrey and Saenger [3], Jeffrey [4], Desiraju and Steiner [5], Nishio, Hirota and Umezawa [6] or Gilli and Gilli [7]. It is important that the monographs describing mostly the theoretical studies on the hydrogen bond refer to the experimental results, among them crystal structures [8,9]. Furthermore, even important monographs are mentioned here since the hydrogen bond interaction was and is the subject of a huge number of studies.

2. From a Variety of Hydrogen Bond Interactions to New Definitions of the Hydrogen Bond

The Pauling definition of a hydrogen bond (HB) was cited in the previous section [1]; it is worth mentioning that further modifications of this definition were mainly inspired by findings concerning crystal structures. For example, the article introducing and explaining the recent definition recommended by IUPAC often refers to properties of the hydrogen bond that are revealed in crystal structures, and one may mention the directionality of the hydrogen bond [10]. This definition states that: *"The hydrogen bond is an attractive interaction between a hydrogen atom from a molecule or a molecular fragment X–H in which X is more electronegative than H, and an atom or a group of atoms in the same or a different molecule, in which there is evidence of bond formation."*

One of the first debates on hydrogen bonds was based on the nature of interactions in crystal structures [11,12]. It was contested that for HB, the hydrogen atom has to be located only between electronegative centers; the C-H ... O interactions were classified as HBs [11]. Several years later, Taylor and Kennard applied subtle statistical methods to analyze numerous crystal structures and they proved that the C-H ... Y (Y designates the Lewis base center) interactions possess characteristics of typical hydrogen bonds [13]. After that it was commonly accepted that the carbon atom may be the proton-donating center in HB systems. Also π-electrons of these systems such as acetylene, ethylene and their derivatives or π-electrons of aromatic systems were classified as possible proton acceptors (Lewis bases) in HBs [6].

The detailed analysis and description of a new kind of hydrogen bond, the dihydrogen bond (DHB), was based on the detection of such interactions in crystal structures [14]. This interaction is characterized by the contact between two H atoms of opposite charges, *i.e.*, one H atom plays the role of the Lewis acid and the other one acts as the Lewis base; Figure 1 shows the fragment of the structure with the Re-H ... H-N dihydrogen bond.

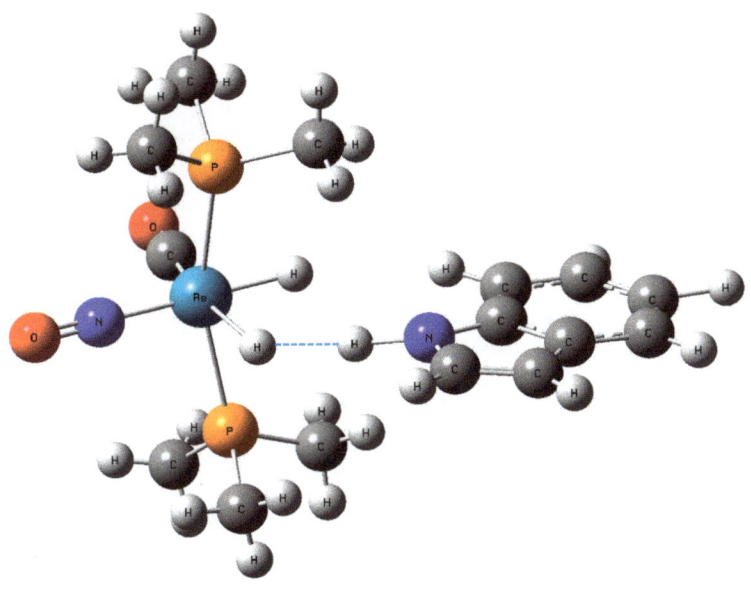

Figure 1. The fragment of the crystal structure of the ReH$_2$(CO)(NO)(PMe$_3$)$_2$ complex with indole; the H ... H contact corresponding to the Re-H ... H-N dihydrogen bond is designated by the blue broken line. The structure was taken from the Cambridge Structural Database [15]; refcode: XATFAZ, following Reference [16].

The above-mentioned IUPAC definition of the hydrogen bond covers the C-H ... Y interactions and those where π-electrons play the role of the Lewis base and the dihydrogen bonds; other interactions possessing numerous characteristics of "the typical Pauling-style HBs" are classified as HBs according to this definition. It is mentioning that the DHB interaction may be treated as a preliminary stage of the reaction leading to the release of the molecular hydrogen; it is important that this process was analyzed in crystal structures, *i.e.*, the evidence of the topochemical control of this reaction was reported [17]. In general, HBs may be treated as the preliminary stage of the proton transfer process; the proton transfer reactions in solid-state structures are often the subject of analysis, especially the double proton transfer which is characteristic for carboxylic acids linked in dimers by two, often equivalent C=O ... H-O hydrogen bonds [18]. In such a way the eight-member ring is formed, characteristic not only for carboxylic acids but also for other species such as, for example, amides. Figure 2 presents examples of crystal structures where such motifs exist.

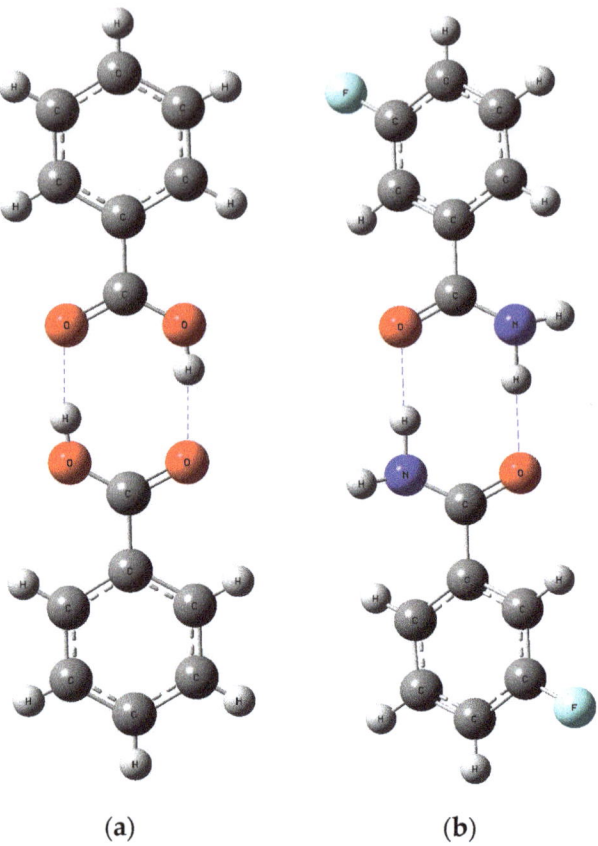

(a) (b)

Figure 2. The fragments of the crystal structures of: (**a**) benzoic acid [19] where the dimer is linked by O-H . . . O HBs; (**b**) m-fluorobenzamide [20] where the dimer is linked by N-H . . . O HBs. The structures were taken from the Cambridge Structural Database [15]; refcodes: BENZAC02 and BENAFM10, respectively.

The method based on the graph theory was introduced by Etter *et al.* [21] and developed later by Bernstein *et al.* [22]; it allows the categorization of HB motifs. The motifs presented in Figure 2 are characterized by eight-member rings closed by two proton donors and two proton acceptors. The other approach proposed was where the supramolecular synthons as the basic molecular entities may form, in a predictable manner, large assemblies such as those in crystal structures [23].

New techniques and approaches used to determine crystal structures require additional comments. For example, the electron density can be reconstructed from diffraction experiments [24] with the use of X-rays or by more recently introduced γ-ray and synchrotron radiation techniques. Numerous interesting studies were performed where the electron density distribution analysis allows the description of inter- and intramolecular interactions, among them hydrogen bonds. It is also important that numerous theoretical approaches may be applied here, such as, for example, the Quantum Theory of Atoms in Molecules which is a powerful approach to deepening the understanding of the nature of interactions [25].

3. Conclusions

One can mention numerous topics connected both with the HB interaction and with crystal structures; it is difficult to briefly mention all of the most important studies and findings. However, it is very important that several topics mentioned here earlier, which are the subject of extensive investigations, are represented in this special issue. This issue is a collection of important scientific contributions where new crystal structures are presented, where the physicochemical phenomena dependent on H-bond interactions are described and where the experimental results are supported by theoretical calculations; also, the hydrogen bonds are compared with other interactions that steer the arrangement of molecules in crystals.

Acknowledgments: The author is thankful for the technical and human support provided by IZO-SGI SGIker of UPV/EHU and European funding (ERDF and ESF). Financial support comes from Eusko Jaurlaritza (Basque Government) through Project No. IT588-13, and from the Spanish Office of Scientific Research through Projects Nos. CTQ2012- 38496-C05-01 and CTQ2015-67660-P.

Conflicts of Interest: The author declares no conflict of interest.

References

1. Pauling, L. *The Nature of the Chemical Bond*; Cornell University Press: New York, NY, USA, 1960.
2. Pimentel, G.C.; McClellan, A.L. *The Hydrogen Bond*; W. H. Freeman and Company: San Fransisco, CA, USA; London, UK, 1960.
3. Jeffrey, G.A.; Saenger, W. *Hydrogen Bonding in Biological Structures*; Springer: Berlin, Germany, 1991.
4. Jeffrey, G.A. *An Introduction to Hydrogen Bonding*; Oxford University Press Inc.: New York, NY, USA, 1997.
5. Desiraju, G.R.; Steiner, T. *The Weak Hydrogen Bond in Structural Chemistry and Biology*; Oxford University Press Inc.: New York, NY, USA, 1999.
6. Nishio, M.; Hirota, M.; Umezawa, Y. *The CH/π Interaction: Evidence, Nature and Consequences*; Wiley: New York, NY, USA, 1998.
7. Gilli, G.; Gilli, P. *The Nature of the Hydrogen Bond*; Oxford University Press Inc.: New York, NY, USA, 2009.
8. Scheiner, S. *Hydrogen Bonding a Theoretical Perspective*; Oxford University Press Inc.: New York, NY, USA, 1997.
9. Grabowski, S.J., Ed.; *Hydrogen Bonding—New Insights*; Springer: Dordrecht, The Netherlands, 2006.

10. Arunan, E.; Desiraju, G.R.; Klein, R.A.; Sadlej, J.; Scheiner, S.; Alkorta, I.; Clary, D.C.; Crabtree, R.H.; Dannenberg, J.J.; Hobza, P.; et al. Definition of the hydrogen bond (IUPAC Recommendations 2011). *Pure Appl. Chem.* **2011**, *83*, 1637–1641.
11. Sutor, D.J. The C-H . . . O hydrogen bond in crystals. *Nature* **1962**, *195*, 68–69.
12. Donohue, J. Selected topics in hydrogen bonding. In *Structural Chemistry and Molecular Biology*; Rich, A., Davidson, N., Eds.; Freeman: San Francisco, CA, USA, 1968.
13. Taylor, R.; Kennard, O. Crystallographic evidence for the existence of C-H . . . O, C-H . . . N and C-H . . . Cl hydrogen bonds. *J. Am. Chem. Soc.* **1982**, *104*, 5063–5070.
14. Richardson, T.B.; de Gala, S.; Crabtree, R.H. Unconventional Hydrogen Bonds: Intermolecular B-H . . . H-N Interactions. *J. Am. Chem. Soc.* **1995**, *117*, 12875–12876.
15. Allen, F.H. The Cambridge Structural Database: A quarter of a million crystal structures and rising. *Acta Crystallogr. Sect. B—Struct. Sci.* **2002**, *58*, 380–388.
16. Belkova, N.V.; Shubina, E.S.; Gutsul, E.I.; Epstein, L.M.; Eremenko, I.L.; Nefedov, S.E. Structural and energetic aspects of hydrogen bonding and proton transfer to $ReH_2(CO)(NO)(PR_3)_2$ and $ReHCl(CO)(NO)(PMe_3)_2$ by IR and X-ray studies. *J. Organomet. Chem.* **2000**, *610*, 58–70.
17. Custelcean, R.; Jackson, J.E. Topochemical Control of Covalent Bond Formation by Dihydrogen Bonding. *J. Am. Chem. Soc.* **1998**, *120*, 12935–12941.
18. Grabowski, S.J. What Is the Covalency of Hydrogen Bonding? *Chem. Rev.* **2011**, *11*, 2597–2625.
19. Feld, R.; Lehmann, M.S.; Muir, K.W.; Speakman, J.C. The crystal structure of benzoic acid—A redetermination with X-rays at room temperature—A summary of neutron diffraction work at temperatures down to 5K. *Zeitschrift für Kristallographie—Crystalline Materials* **1981**, *157*, 215–231.
20. Kato, Y.; Sakurai, K. The Crystal Structure of o-Fluorobenzamide. *Bull. Chem. Soc. Jpn.* **1982**, *55*, 1643–1644.
21. Etter, M.C.; MacDonald, J.C.; Bernstein, J. Graph-set analysis of hydrogen-bond patterns in organic crystals. *Acta Crystallogr. Sect. B—Struct. Sci.* **1990**, *46*, 256–262.
22. Bernstein, J.; Davis, R.E.; Shimoni, L.; Cheng, N.-L. Patterns in Hydrogen Bonding: Functionality and Graph Set Analysis in Crystals. *Angew. Chem. Int. Ed.* **1995**, *34*, 1555–1573.
23. Desiraju, G.R. Supramolecular Synthons in Crystal Engineering—A New Organic Synthesis. *Angew. Chem. Int. Ed.* **1995**, *34*, 2311–2327.
24. Tsirelson, V.G.; Ozerov, R.P. *Electron Density and Bonding in Crystals*; Taylor & Francis Group: New York, NY, USA, 1996.
25. Bader, R.F.W. *Atoms in Molecules: A Quantum Theory*; Oxford University Press: Oxford, UK, 1990.

A Cadmium Anionic 1-D Coordination Polymer $\{[Cd(H_2O)_6][Cd_2(atr)_2(\mu_2\text{-btc})_2(H_2O)_4]\ 2H_2O\}_n$ within a 3-D Supramolecular Charge-Assisted Hydrogen-Bonded and π-Stacking Network

Anas Tahli, Ümit Köc, Reda F. M. Elshaarawy, Anna Christin Kautz and Christoph Janiak

Abstract: The hydrothermal reaction of 4,4'-bis(1,2,4-triazol-4-yl) (btr) and benzene-1,3,5-tricarboxylic acid (H_3btc) with Cd(OAc)$_2 \cdot$ 2H$_2$O at 125 °C *in situ* forms 4-amino-1,2,4-triazole (atr) from btr, which crystallizes to a mixed-ligand, poly-anionic chain of $[Cd_2(atr)_2(\mu_2\text{-btc})_2(H_2O)_4]^{2-}$. Together with a hexaaquacadmium(II) cation and water molecules the anionic coordination-polymeric forms a 3-D supramolecular network of hexaaquacadmium(II)-*catena*-[bis(4-amino-1,2,4-triazole) tetraaquabis(benzene-1,3,5-tricarboxylato)dicadmate(II)] dihydrate, 1-D- $\{[Cd(H_2O)_6][Cd_2(atr)_2\ (\mu_2\text{-btc})_2(H_2O)_4]\ 2H_2O\}_n$ which is based on hydrogen bonds (in part charge-assisted) and π–π interactions.

Reprinted from *Crystals*. Cite as: Tahli, A.; Köc, Ü.; Elshaarawy, R.F.M.; Kautz, A.C.; Janiak, C. A Cadmium Anionic 1-D Coordination Polymer $\{[Cd(H_2O)_6][Cd_2(atr)_2(\mu_2\text{-btc})_2(H_2O)_4]\ 2H_2O\}_n$ within a 3-D Supramolecular Charge-Assisted Hydrogen-Bonded and π-Stacking Network. *Crystals* **2016**, *6*, 23.

1. Introduction

Metal-organic frameworks (MOFs) which are porous coordination polymers (PCPs) attract great interest for their potential applications in separation processes [1], sensor technology [2], luminescence [3], ionic or electrical conductivity [4,5], magnetism [6], and heat transformation through reversible water de- and adsorption [7,8]. Benzene carboxylic acid ligands, such as terephthalic acid or trimesic acid (H_3btc), are common rigid ligands for porous coordination polymers (PCPs). 1,2,4-Triazol-4-yl derivatives match the coordination geometry of pyrazoles with their N1 and N2 donor atoms, and can form different secondary building units (SBUs). The amino-functionalized triazole ligand 4-amino-1,2,4-triazole (atr) can coordinate to metal atoms to build molecular complexes [9,10], polynuclear complexes [11,12], inorganic-organic coordination polymers [13–16] and different dimensional metal-organic networks of mixed ligands [17,18].

Coordination polymers based on mixed-linkers allow for a fine-tuning of MOF properties and can show additional characteristics such as crystal-to-crystal transformations [19], short and long-range magnetic ordering [20], luminescence [21], etc. [22]. The combination of neutral nitrogen donor ligands with anionic carboxylate ligands are frequent choices for the synthesis of mixed-ligand networks [22]. The linker 1,2-bis(1,2,4-triazol-4-yl)ethane (abbreviated as btre, Scheme 1) has recently been intensely studied in mixed-linker MOFs [19–24] and single-linker networks [25]. Herein, we report an attempt to construct a mixed-linker network with the 4,4′-bis(1,2,4-triazol-4-yl) (btr) ligand which was not hydrothermally stable under the synthesis conditions of 125 °C in water, so that the hydrolysis product 4-amino-1,2,4-triazole (atr) was incorporated instead (Scheme 1).

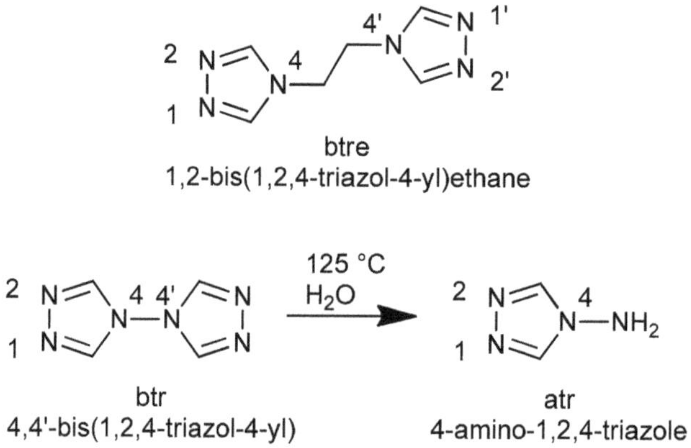

Scheme 1. Triazole ligands relevant in this work and indication of the hydrolysis of btr to atr.

2. Results and Discussion

Colorless crystals (Figure S1, Supplementary Information) were obtained from the hydrothermal reaction (125 °C) of $Cd(OAc)_2 \cdot 2H_2O$, 4,4′-bis(1,2,4-triazol-4-yl) (btr) and benzene-1,3,5-tricarboxylic acid (H_3btc) in approximately 1:3:1 molar ratio in the presence of three equivalents of triethylamine as a base to deprotonate the carboxylate groups of H_3btc. The reaction was repeated several times and found reproducible. The crystals are soluble in water and ethanol.

Comparison between the FT-IR spectra (attenuated total reflection, ATR) of the crystalline product and the mixture of atr and H_3btc ligands (Figure S2, Supplementary Information) shows significant differences in the fingerprint region which suggests that the ligands became coordinated to the metal ion. Multiple weak broad peaks between 3300 and 3100 cm^{-1} can be assigned to the O–H/N–H stretching vibrations of aqua ligands and the amino group of the atr ligand [26]. The absence of a peak at *ca* 1715 cm^{-1} indicates the full deprotonation of the H$_3$btc ligand. Additionally, the asymmetric and symmetric stretching vibrations of the carboxylate group [27,28] are observed at 1607 cm^{-1}, 1204 cm^{-1}, and 1527 cm^{-1}, 1110 cm^{-1}, which reveal different binding modes of the carboxylate group. Bands at 752 cm^{-1} and 731 cm^{-1} in the fingerprint region are due to 1,3,5-trisubstituted benzene [27]. Furthermore, a band at 612 cm^{-1} can be assigned to the vibrational mode of the triazole ring of the atr ligand [29].

The sample was dissolved in DMSO-d$_6$ via heating in an ultrasonic bath at 50 °C. For the NMR analysis of the crystalline product an excess NaCN was added to the sample in order to bind the Cd^{2+} ions as stable cyanido complexes and to free the ligands so that a ligand ratio can be determined. After centrifugation, the pipette-separated supernatant was measured. The ^{1}H NMR spectrum (Figure S3, Supplementary Information) then shows both atr and btc ligand signals. The signal of the two protons of the triazole ring in the atr ligand appears at 9.17 ppm and the signal for the (protonated) amino group is observed at 6.30 ppm. The signal at 8.36 is assigned to the three protons of the btc benzene ring. Signals for residual Et$_3$N appeared at ~0.89–0.92 and ~2.37–2.42 ppm. The integration ratio between atr: btc and NH$_3^+$ is 1.0:1.7:1.7 which agrees with two H-atoms for the triazole ring, three H-atoms of btc (1:1.5) and three H-atoms for the protonated amino group (1:1.5).

The title compound crystallizes in the monoclinic crystal system with the P2$_1$/c space group. The crystallographic asymmetric unit (Figure 1a) consists of a Cd(II) ion with benzene-tricarboxylate (btc^{3-}), 4-amino-1,2,4-triazole (atr), and two aqua ligands, plus a half-occupied cationic hexaaquacadmium(II) complex and a lattice water molecule. The original precursor compound 4,4'-bis(1,2,4-triazol-4-yl) (btr) is not incorporated as a ligand into the structure due to the *in situ* hydrolysis into 4-amino-1,2,4-triazole (atr), which was then found instead. Yet, the slow delivery of atr in the course of the reaction is apparently crucial for the product formation. When the reaction was carried out with atr directly instead of btr no crystals or precipitate formed. The charge-neutral product formula is 1-D {[Cd(H$_2$O)$_6$][Cd$_2$(atr)$_2$(µ$_2$-btc)$_2$(H$_2$O)$_4$] 2H$_2$O}$_n$, **1**, named hexaaqua-cadmium(II)-*catena*-[bis(4-amino-1,2,4-triazole)tetraaquabis(benzene-1,3,5-tricarboxylato) dicadmate(II)] dihydrate.

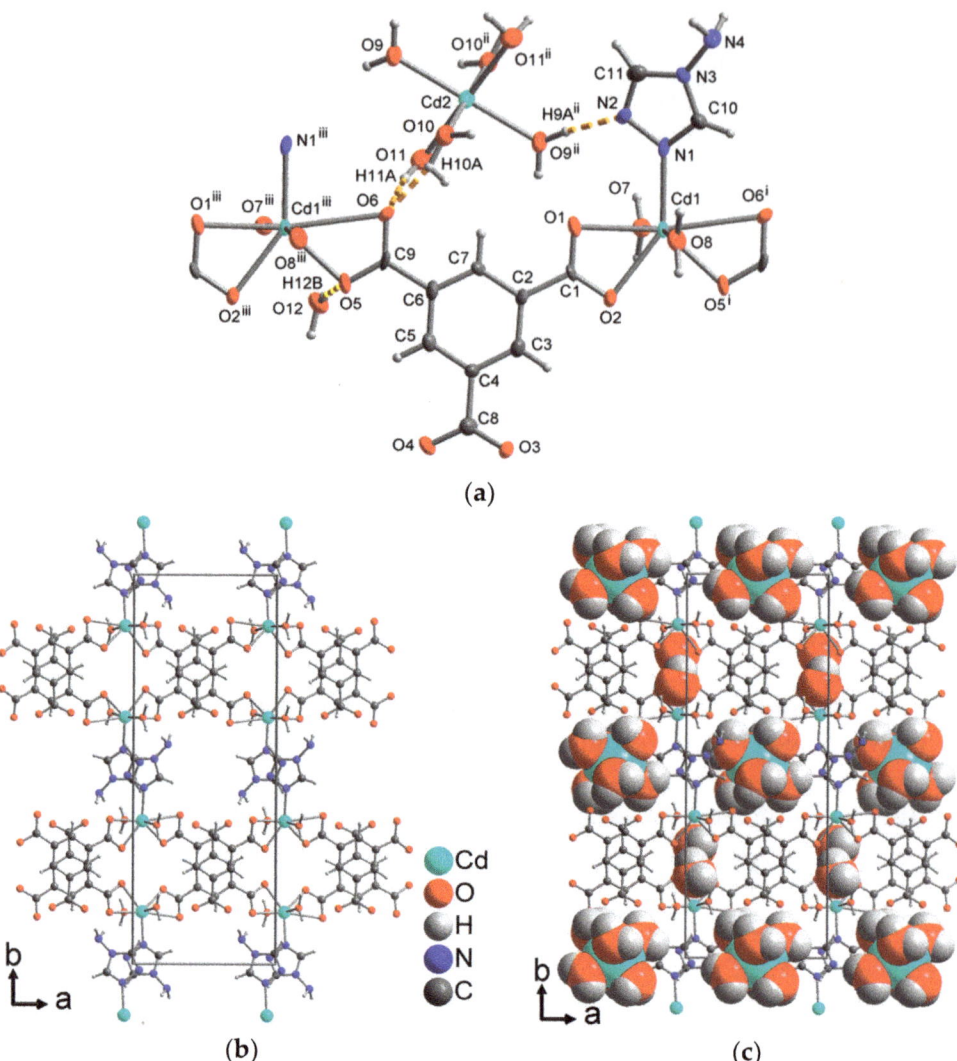

Figure 1. (a) Extended asymmetric unit of **1** (70% thermal ellipsoids, hydrogen atoms with arbitrary radii), showing also part of the hydrogen bonds (orange dashed lines). Symmetry transformations $i = x + 1, y, z$; $ii = -x + 1, -y + 1, -z + 1$; $iii = x - 1, y, z$. Section of the packing diagram of (b) the anionic chains and (c) the full structure with the $[Cd(H_2O)_6]^{2+}$ cations and the crystal water molecules highlighted in space-filling mode. Selected distances and angles are given in Table 1 and details of H-bonds in Table 2.

Table 1. Selected bond lengths [Å] and angles [°] in **1**.

Cd1–N1	2.253(6)	O8–Cd1–O7	171.88(16)
Cd1–O1	2.412(5)	N1–Cd1–O2	142.95(16)
Cd1–O2	2.355(5)	O5 *i*–Cd1–O2	84.36(16)
Cd1–O5 *i*	2.258(5)	O8–Cd1–O2	89.48(18)
Cd1–O6 *i*	2.766(5)	O7–Cd1–O2	82.40(17)
Cd1–O7	2.325(5)	N1–Cd1–O1	88.09(17)
Cd1–O8	2.302(5)	O5 *i*–Cd1–O1	139.22(16)
–	–	O8–Cd1–O1	90.35(17)
Cd2–O9	2.275(5)	O7–Cd1–O1	84.81(17)
Cd2–O10	2.304(6)	O2–Cd1–O1	54.86(15)
Cd2–O11	2.266(5)	–	–
–	–	O9–Cd2–O10	86.65(18)
N1–Cd1–O5 *i*	132.67(18)	O9–Cd2–O10 *ii*	93.35(18)
N1–Cd1–O8	90.45(19)	O10-Cd2–O11	86.1(2)
O5 *i*–Cd1–O8	88.47(18)	O10–Cd2–O11 *ii*	93.9(2)
N1–Cd1–O7	95.88(18)	O9–Cd2–O11	87.50(19)
O5 *i*–Cd1–O7	90.94(17)	O9–Cd2–O11 *ii*	92.50(19)

Symmetry transformations used to generate equivalent atoms: $i = x + 1, y, z$; $ii = -x + 1, -y + 1, -z + 1$.

Table 2. Details of the hydrogen bonding interactions in **1** [a].

D-H···A	D-H [Å]	H···A [Å]	D···A [Å]	D-H···A [°]	Symmetry Transformations
O7–H7A···N3 *iv*	0.95	2.33	3.280(7)	176	$iv = -x + 2, -y + 1, -z + 1$
O7–H7A···N4 *iv*	0.95	2.33	3.138(8)	143	$iv = -x + 2, -y + 1, -z + 1$
O7–H7B···O12 *i*	0.95	2.07	2.684(7)	121	$i = x + 1, y, z$
O8–H8A···O11 *v*	0.95	2.41	3.284(7)	152	$v = x + 1, y, z + 1$
O8–H8B···O12 *v*	0.95	2.05	2.695(7)	124	$v = x + 1, y, z + 1$
O9–H9A···O3 *vi*	0.92(4)	1.88(5)	2.785(7)	167(8)	$vi = -x + 1, y + 1/2, -z + 3/2$
O9–H9B···N2 *ii*	0.89(9)	2.02(9)	2.897(8)	171(8)	$ii = -x + 1, -y + 1, -z + 1$
O10–H10A···O6	0.95(9)	1.84(9)	2.742(7)	158(9)	–
O10–H10B···N4 *vii*	0.98(5)	1.91(6)	2.803(8)	151(7)	$vii = -x + 2, -y + 1, -z + 2$
O11–H11A···O6	1.00(9)	1.93(9)	2.906(8)	167(8)	–
O11–H11B···O4 *viii*	0.92(5)	1.78(6)	2.641(7)	155(8)	$viii = x, -y + 1/2, z - 1/2$
O12–H12A···O2 *x*	0.89(5)	1.84(6)	2.661(7)	152(8)	$x = x - 1, -y + 1/2, z - 1/2$
O12–H12B···O5	0.92	1.85	2.763(7)	172	–
N4–H4A···O3 *ix*	0.85(8)	2.24(8)	3.081(8)	170(7)	$ix = 2 - x, y + 1/2, -z + 3/2$
N4–H4B···O1 *vii*	0.83(4)	2.04(5)	2.857(8)	166(8)	$vii = -x + 2, -y + 1, -z + 2$
C10–H10···O10 *i*	0.95	2.54	3.381(8)	147	$i = x + 1, y, z$

Notes: [a] D = donor, A = acceptor.

The Cd1 ion forms a coordination polymeric chain with benzene-tricarboxylate (btc^{3-}), 4-amino-1,2,4-triazole (atr), and two aqua ligands. The Cd1 atom is seven-fold coordinated in a distorted pentagonal-bipyramidal fashion by a triazole nitrogen atom of atr and six oxygen atoms; two of them belong to axial aqua ligands. The other four O-atoms come from the carboxylate groups of two fully deprotonated btc^{3-} ligands, which coordinate in a bidentate chelating mode. The O-atoms of btc^{3-} and the N-atom of atr form the equatorial plane of the pentagonal bipyramid. One of these chelated Cd–O bonds is slightly longer (Cd1–O6 = 2.766(5) Å) than the range of the other Cd–O bonds (Cd1–O = 2.258(5)–2.412(5) Å). The atr ligand coordinates to Cd1 through the imine N1-atom of the triazole ring as a terminal ligand; the other imine N atom (N2) and the amino group remain without Cd coordination but engage in hydrogen bonding (see below). The tri-anionic btc^{3-} ligands coordinate as bridges between two Cd1 atoms to form the one-dimensional mixed-ligand chain (Figure 1b); the third carboxylate group remains uncoordinated but is part of the hydrogen-bonding network. This chain is a polyanion and has the formula of [Cd$_2$(atr)$_2$(μ_2-btc)$_2$(H$_2$O)$_4$]$_n$$^{2-}$. Charge neutrality is reached by one hexaaquacadmium(II) cation, [Cd(H$_2$O)$_6$]$^{2+}$ as equivalent for each of the two Cd1 atoms. The Cd2 atom in [Cd(H$_2$O)$_6$]$^{2+}$ sits on an inversion center as a special position (Figure 1a). Hexa-coordinated cadmium can have a coordination environment in-between octahedral and trigonal prismatic [30,31]. Here, the Cd2 atom has a slightly distorted octahedral environment of six water molecules. Two crystal water molecules per formula unit of 1-D {[Cd(H$_2$O)$_6$][Cd$_2$(atr)$_2$(μ_2-btc)$_2$(H$_2$O)$_4$] 2H$_2$O}$_n$ complete the packing (Figure 1c).

Anionic coordination polymers are not very frequent in view of the several thousand publications on coordination polymers [32]. Only a few of the ones which are reported feature the metal cadmium, as in [Cd((P$_6$O$_{18}$)(H$_2$O)$_2$)]$^{2-}$ [33], [Cd(OABDC)(H$_2$O)$_2$]$^-$ · (OABDC = 5-(carboxylatomethoxy)benzene-1,3-dicarboxylato) [34], and in [Cd(P$_6$O$_{18}$)]$_n$$^{4-}$ [35].

The infinite anionic chains are stacked parallel to each other along the c direction (Figure 1b) through significant π–π interactions [36] between the adjacent benzene rings of the btc^{3-} ligands and the adjacent triazole rings of the atr ligands, respectively (Figure 2a). Strong π-stacking interactions have rather short centroid-centroid contacts (<3.8 Å) and near parallel ring planes which translate into a sizable overlap of the aryl-plane areas. The centroid–centroid distance of adjacent btc-benzene rings is 3.726(4) Å and for neighboring triazole rings it is 3.598(4) and 3.997(4) Å (Figure 2a). The interplanar angle of the benzene rings is 4°; the triazole rings are exactly parallel by symmetry; (see Supplementary Information for further details, Scheme S1 and Table S1). In addition to π-stacking the inter-chain packing is controlled by charge-assisted amino-N–H···$^{(-)}$O$_2$C- hydrogen bonds (N4–H···O1 and O3, Figure 2b and Table 2). The carboxylate groups of btc^{3-} as hydrogen bond acceptors carry negative ionic charges. Such charge-assisted H-bonds are usually stronger and shorter than neutral H-bonds [37–42].

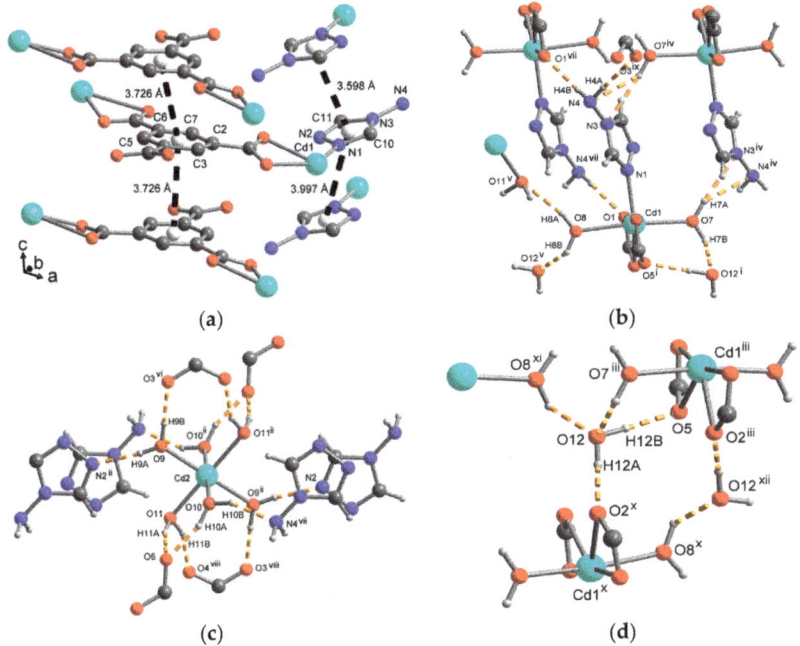

Figure 2. Supramolecular packing interactions in **1**: (**a**) π-stacking interactions with centroid-centroid contacts given. Hydrogen-bonding interactions shown separately for the different H-bond donors for clarity: (**b**) around the [Cd(H$_2$O)$_6$]$^{2+}$ cations, (**c**) around the 4-amino-1,2,4-triazole (atr) ligands, and (**d**) around the crystal water molecule. Details of π-stacking interactions are given in Supplementary Information, details of H-bonds together with symmetry transformations are listed in Table 2. Additional symmetry transformations used in (d): $iii = x - 1, y, z$; $xi = x - 1, y, z - 1$; $xii = x, -y + 1/2, z + 1/2$.

The space which is created between the parallel anion chains is occupied by the cationic [Cd(H$_2$O)$_6$]$^{2+}$ complexes and lattice water molecules (Figure 1c). Both [Cd(H$_2$O)$_6$]$^{2+}$ and the crystal water molecules function as H-bond donors towards the carboxylate oxygen atoms, again with charge-assisted O–H$\cdots$$^{(-)}O_2$C-hydrogen bonds (Figure 2c,d). The O9 atom of [Cd(H$_2$O)$_6$]$^{2+}$ also forms a hydrogen bond towards the N2-triazole atom (Figure 2c). The crystal water oxygen atom O12 and O11 of [Cd(H$_2$O)$_6$]$^{2+}$ act further as H-bond acceptors from the aqua ligands (O7, O8) on Cd1 in the anionic chain (Figure 2b,d). Overall this gives a tight hydrogen bonding network which, apparently, prevents any disorder in the crystal water and allowed for the finding and refining of most protic H-atom positions (see X-ray crystallography section).

Together, the π–π interactions between adjacent anionic chains and the hydrogen bonds between cationic complexes and anionic chains and lattice water molecules build a 3-D supramolecular network (Figure 1c).

3. Materials and Methods

Reagents and solvents were obtained from commercial sources and used without any further purification. The bis(1,2,4-triazol-4-yl) ligand (btr) was synthesized under inert conditions according to previous work [43] from hydrazine monohydrate, $N_2H_4 \cdot H_2O$, N,N'-dimethylformamide azine [44], and p-toluenesulfonic acid monohydrate in dry toluene. A programmable oven type (UFP 400) from Memmert GmbH (Schwabach, Germany) was used for the hydrothermal synthesis. The reactions were carried out in DURAN® (DURAN Group GmbH, Wertheim, Germany) culture glass tubes with PTFE-faced sealing wad, diameter 12 mm, height 100 mm, and DIN thread 14 GL, closed with a red screw cap (Figure S4, Supplementary Information), suitable for hydro-/solvothermal synthesis for coordination polymer synthesis up to 150 °C instead of an autoclave. The contents only come into contact with the glass and polytetrafluoroethylene (PTFE) seal. Elemental analyses were performed on a Vario MicroCube from Elementar GmbH. The light microscopy images were observed with a Leica MS5 binocular eyepiece with transmitted light and polarization filter. The images of isolated crystals were taken with a Nikon *COOLPIX 4500* (Tokyo, Japan) digital camera through a special ocular connection. Infrared spectra were recorded with a Bruker Optik TENSOR 37 spectrophotometer (Bruker Optik GmbH, Ettlingen, Germany) using a Diamond ATR (Attenuated Total Reflection) unit from 4000 to 500 cm^{-1}. The following abbreviations were used to classify spectral bands: br (broad), sh (shoulder), very weak (vw), w (weak), m (medium), s (strong), and vs (very strong). The ^{1}H-NMR spectra were recorded on a Bruker Advance DRX 500 MHz NMR spectrometer with calibration against the residual protonated solvent signal DMSO-d_6 (2.50 ppm).

1D {[Cd(H$_2$O)$_6$][Cd$_2$(atr)$_2$(μ$_2$-btc)$_2$(H$_2$O)$_4$] 2H$_2$O}$_n$: A portion of Cd(OAc)$_2 \cdot$ 2H$_2$O (10.7 mg, 0.04 mmol) and 4,4′-bis(1,2,4-triazol-4-yl) (btr) (16.3 mg, 0.12 mmol) was combined in 1 mL of water in a DURAN glass tube and shaken for about 3 min. Then a solution of benzene-1,3,5-tricarboxylic acid (H$_3$btc) (8.4 mg, 0.04 mmol) and Et$_3$N/H$_2$O (0.75 mL of 0.16 mol/L, 3eq. to H$_3$btc) in 2.5 mL of water prepared in an ultrasonic bath at 45–50 °C was added to the previous mixture. The sealed glass tube was shaken for about 5 min and then placed in a programmable furnace, heated to 125 °C for 3 h and held at that temperature for 48 h, then cooled at a rate of 5 °C/h to ambient temperature. The resulting colorless crystals were separated from powdery precipitate and washed with the mother liquor. Yield: (14 mg, 30% based on metal salt). FT-IR (ATR, cm^{-1}): 3100 (w, ν (C–H)), 1607 (s, ν$_{asym}$ C=O, C=N), 1527 (m, ν$_{sym}$ C=O, C=N), 1204 (s, ν$_{asym}$ C–O, C–N),

1110 (s, ν_{sym} C–O, C–N), 871 (m, sh, γ (C–H)), 752, 731 (s, γ, 1,3,5-trisubstituted benzene ring of Hbtc), 612 (m, γ, triazole ring). Calcd. for $C_{22}H_{38}Cd_3N_8O_{24}$ (M_w = 1135.8 g mol^{-1}): C 23.27, H 3.37, N 9.87; found: C 23.35, H 2.99, N 10.35%.

Single Crystal X-Ray Structure

A suitable single crystal (Figure S1, Supplementary Information) was carefully selected under a polarizing microscope and mounted in oil in a glass loop. Data collection: Bruker AXS APEX II CCD area-detector diffractometer with multilayer mirror monochromator, Mo-Kα radiation (λ = 0.71073 Å) from microsource, double-pass method with φ- and ω-scans; data collection with APEX2, cell refinement and data reduction with SAINT [45], experimental absorption correction with SADABS [46]. Structure analysis and refinement: All structures were solved by direct methods using SHELXL2014; refinement was done by full-matrix least squares on F^2 using the SHELX-97 program suite [47]. Non-hydrogen atoms were refined with anisotropic displacement parameters. All non-hydrogen positions were found and refined with anisotropic temperature factors. Hydrogen atoms for aromatic CH were positioned geometrically (CH = 0.95 Å) and refined using a riding model (AFIX 43) with U_{iso}(H) = 1.2U_{eq}(C). Hydrogen atoms on aqua ligands and on the amino, NH_2 group were treated in a mixed refinement. On O7 and O8 the H atoms were positioned geometrically (O–H = 0.95 Å, N–H = 0.88 Å) and refined using a riding model (AFIX 93) with U_{iso}(H) = 1.5U_{eq}(O,N). On O9, O10, O11, O12 and N4 the H atoms could be found and refined with U_{iso}(H) = 1.5U_{eq}(O,N) and DFIX constraints (0.95, 0.05) for H9B, O10B, H11B, N4B, and O12A. H12B was found but had to be kept fixed upon further refinement (AFIX 1). The two highest peaks in the electron density map are within 1.52 Å of the Cd1 atom, the next two highest peaks are within 1.5 Å of Cd2. Details of the X-ray crystal data structure determination and refinement are provided in Table 3. Graphics were drawn with *DIAMOND* (Version 3.2) [48] Analyses on the supramolecular π–π -stacking interactions were done with *PLATON* for Windows [49]. CCDC No. 1451376 contains the supplementary crystallographic data for this paper. These data can be obtained free of charge via http://www.ccdc.cam.ac.uk/conts/retrieving.html.

Table 3. Crystal data and refinement details for 1-D {[Cd(H$_2$O)$_6$][Cd$_2$(atr)$_2$(μ_2-btc)$_2$(H$_2$O)$_4$] 2H$_2$O}$_n$, **1**.

Chemical Formula	C$_{22}$H$_{38}$Cd$_3$N$_8$O$_{24}$
Mr	1131.80
Crystal system, space group	Monoclinic, $P\,2_1/c$
Temperature (K)	100(2)
a (Å)	10.1435(18)
b (Å)	26.471(5)
c (Å)	7.0263(13)
β (°)	106.213(13)
V (Å^3)	1811.6(6)
Z	2
Density (calculated), Mg/m^3	2.082
Absorpt. coefficient, μ (mm^{-1})	1.850
Crystal size (mm)	0.3 × 0.3 × 0.1
F(000)	1124
Theta range for data collection, (°)	2.09–25.34
h, k, l ranges	±12, ±31, ±8
Reflections collected	15909
Independent reflections	3279 [R(int) = 0.1114]
Completeness to theta 25.34°	98.9%
Data/restraints/parameters	3279/5/286
Final R indices [$I > 2\mathrm{sigma}(I)$] [a]	R1 = 0.0519, wR2 = 0.1284
R indices (all data) [a]	R1 = 0.0734, wR2 = 0.1418
Goodness-of-fit on F^2 [b]	1.020
Weighting scheme w; a/b [c]	0.0521/0.000
$\Delta\rho_{max}, \Delta\rho_{min}$ (e Å^{-3}) [d]	0.30, −0.34

[a] $R_1 = [\sum(||F_o| - |F_c||)/\sum|F_o|]$; $wR_2 = [\sum[w(F_o^2 - F_c^2)^2]/\sum[w(F_o^2)^2]]^{1/2}$;
[b] Goodness-of-fit, $S = [\sum[w(F_o^2 - F_c^2)^2]/(n-p)]^{1/2}$; [c] $w = 1/[\sigma^2(F_o^2) + (aP)^2 + bP]$ where $P = (\max(F_o^2 \text{ or } 0) + 2F_c^2)/3$; [d] Largest difference peak and hole.

Supplementary Materials: The supplementary files are available online at http://www.mdpi.com/2073-4352/6/3/23/s001.

Acknowledgments: Anas Tahli sincerely thanks the Al-Furat University and the Ministry of Higher Education in Syria for the financial support (scholarship) during his study in Germany (language course, Master and PhD degree). The work was supported by the German Science Foundation (DFG) through grant Ja466/25-1.

Author Contributions: Anas Tahli and Ümit Köc designed the experiments, synthesized the btr ligand and compound **1**. X-ray measurements by Anna Christin Kautz, data analysis by Anas Tahli. Anas Tahli, Reda Elshaarawy and Christoph Janiak have written the manuscript.

Conflicts of Interest: The authors declare no conflict of interest.

References

1. Li, J.-R.; Sculley, J.; Zhou, H.-C. Metal–Organic Frameworks for Separations. *Chem. Rev.* **2012**, *112*, 869–932.
2. Hu, Z.; Deibert, B.J.; Li, J. Luminescent metal–organic frameworks for chemical sensing and explosive detection. *Chem. Soc. Rev.* **2014**, *43*, 5815–5840.
3. Cui, Y.-J.; Yue, Y.-F.; Qian, G.-D.; Chen, B.-L. Luminescent Functional Metal–Organic Frameworks. *Chem. Rev.* **2012**, *112*, 1126–1162.
4. Horike, S.; Umeyama, D.; Kitagawa, S. Ion Conductivity and Transport by Porous Coordination Polymers and Metal–Organic Frameworks. *Acc. Chem. Res.* **2013**, *46*, 2376–2384.
5. Givaja, G.; Amo-Ochoa, P.; Gómez-García, C.J.; Zamora, F. Electrical conductive coordination polymers. *Chem. Soc. Rev.* **2012**, *41*, 115–147.
6. Kurmoo, M. Magnetic metal–organic frameworks. *Chem. Soc. Rev.* **2009**, *38*, 1353–1379.
7. Fröhlich, D.; Henninger, S.K.; Janiak, C. Multicyle water vapour stability of microporous breathing MOF aluminium isophthalate CAU-10-H. *Dalton Trans.* **2014**, *43*, 15300–15304.
8. Jeremias, F.; Fröhlich, D.; Janiak, C.; Henninger, S.K. Water and methanol adsorption on MOFs for fast cycling heat transformation processes. *New J. Chem.* **2014**, *38*, 1846–1852.
9. Wang, S.-W.; Yang, L.; Feng, J.-L.; Wu, B.-D.; Zhang, J.-G.; Zhang, T.-L.; Zhou, Z.-N. Synthesis, Crystal Structure, Thermal Decomposition, and Sensitive Properties of Two Novel Energetic Cadmium(II) Complexes Based on 4-Amino-1,2,4-triazole. *Z. Anorg. Allg. Chem.* **2011**, *637*, 2215–2222.
10. Carpenter, J.A.; Dshemuchadse, J.; Busato, S.; Braeunlich, I.; Pöthig, A.; Caseri, W. Tetrakis(4-amino -1,2,4-triazole)platinum(II) Salts: Syntheses, Crystal Structures, and Properties. *Z. Anorg. Allg. Chem.* **2014**, *640*, 724–732.
11. Zhou, J.-H.; Cheng, R.-M.; Song, Y.; Li, Y.-Z.; Yu, Z.; Chen, X.-T.; Xue, Z.-L.; You, X.-Z. Syntheses, Structures, and Magnetic Properties of Unusual Nonlinear Polynuclear Copper(II) Complexes Containing Derivatives of 1,2,4-Triazole and Pivalate Ligands. *Inorg. Chem.* **2005**, *44*, 8011–8022.
12. Yoo, H.-S.; Lim, J.-H.; Kang, J.-S.; Koh, E.-K.; Hong, C.-S. Triazole-bridged magnetic M(II) assemblies (M = Co, Ni) capped with the end-on terephthalate dianion involving multi-intermolecular contacts. *Polyhedron* **2007**, *26*, 4383–4388.
13. Yeh, C.-W.; Chang, W.-J.; Suen, M.-C.; Lee, H.-T.; Tsai, H.-A.; Tsou, C.-H. Roles of the anion in the self-assembly of silver(I) complexes containing 4-amino-1,2,4-triazole. *Polyhedron* **2013**, *61*, 151–160.
14. Wu, X.-Y.; Kuang, X.-F.; Zhao, Z.-G.; Chen, S.-C.; Xie, Y.-M.; Yu, R.-M.; Lu, C.-Z. A series of POM-based hybrid materials with different copper/aminotriazole motifs. *Inorg. Chim. Acta* **2010**, *363*, 1236–1242.
15. Grosjean, A.; Daro, N.; Kaufmann, B.; Kaiba, A.; Létard, J.-F.; Guionneau, P. The 1-D polymeric structure of the $[Fe(NH_2trz)](NO)_2 \cdot nH_2O$ (with $n = 2$) spin crossover compound proven by single crystal investigations. *Chem. Commun.* **2011**, *47*, 12382–12384.

16. Gou, Y.-T.; Yue, F.; Chen, H.-M.; Liu, G.; Sun, D.-C. Syntheses and structures of 3d–4d heterometallic coordination polymers based on 4-amino-1,2,4-triazole. *J. Coord. Chem.* **2013**, *66*, 1889–1896.
17. Yang, E.-C.; Zhang, C.-H.; Liu, Z.-Y.; Zhang, N.; Zhao, L.-N.; Zhao, X.-J. Three copper(II) 4-amino-1,2,4-triazole complexes containing differently deprotonated forms of 1,3,5-benzenetricarboxylic acid: Synthesis, structures and magnetism. *Polyhedron* **2012**, *40*, 65–71.
18. Wang, X.-L.; Zhao, W.; Zhang, J.-W.; Lu, Q.-L. Three tetranuclear copper(II) cluster-based complexes constructed from 4-amino-1,2,4-triazole and different aromatic carboxylates: Assembly, structures, electrochemical and magnetic properties. *J. Solid State Chem.* **2013**, *198*, 162–168.
19. Habib, H.A.; Sanchiz, J.; Janiak, C. Mixed-ligand coordination polymers from 1,2-bis(1,2,4-triazol-4-yl)-ethane and benzene-1,3,5-tricarboxylate: Trinuclear nickel or zinc secondary building units for three dimensional networks with crystal-to-crystal transformation upon dehydration. *Dalton Trans.* **2008**, 1734–1744.
20. Habib, H.A.; Sanchiz, J.; Janiak, C. Magnetic and luminescence properties of Cu(II), Cu(II)$_4$O$_4$ core, and Cd(II) mixed-ligand metal–organic frameworks constructed from 1,2-bis(1,2,4-triazol-4-yl)ethane and benzene-1,3,5-tricarboxylate. *Inorg. Chim. Acta* **2009**, *362*, 2452–2460.
21. Habib, H.A.; Hoffmann, A.; Höppe, H.A.; Janiak, C. Crystal structures and solid-state CPMAS ^{13}C NMR correlations in luminescent zinc(II) and cadmium(II) mixed-ligand coordination polymers constructed from 1,2-bis(1,2,4-triazol-4-yl)ethane and benzenedicarboxylate. *Dalton Trans.* **2009**, 1742–1751.
22. Yin, Z.; Zhou, Y.-L.; Zeng, M.-H.; Kurmoo, M. The concept of mixed organic ligands in metal–organic frameworks: design, tuning and functions. *Dalton Trans.* **2015**, *44*, 5258–5275.
23. Ding, J.-G.; Zhu, X.; Cui, Y.-F.; Liang, N.; Sun, P.-P.; Chen, Q.; Li, B.-L.; Li, H.-Y. Structurally versatile cadmium coordination polymers based on bis(1,2,4-triazol)ethane and rigid aromatic multicarboxylates: Syntheses, structures and properties. *CrystEngComm* **2014**, *16*, 1632–1644.
24. Habib, H.A.; Janiak, C. Benzene-1,3,5-tricarboxylic acid-1,2-bis-(1,2,4-triazol-4-yl) ethane–water (4/1/2). *Acta Crystallogr. Sect. E Struct. Rep. Online* **2008**, *64*, o1199.
25. Habib, H.A.; Hoffmann, A.; Höppe, H.A.; Steinfeld, G.; Janiak, C. Crystal Structure Solid-State Cross Polarization Magic Angle Spinning ^{13}C NMR Correlation in Luminescent d^{10} Metal-Organic Frameworks Constructed with the 1,2-Bis(1,2,4-triazol-4-yl)ethane Ligand. *Inorg. Chem.* **2009**, *48*, 2166–2180.
26. Nakamoto, K. *Infrared and Raman Spectra of Inorganic and Coordination Compounds*; Wiley: New York, NY, USA, 1970.
27. Shi, Z.; Li, G.; Wang, L.; Gao, L.; Chen, X.; Hua, J.; Feng, S. Two Three-Dimensional Metal−Organic Frameworks from Secondary Building Units of Zn$_8$(OH)$_4$(O$_2$C−)$_{12}$ and Zn$_2$((OH)(O$_2$C−)$_3$: [Zn$_2$(OH)(btc)]$_2$(4,4′-bipy) and Zn$_2$(OH)(btc)(pipe). *Cryst. Growth Des.* **2004**, *4*, 25–27.

28. Dai, J.-C.; Wu, X.-T.; Fu, Z.-Y.; cui, C.-P.; Hu, S.-M.; Du, W.-X.; Wu, L.-M.; Zhang, H.-H.; Sun, R.-Q. Synthesis, Structure, and Fluorescence of the Novel Cadmium(II)−Trimesate Coordination Polymers with Different Coordination Architectures. *Inorg. Chem.* **2002**, *41*, 1391–1396.
29. Slangen, P.M.; van Koningsbruggen, P.J.; Goubitz, K.; Haasnoot, J.G.; Reedijk, J. Isotropic Magnetic Exchange Interaction through Double µ-1,2,4-Triazolato-N1,N2 Bridges: X-ray Crystal Structure, Magnetic Properties, and EPR Study of Bis(µ-3-pyridin-2-yl-1,2,4-triazolato-N′,N1,N2)(sulfato-O)aquacopper(II)-diaquacopper(II) Trihydrate. *Inorg. Chem.* **1994**, *33*, 1121–1126.
30. Stiefel, E.I.; Brown, G.F. Detailed nature of the six-coordinate polyhedra in tris(bidentate ligand) complexes. *Inorg. Chem.* **1972**, *11*, 434–436.
31. Banerjee, S.; Ghosh, A.; Wu, B.; Lassahn, P.-G.; Janiak, C. Polymethylene spacer regulated structural divergence in cadmium complexes: Unusual trigonal prismatic and severely distorted octahedral coordination. *Polyhedron* **2005**, *24*, 593–599.
32. Janiak, C.; Vieth, J.K. MOFs, MILs and more: Concepts, properties and applications for porous coordination networks (PCNs). *New J. Chem.* **2010**, *34*, 2366–2388.
33. Ameur, I.; Dkhili, S.; Sbihi, H.; Besbes-Hentati, S.; Rzaigui, M.; Abid, S. A Cadmium Coordination Polymer Based on Phosphate Clusters: Synthesis, Crystal Structure, Electrochemical Investigation and Antibacterial Activity. *J. Cluster Sci.* **2016**.
34. Wang, P.; Zhao, Y.; Chen, Y.; Kou, X.-Y. A novel one-dimensional tubular cadmium(II) coordination polymer based on 5-(carboxylatomethoxy)benzene-1,3-dicarboxylate containing 4-aminopyridinium ions. *Acta Crystallogr. Sect. C Cryst. Struct. Commun.* **2013**, *69*, 1340–1343.
35. Abid, S.; Al-Deyab, S.S.; Rzaigui, M. The one-dimensional coordination polymer poly[tetrakis [(4-chloro-phenyl)methanaminium] [cadmate-µ-cyclohexaphosphorato]]. *Acta Crystallogr. Sect. E Struct. Rep. Online* **2011**, *67*, m1549–m1550.
36. Janiak, C. A critical account on π–π stacking in metal complexes with aromatic nitrogen-containing ligands. *J. Chem. Soc. Dalton Trans.* **2000**, 3885–3896.
37. Ward, M.D. Design of crystalline molecular networks with charge-assisted hydrogen bonds. *Chem. Commun.* **2005**, 5838–5842.
38. Heering, C.; Nateghi, B.; Janiak, C. Charge-Assisted Hydrogen-Bonded Networks of NH_4^+ and $[Co(NH_3)_6]^{3+}$ with the New Linker Anion of 4-Phosphono-Biphenyl-4′-Carboxylic Acid. *Crystals* **2016**, *6*, 22.
39. Chamayou, A.-C.; Neelakantan, M.A.; Thalamuthu, S.; Janiak, C. The first vitamin B6 zinc complex, pyridoxinato-zinc acetate: A 1D coordination polymer with polar packing through strong inter-chain hydrogen bonding. *Inorg. Chim. Acta* **2011**, *365*, 447–450.
40. Drašković, B.M.; Bogdanović, G.A.; Neelakantan, M.A.; Chamayou, A.-C.; Thalamuthu, S.; Avadhut, Y.S.; Schmedt auf der Günne, J.; Banerjee, S.; Janiak, C. N-o-Vanillylidene-L-histidine: Experimental charge density analysis of a double zwitterionic amino acid Schiff-base compound. *Cryst. Growth Des.* **2010**, *10*, 1665–1676.

41. Dorn, T.; Chamayou, A.-C.; Janiak, C. Hydrophilic interior between hydrophobic regions in inverse bilayer structures of cation–1,1′-binaphthalene-2,2′-diyl phosphate salts. *New J. Chem.* **2006**, *30*, 156–167.
42. Maclaren, J.K.; Sanchiz, J.; Gili, P.; Janiak, C. Hydrophobic-exterior layer structures and magnetic properties of trinuclear copper complexes with chiral amino alcoholate ligands. *New J. Chem.* **2012**, *36*, 1596–1609.
43. Naik, A.; Marchand-Brynaert, J.; Garcia, Y. A Simplified Approach to *N*-and *N,N*′-Linked 1,2,4-Triazoles by Transamination. *Synthesis* **2008**, *1*, 149–154.
44. Bartlett, R.K.; Humphrey, I.R. Transaminations of *NN*-dimethylformamide azine. *J. Chem. Soc. C* **1967**, 1664–1666.
45. APEX2. *SAINT, Data Reduction and Frame Integration Program for the CCD Area-Detector System, Bruker Analytical X-ray Systems, Data Collection program for the CCD Area-Detector System*; Brucker: Madison, WI, USA, 1997–2006.
46. Sheldrick, G. *SADABS: Area-Detector Absorption Correction*; University of Göttingen: Göttingen, Germany, 1996.
47. Hübschle, C.B.; Sheldrick, G.M.; Dittrich, B. ShelXle: A graphical user interface for SHELXL. *J. Appl. Cryst.* **2011**, *44*, 1281–1284.
48. Brandenburg, K. Crystal and Molecular Structure Visualization, Crystal Impact—K. In *Diamond*; version 3.2; Brandenburg, H., Ed.; Putz Gbr: Bonn, Germany, 2009.
49. Spek, A.L. Structure validation in chemical crystallography. *Acta Crystallogr. Sect. D Biol. Crystallogr.* **2009**, *65*, 148–155.

Charge-Assisted Hydrogen-Bonded Networks of NH_4^+ and $[Co(NH_3)_6]^{3+}$ with the New Linker Anion of 4-Phosphono-Biphenyl-4'-Carboxylic Acid

Christian Heering, Bahareh Nateghi and Christoph Janiak

Abstract: The new linker molecule 4-phosphono-biphenyl-4'-carboxylic acid (H_2O_3P-$(C_6H_4)_2$-COOH, H_3BPPA) has been structurally elucidated in hydrogen-bonded networks with the ammonium cation $NH_4(H_2BPPA)(H_3BPPA)$ (**1**) and the hexaamminecobalt(III) cation $[Co(NH_3)_6](BPPA) \cdot 4H_2O$ (**2**). The protic O-H and N-H hydrogen atoms were found and refined in the low-temperature single-crystal X-ray structures. The hydrogen bonds in both structures are so-called charge-assisted; that is, the H-bond donor and/or acceptor carry positive and/or negative ionic charges, respectively. The H-bonded network in **1** consists of one formally mono-deprotonated 4-phosphonato-biphenyl-4'-carboxylic acid group; that is, a H_2BPPA^- anion and a neutral H_3BPPA molecule, which together form a 3D hydrogen-bonded network. However, an almost symmetric resonance-assisted hydrogen bond (RAHB) bond [O···H = 1.17 (3) and 1.26 (3) Å, O···H···O = 180 (3)°] signals charge delocalization between the formal H_2BPPA^- anion and the formally neutral H_3BPPA molecule. Hence, the anion in **1** is better formulated as $[H_2BPPA···H···H_2BPPA]^-$. In the H-bonded network of **2** the 4-phosphonato-biphenyl-4'-carboxylic acid is triply deprotonated, $BPPA^{3-}$. The $[Co(NH_3)_6]^{3+}$ cation is embedded between H-bond acceptor groups, $-COO^-$ and $-PO_3^-$ and H_2O molecules. The incorporation of sixteen H_2O molecules per unit cell makes **2** an analogue of the well-studied guanidinium sulfonate frameworks.

Reprinted from *Crystals*. Cite as: Heering, C.; Nateghi, B.; Janiak, C. Charge-Assisted Hydrogen-Bonded Networks of NH_4^+ and $[Co(NH_3)_6]^{3+}$ with the New Linker Anion of 4-Phosphono-Biphenyl-4'-Carboxylic Acid. *Crystals* **2016**, *6*, 22.

1. Introduction

The organophosphonic acid function, which has a pK_{a1} of 2.0 for the first and a pK_{a2} of 6.59 for the second proton, is capable of forming strong metal-to-ligand coordinative bonds in thermodynamically stable complexes with high stability constants [1]. Metal organophosphonate compounds are multifunctional organic-inorganic hybrid materials and as open frameworks can be regarded in between zeolite-like [1,2] and metal-organic framework materials [3,4], whereas phosphonate metal-organic frameworks (MOFs) are considerably rarer than MOFs with carboxylate linkers, with phosphonates forming stronger bonds to metals than carboxylate groups [3]. Metal organophosphonates are stable in water or aqueous environment [5]. The use of metal phosphonates in catalysis, luminescence [6], ion or proton exchange or conductivity [7,8] and in separation is discussed and investigated [9]. Further, cobalt and iron organophosphonates are investigated for their magnetic properties [10–13]. Metal phosphonates are also promising porous materials [14,15], and can be reversibly hydrated and dehydrated [16].

Organophosphonates can contain additional functional groups such as carboxylate, hydroxyl or amino in the organo-moiety which presents a tunable functionality with a wide variety of structural motifs and properties [1,3,17]. Carboxy-phosphonates (Scheme 1) can be seen as intermediates between pure carboxylates and pure phosphonates, sharing synergies of both ligand classes. Carboxy-phosphonates can form porous or 3D metal-ligand networks [18–20]. Weng et al. described a 3D zinc carboxy-phosphonate, ZnPC-2, as a material for CO_2 adsorption [21].

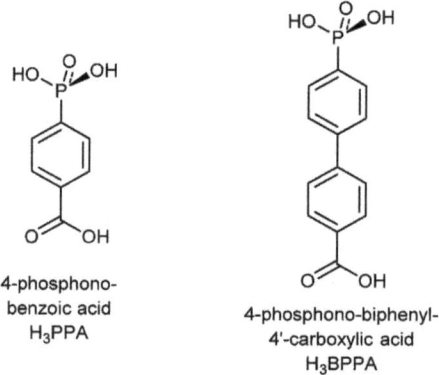

4-phosphono-
benzoic acid
H_3PPA

4-phosphono-biphenyl-
4'-carboxylic acid
H_3BPPA

Scheme 1. Examples of phosphono-carboxylic acids.

Various metal complexes have been synthesized with a ligand from deprotonated 4-phosphono-benzoic acid, including the metals barium [22], cobalt [23], copper [23], europium [24], lead [20], lithium [25], silver [26], strontium [27], thorium [28], titanium [29], uranium [28] and zinc [30,31]. However, the extended biphenyl-based variant 4-phosphono-biphenyl-4'-carboxylic acid (H_3BPPA) was unknown so far (Scheme 1).

Herein, we present the new linker 4-phosphono-biphenyl-4'-carboxylic acid, H_3BPPA, and its deprotonated structure in hydrogen-bonded networks with NH_4^+ and $[Co(NH_3)_6]^{3+}$ cations.

2. Results and Discussion

4-Phosphono-biphenyl-4′-carboxylic acid H$_3$BPPA has been synthesized, following a known procedure by Merkushev *et al.* from 4-biphenyl carboxylic acid through the intermediates 4′-iodo-biphenyl-4-carboxylic acid [32] and its methyl ester, followed by the nickel(II)-catalyzed conversion to a phosphonate ester, which after hydrolysis gave H$_3$BPPA (Scheme 2).

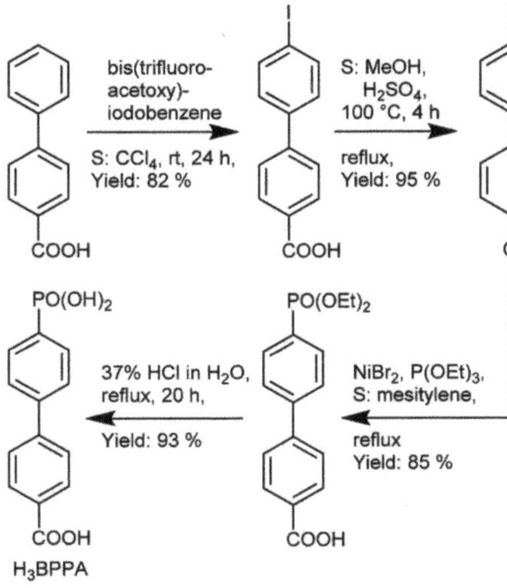

Scheme 2. Reaction sequence for the synthesis of H$_3$BPPA from 4-biphenyl carboxylic acid.

Neutralization of H$_3$BPPA with one equivalent of ammonium acetate yielded colorless crystals of formula NH$_4$(HO$_3$P-(C$_6$H$_4$)$_2$-COOH)(H$_2$O$_3$P-C$_6$H$_4$-C$_6$H$_4$-COOH) (**1**). The ammonium monohydrogenphosphonato-biphenyl–carboxylic acid crystallized with one molecule of the free H$_3$BPPA acid. The best results were obtained using a 1:1 ratio, though less ammonium acetate also led to product formation of lower quality. When the neutralization of H$_3$BPPA was carried out with excess conc. aqueous NH$_3$ instead of stoichiometric ammonium acetate, the same product, **1**, was formed, albeit of lower purity. Importantly, no complete or even twofold deprotonation of H$_3$BPPA was achieved in that way.

The asymmetric unit of **1** consists of the ammonium-cation, and formally a H$_2$BPPA$^-$ anion and a neutral H$_3$BPPA molecule (see below) (Figure 1a). The H$_2$BPPA$^-$ anion is derived by mono-deprotonation of the phosphonic acid group. The protic O–H and N–H hydrogen atoms were found and refined with U$_{eq}$ = 1.5 U$_{eq}$(O,N). The three building blocks form a three-dimensional (3D) hydrogen bonded network.

The carboxylic acid groups are oriented towards each other with the typical tail-to-tail arrangement, also known as $R_2^2(8)$-motif in the Etter-notation (Figure 1b) [33].

The biphenyl systems of the BPPA molecules are in nearly planar geometry with 0.31 (14) and 2.79 (14)° for the dihedral angles between the aryl ring planes, and 1.7 (2)° and 3.3 (2)° for the dihedral angle between −COOH and its aryl ring in the P1 and P2 molecule, respectively. The shape of the thermal ellipsoids of the carboxyl oxygen atoms O1, O2, O6, and O7 is indicative of some rotational movement (vibration) around the (carboxyl)C–C(aryl) bond (yet, no split refinement was suggested by SHELX from the principal mean square atomic displacements). Despite the presence of the biphenyl π-systems in **1**, there are no π-π interactions [34] and only few intermolecular C–H···π [35] are evident. The angle is 57° for the plane formed by one biphenyl system to its neighbor.

The ammonium cation engages all of its four (found and refined) N–H bonds in the hydrogen network to four different phosphono groups. The ammonium cations and phosphono groups form hydrogen-bonded layers parallel to the *ab*-plane, separated by the biphenyl-carboxylic acid parts (Figure 1c). Ammonium benzenephosphonate, $NH_4(HO_3PC_6H_5)$ [36], consists of a layered structure due to hydrogen bonds, with a similar motif to that of **1**.

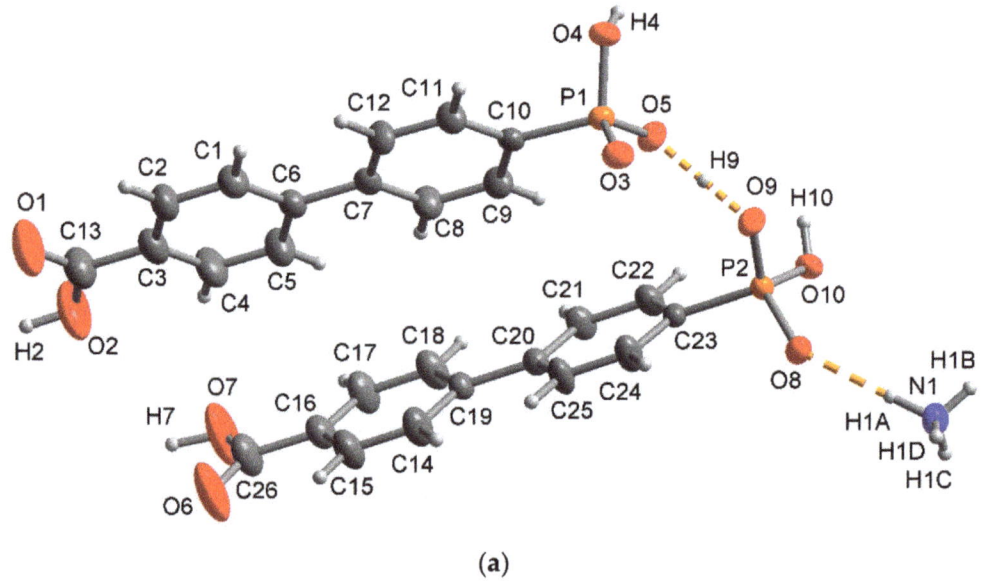

(a)

Figure 1. *Cont.*

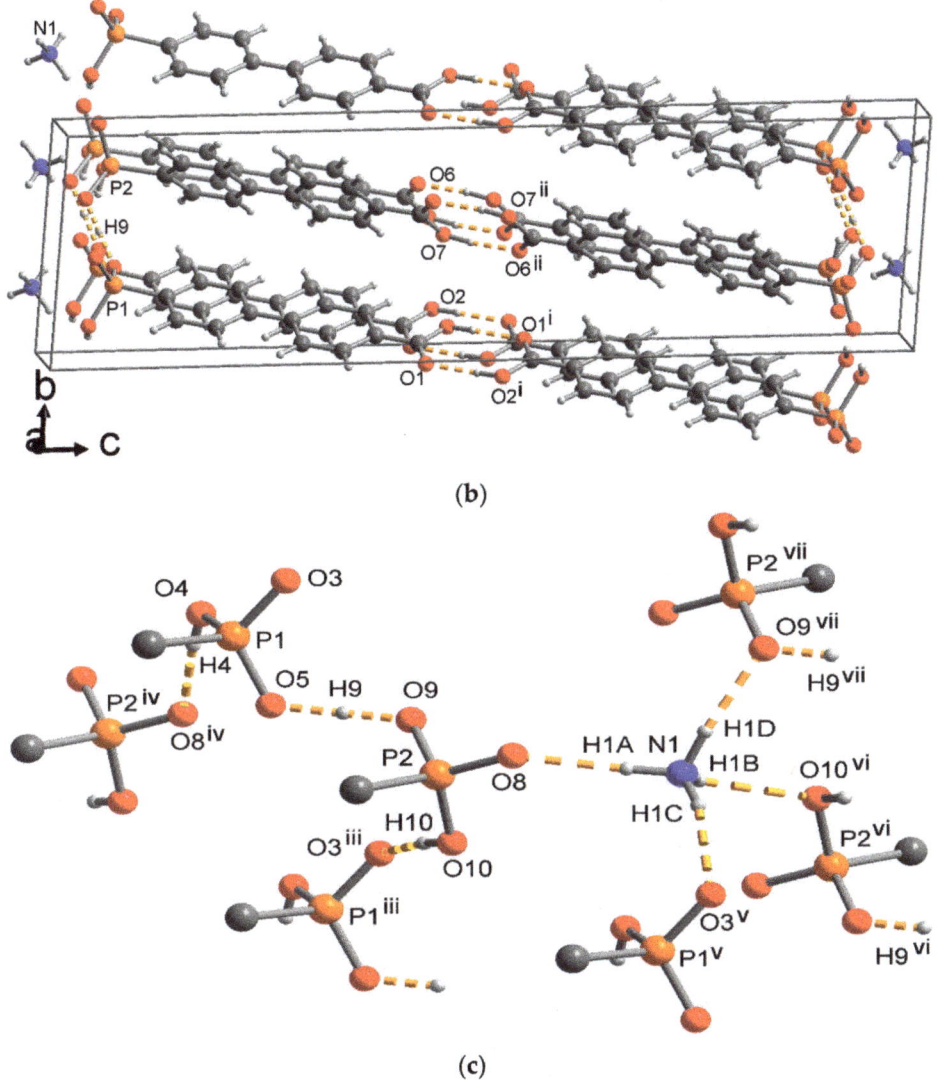

Figure 1. (a) Asymmetric unit of **1** (50% thermal ellipsoids); (b) unit-cell packing diagram with tail-to-tail arrangement of the carboxylic acid groups (showing only the carboxyl and the O9-H-O5 H bonds for clarity); and (c) full hydrogen-bonding arrangement around the NH_4^+ cation and the phosphonate and phosphonic acid groups. Details of the H-bonding interactions (orange dashed lines) are given in Table 1, selected non-hydrogen bonds and angles in Table 2. Symmetry transformations: i = −1 − x, −y, 1 − z; ii = −x, 1 − y, 1 − z; iii = 1 + x, y, z; iv = x, −1 + y, z; v = 1 + x, 1 + y, z; vi = 2 − x, 2 − y, −z; vii = 1 − x, 2 − y, −z.

Table 1. Details of the hydrogen bonding interactions in **1** [a].

D–H···A	D–H [Å]	H···A [Å]	D···A [Å]	D–H···A [°]	Symmetry Transformations
N1–H1A···O8	0.94 (3)	1.90 (3)	2.840 (3)	177 (3)	
N1–H1B···O10 vi	0.86 (3)	2.25 (3)	2.945 (3)	138 (3)	vi = 2 − x, 2 − y, −z
N1–H1C···O3 v	0.84 (3)	2.00 (3)	2.817 (3)	165 (3)	v = 1 + x, 1 + y, z
N1–H1D···O9 vii	0.95 (3)	1.92 (3)	2.845 (3)	164 (2)	vii = 1 − x, 2−y, −z
O2–H2···O1 i	0.94 (5)	1.71 (5)	2.623 (3)	164 (4)	i = −1 − x, −y, 1 − z
O4–H4···O8 iv	0.78 (3)	1.78 (3)	2.563 (2)	175 (3)	iv = x, −1 + y, z
O7–H7···O6 ii	0.99 (5)	1.66 (5)	2.642 (3)	172 (4)	ii = −x, 1 − y, 1 − z
O9–H9···O5	1.17 (3)	1.26 (3)	2.428 (2)	180 (3)	
O10–H10···O3 iii	0.83 (3)	1.72 (3)	2.537 (2)	170 (3)	iii = 1 + x, y, z

Notes: [a] D = donor, A = acceptor.

Table 2. Selected bond lengths [Å] and angles [°] in **1**.

P1–O3	1.4975 (18)	P2–O8	1.5045 (18)
P1–O5	1.5174 (18)	P2–O9	1.5190 (18)
P1–O4	1.5583 (19)	P2–O10	1.5553 (19)
P1–C10	1.787 (3)	P2–C23	1.796 (3)
C13–O1	1.235 (4)	C26–O6	1.230 (4)
C13–O2	1.283 (4)	C26–O7	1.275 (4)
O3–P1–O5	115.63 (11)	O8–P2–O9	112.38 (11)
O3–P1–O4	107.83 (10)	O8–P2–O10	109.16 (10)
O5–P1–O4	108.78 (11)	O9–P2–O10	110.04 (10)
O3–P1–C10	109.35 (11)	O8–P2–C23	109.32 (11)
O5–P1–C10	107.54 (11)	O9–P2–C23	108.76 (11)
O4–P1–C10	107.44 (11)	O10–P2–C23	107.04 (11)

Noteworthy, the H-bond O9–H9···O5 refined to an almost symmetric O9···H9···O5 hydrogen bridge with very similar distances of the H atom to both oxygen neighbors (1.17 (3) and 1.26 (3) Å) and a 180 (3)° O–H···O bond angle. This symmetric resonance-assisted hydrogen bond (RAHB) [37–41] O···H···O signals charge delocalization between the formal H_2BPPA^- anion (of P2) and the formally neutral H_3BPPA molecule (of P1). Hence, the anion is better formulated as $[H_2BPPA···H···H_2BPPA]^-$.

The interpretation of delocalized anion charge over the two phosphono groups is in agreement with the P–O bond lengths (Scheme 2). In each phosphonato group, there is a longer P–O bond of ~1.56 Å and two shorter P–O bonds between 1.50–1.52 Å. The P–O(H) bonds are 1.5583 (19) Å and 1.5553 (19) Å. One cannot clearly distinguish between a formally P=O double bond and a formally deprotonated P–O⁻ bond. The P–O bond lengths of the symmetric O9···H9···O5 hydrogen bridge are only slightly longer (~1.52 Å) then what should be P=O double bonds (~1.50 Å). The negative charge is delocalized over the P–O⁻ and P=O bonds, giving both of them a partial double bond character with P–O bond lengths between 1.50–1.52 Å (Scheme 3).

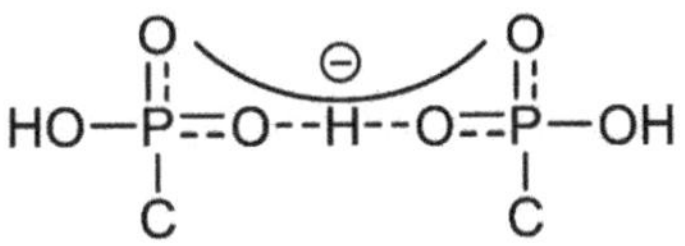

$$HO-\overset{\overset{O}{\|}}{\underset{C}{P}}=O--H--O=\overset{\overset{O}{\|}}{\underset{C}{P}}-OH$$

Scheme 3. Lewis valence structure for the bond order and charge-delocalization in the phosphonate groups in **1**.

Thermogravimetric analysis (TGA) of **1** shows a first a mass loss of ~4% up to 240 °C (Figure 2), which can be assigned to one molecule of ammonia (~3%), which is in agreement with literature values [42]. In a second step decarboxylation of one mol CO_2 (44 g/mol) leads to a mass loss of ~8%. With a third step of ~24% rapid dephosphonation of one mol $PO(OH)_2$ and final decarboxylation of another mol CO_2 takes place, which is followed by steady decomposition up to 700 °C.

The hydrated salt $[Co(NH_3)_6](O_3P\text{-}(C_6H_4)_2\text{-}COO) \cdot 4H_2O$, **2** could be crystallized from an aqueous ammonia solution 4-phosphono-biphenyl-4'-carboxylic acid. In a similar approach, 4-phosphono benzoic acid was crystallized with hexaaquacobalt(II) [43], and sulfonate ligands were crystallized with $[Co(NH_3)_6]^{3+}$ cations, resulting in hydrogen-bonded networks [44].

In the asymmetric unit of **2** there is one trivalent hexaamminecobalt cation, one completely deprotonated $BPPA^{3-}$ trianion and four water molecules (Figure 3a). The coordination sphere of Co^{3+} with six crystallographically different ammine ligands results in the well-known $[Co(NH_3)_6]^{3+}$ octahedron [45]. Despite its high symmetry, the $Co(NH_3)_6]^{3+}$ octahedron does not reside on a special position. The Co-N distances (Table 3) are comparable with that of $[Co(NH_3)_6]^{3+}$ in related complexes (Co-N = 1.951 (2) – 1.976 (2) Å, av. 1.956 (2) Å) [43,44,46].

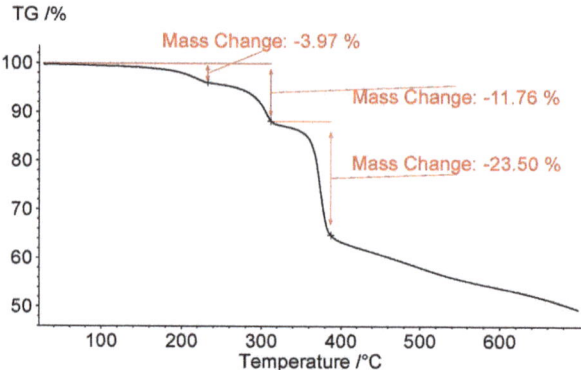

Figure 2. Thermogravimetric analysis (TGA) of **1** in the temperature range 20–700 °C.

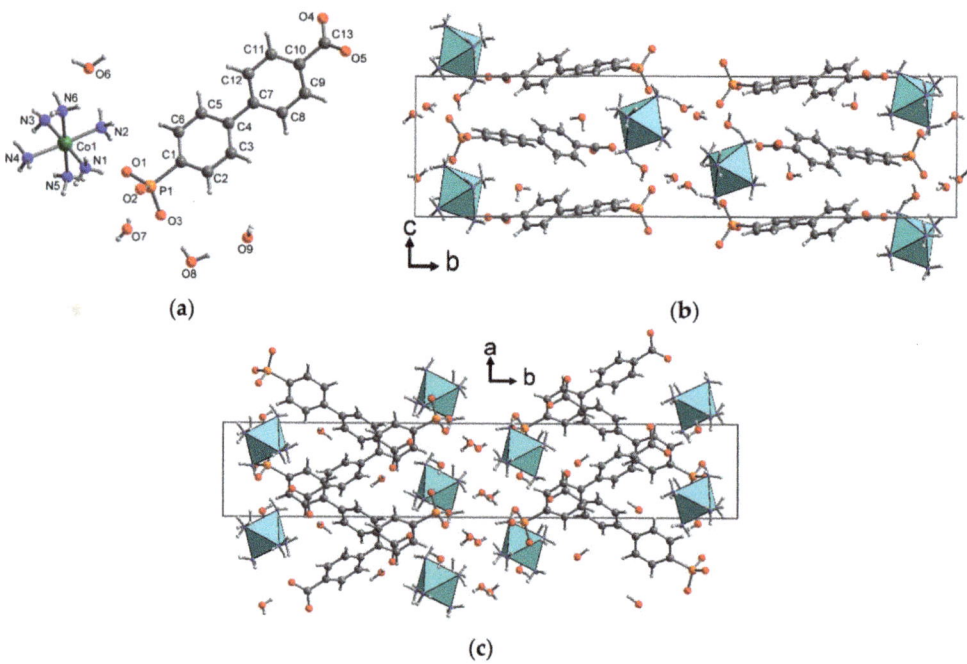

Figure 3. (a) Asymmetric unit of **2**; and (**b,c**) projections of the unit-cell packing on different planes. The [Co(NH$_3$)$_6$]$^{3+}$ cations are illustrated as octahedra; hydrogen bonds are not shown in (**a–c**) for clarity. Selected bond distances and angles are given in Table 3.

Again, the hydrophilic groups, [Co(NH$_3$)$_6$]$^{3+}$, –COO$^-$, and –PO$_3$$^{2-}$, and the crystal water molecules are arranged in slabs (parallel to the *ac* plane) with slabs of the hydrophobic biphenyl part in-between (Figure 3b,c). Such a separation of hydrophilic and hydrophobic parts of molecules is a common packing motif [47–49]. Here the hydrophilic region is organized by hydrogen bonding, the biphenyl rings are arranged by singular N–H···π, O–H···π, C–H···π or van-der-Waals interactions (see Supplementary Information). The dihedral angles within the biphenyl-carboxylate are 26.8 (4)° (ring to ring) and 42.0 (4)° (–CCOO⁻ to aryl ring).

Table 3. Selected bond lengths and angles (Å, °) in 2.

Co1–N1	1.957 (2)	P1–O1	1.5235 (18)
Co1–N2	1.965 (2)	P1–O2	1.5283 (18)
Co1–N3	1.951 (2)	P1–O3	1.5247 (18)
Co1–N4	1.959 (2)	P1–C1	1.820 (2)
Co1–N5	1.976 (2)		
Co1–N6	1.961 (2)		
N5–Co1–N1	87.17 (9)	N3–Co1–N4	89.51 (9)
N4–Co1–N1	91.46 (9)	N3–Co1–N2	90.46 (9)
N4–Co1–N5	90.01 (9)	N6–Co1–N1	91.82 (9)
N2–Co1–N1	88.66 (9)	N6–Co1–N5	178.81 (9)
N2–Co1–N5	92.53 (9)	N6–Co1–N2	88.07 (9)
N2–Co1–N4	177.47 (9)	N6–Co1–N3	90.31 (9)
N3–Co1–N1	177.66 (9)	N6–Co1–N4	89.40 (9)
N3–Co1–N5	90.71 (9)		

The BPPA^{3-} anions and the [Co(NH$_3$)$_6$] octahedra are connected to each other by hydrogen bonding (Table 4, Figure 4). The fully deprotonated phosphonate-carboxylate is solely an H-acceptor for the N–H and water O–H bonds. The carboxylate group is acceptor to O–H from water molecules. The four water molecules are held by hydrogen bonding from N–H and O–H-donors and –COO⁻ and –P(O)$_2$O^{2-} acceptors.

Finally, we note that in both structures, 1 and 2, the H-bonds to the phosphonato groups are so-called charge-assisted hydrogen bonds. The hydrogen bond donor and/or acceptor carry positive and negative ionic charges, respectively, hence are usually stronger and shorter than neutral H-bonds [12,46,47,50–54]. In 1 these are bonds NH$_4^{(+)}$···$^{(-)}$O-P and NH$_4^{(+)}$···(H)O-P, -P-OH···$^{(-)}$O-P (Figure 1c). In 2 these are bonds Co-NH$_3^{(+)}$···$^{(-)}$O-P and HOH···$^{(-)}$O-P (Figure 4).

Table 4. Details of the hydrogen bonding interactions in **2** [a].

D–H···A	D–H [Å]	H···A [Å]	D···A [Å]	D–H···A [°]	Symmetry Transformations
N1–H1A···O8 [i]	0.84 (5)	2.71 (4)	3.327 (3)	132 (4)	i = x, y, z − 1
N1–H1B···O2	0.84 (4)	2.14 (4)	2.976 (3)	173 (3)	
N1–H1C···O8 [ii]	0.87 (4)	2.15 (4)	2.971 (3)	156 (3)	ii = −x + 1, −y, −z + 1
N2–H2A···O4 [vi]	0.86 (4)	2.28 (4)	3.086 (3)	157 (3)	vi = x + 1/2, −y + 1/2, z − 1/2
N2–H2B···O6	0.86 (4)	2.07 (4)	2.909 (4)	163 (3)	
N2–H2C···O1	0.89 (3)	2.00 (3)	2.864 (3)	165 (3)	
N3–H3A···O6	0.82 (5)	2.60 (5)	3.177 (4)	129 (4)	
N3–H3B···O2 [iii]	0.90 (4)	1.90 (4)	2.791 (3)	174 (3)	iii = x + 1, y, z
N3–H3C···O5 [viii]	0.69 (5)	2.56 (4)	3.048 (3)	130 (4)	viii = x + 3/2, −y + 1/2, z − 1/2
N3–H3C···O4 [viii]	0.69 (5)	2.56 (4)	3.180 (3)	152 (4)	viii = x + 3/2, −y + 1/2, z − 1/2
N4–H4A···O9 [iv]	0.90 (4)	2.05 (4)	2.937 (3)	171 (3)	iv = x + 1, y, z − 1
N4–H4B···O7 [v]	0.83 (4)	2.77 (3)	3.270 (4)	120 (3)	v = −x + 1, −y, −z
N4–H4C···O7 [iii]	0.86 (4)	2.00 (4)	2.852 (3)	171 (3)	iii = x + 1, y, z
N5–H5A···O8 [ii]	0.90 (4)	2.19 (4)	3.035 (3)	157 (3)	ii = −x + 1, −y, −z + 1
N5–H5B···O2 [iii]	0.88 (4)	2.63 (4)	3.437 (3)	153 (3)	iii = x + 1, y, z
N5–H5C···O1	0.83 (4)	2.13 (4)	2.939 (3)	162 (3)	
N6–H6A···O3 [i]	0.82 (4)	2.09 (5)	2.904 (3)	175 (3)	i = x, y, z − 1
N6–H6B···O5 [viii]	0.93 (4)	2.26 (4)	3.170 (3)	167 (3)	viii = x + 3/2, −y + 1/2
N6–H6C···O4 [vi]	0.85 (4)	2.13 (4)	2.948 (3)	160 (3)	vi = x + 1/2, −y + 1/2, z − 1/2
O6–H6E···O5 [viii]	0.87	1.82 (1)	2.657 (4)	161 (4)	viii = x + 3/2, −y + 1/2, z − 1/2
O7–H7A···O3 [ii]	0.62 (5)	2.12 (5)	2.740 (3)	171 (6)	ii = −x + 1, −y, −z + 1
O7–H7B···O2	0.85 (4)	1.89 (5)	2.690 (3)	156 (4)	
O8–H8A···O9	0.82 (5)	1.99 (5)	2.802 (3)	175 (4)	
O8–H8B···O3	0.73 (5)	1.97 (5)	2.693 (3)	175 (5)	
O9–H9A···O1 [vii]	0.74 (4)	1.96 (4)	2.704 (3)	176 (4)	vii = x − 1, y, z
O9–H9B···O4 [ix]	0.78 (4)	1.95 (4)	2.728 (3)	173 (4)	ix = x + 1/2, −y + 1/2, z + 1/2

Notes: [a] D = donor, A = acceptor.

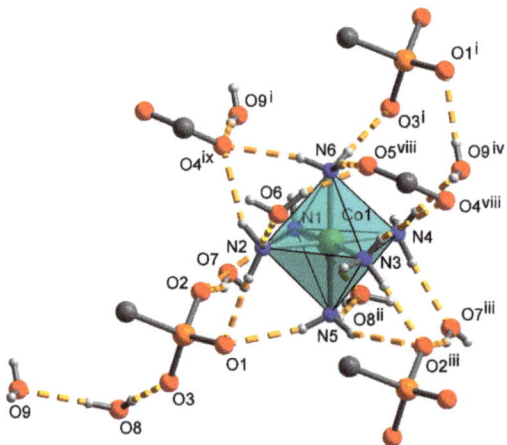

Figure 4. Most relevant H-bonding interactions (orange dashed lines) around a [Co(NH$_3$)$_6$]$^{3+}$ cation in the structure of **2**. Details of the H-bond distances and angles are listed in Table 4. Symmetry transformations: i = x, y, z − 1; ii = −x + 1, −y, −z + 1; iii = x + 1, y, z; iv = x + 1, y, z − 1; v = −x + 1, −y, −z; vi = x + 1/2, −y + 1/2, z − 1/2; vii = x − 1, y, z; viii = x + 3/2, −y + 1/2, z − 1/2; ix = x + 1/2, −y + 1/2, z + 1/2.

The large number of hydrogen-bonds in **2** results in a thermal stability that is higher than that of other supramolecular complexes of $[Co(NH_3)_6]^{3+}$ [55,56]. The thermal stability of $BPPA^{3-}$ is reflected by the TGA measurement (Figure 5). The mass loss in **2** up to (~17%) is due to the evaporation of the four water molecules together with one ammine ligand (17.5%). The next five ammine ligands are removed along with decarboxylation of BPPA in the range from 220 °C to 500 °C, followed by decomposition of the biphenyl system (~42% in total). The remaining mass of ~40% can be assigned to cobalt phosphonate species (~35%). It has been observed that metal phosphonates are stable up to 650 °C and higher [57].

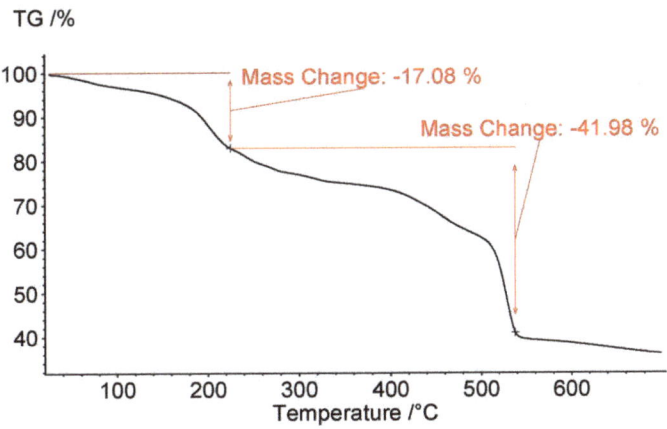

Figure 5. TGA of **2** in the temperature range 20–700 °C.

Comparison of the experimental powder X-ray diffractogram for **2** with the simulation from the single-crystal X-ray dataset (Figure 6) shows that the investigated single crystal was representative of the bulk amount when one takes into account the preferential orientation of the column- or rod-shaped crystals of **2** (Figure S1 in Supplementary Material) on the flat sample holder. Due to the preferred orientation of the rod-shaped crystals on the flat sample holder during the powder X-ray diffraction (PXRD) measurement, and their small quantity, some reflections were not present in the experimental diffraction pattern or their intensity was strongly changed. Such a behavior is discussed in detail in the literature [58–60].

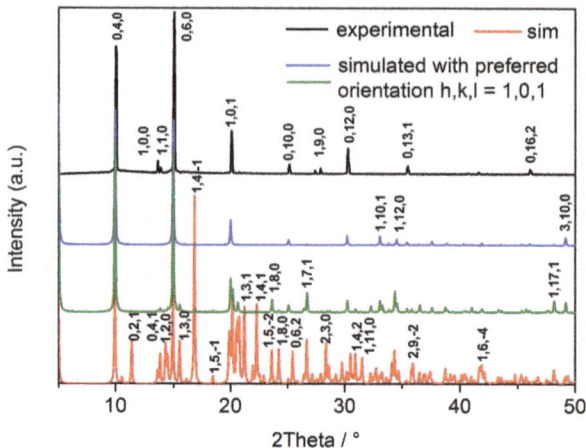

Figure 6. Comparison of the experimental PXRD pattern of **2** (**black**) with the unconstrained simulated pattern from the X-ray data (**red**) and simulated patterns with the preferred orientation of h, k, l = 1, 0, 1 and March–Dollase parameter = 4 (**green**) and = 10 (**blue**). The latter simulations try to take into account the rod-shaped crystal morphology of **2** with their non-random orientation on the flat sample holder. The Miller indices have been assigned to the reflections. Simulations were carried out with *Mercury* [61].

3. Materials and Methods

The chemicals used were obtained from commercial sources. No further purification has been carried out. The ligand has been synthesized starting from 4-biphenyl carboxylic acid in a four-step-synthesis. CHN analysis was performed with a Perkin Elmer CHN 2400. IR-spectra were recorded on a Bruker Tensor 37 IR spectrometer with ATR unit. Thermogravimetric analysis (TGA) was done with a Netzsch TG 209 F3 Tarsus in the range from 20 to 700 °C, equipped with Al-crucible and applying a heating rate of 10 K·min^{-1}. The melting point was determined using a Büchi Melting Point apparatus B540. For powder X-ray diffraction patterns (PXRD), a Bruker D2 Phaser powder diffractometer was used with a flat silicon, low background sample holder, at 30 kV, 10 mA for Cu-Kα radiation (λ = 1.5418 Å), with a scan speed of 0.2 s/step and a step size of 0.02° (2θ). Diffractograms were obtained on flat layer sample holders with a beam scattering protection blade installed, which led to the low relative intensities measured at 2θ < 7°. Details of the synthesis of 4-phosphono-biphenyl-4'-carboxylic acid (H$_2$O$_3$P-(C$_6$H$_4$)$_2$-COOH, H$_3$BPPA) will be given elsewhere [62].

NH$_4$(HO$_3$P-(C$_6$H$_4$)$_2$-COOH)(H$_2$O$_3$P-(C$_6$H$_4$)$_2$-COOH): In a Teflon-lined stainless steel reactor 30 mg (0.108 mmol) of H$_3$BPPA and 8.3 mg (0.108 mmol) of ammonium acetate, NH$_4$(CH$_3$COO), were suspended in 2 mL of doubly de-ionized water. Heating at 180 °C for 24 h and cooling to room temperature within 12 h led to formation of colorless crystals (Figure S1a in Supplementary Information). Yield: 29 mg (91% based on BPPA). Mp > 350 °C. Calc. for $C_{26}H_{25}NO_{10}P_2$ (573.43 g·mol^{-1}): C, 54.46; H, 4.39; N, 2.44%. Found: C, 53.93; H, 4.36; N, 2.02%. FT-IR (ATR) ν/cm^{-1} = 3810 (w), 3196 (w, b) 2999 (w, b), 2859 (w, b), 1672 (m), 1605 (m), 1569 (w), 1446 (m), 1248 (m), 1126 (vs), 1029 (vs), 921 (vs), 824 (vs), 763 (vs), 704 (m), 576 (s), 560 (s) (Figure S2 in Supplementary Material).

[Co(NH$_3$)$_6$](O$_3$P-(C$_6$H$_4$)$_2$-COO)·4H$_2$O: In a glass vial 9.6 mg (0.036 mmol) of [Co(NH$_3$)$_6$]Cl$_3$ and 10 mg (0.036 mmol) of 4-phosphono-biphenyl-4'-carboxylic acid were dissolved in 1.5 mL of 25% aqueous ammonia. The vial was sealed and the crystals were allowed to grow for a period of days at room temperature. After several days deep orange column-shaped crystals had grown (Figure S1b in Supplementary Material). Yield: 17 mg (91%). Mp > 350 °C. Calc. for $C_{13}H_{33}CoN_6O_9P$ (507.34 g·mol^{-1}): C, 30.78; H, 6.56; N, 16.57%. Found: C, 30.92; H, 6.33; N, 17.12%. FT-IR (ATR) ν/cm^{-1} = 3466 (w, b), 3132 (m, b), 3051 (m, b), 2852 (w, b), 1586 (m), 1554 (m), 1528 (m), 1388 (s), 1228 (w), 1138 (m), 1100 (s), 873 (vs), 835 (m), 786 (m), 701 (m), 579 (vs) (Figure S3 in Supplementary Material).

Single Crystal X-ray Structures

Suitable crystals were carefully selected under a polarizing microscope, covered in protective oil and mounted on a 0.05 mm cryo loop. *Data collection.* Bruker Kappa APEX2 CCD X-ray diffractometer with microfocus tube, Mo-Kα radiation (λ = 0.71073 Å), multi-layer mirror system, ω- and θ-scans; data collection with APEX2, cell refinement and data reduction with SAINT [63], experimental absorption correction with SADABS [64]. *Structure analysis and refinement:* All structures were solved by direct methods using SHELXL2014; refinement of **1** was done by full-matrix least squares on F^2 using the SHELX-97 program suite [65], of **2** with OLEX 2 [66, 67]. Non-hydrogen atoms were refined with anisotropic displacement parameters. Hydrogen atoms were positioned geometrically (C–H = 0.95 Å) and refined using riding models (AFIX 43 for aromatic CH with C–H = 0.93 Å and $U_{iso}(H) = 1.2U_{eq}(C)$. In **1** the protic hydrogen atoms (O–H, N–H) were found and freely refined with $U_{iso}(H) = 1.5U_{eq}$ (NH and OH).

In **2**, NH$_3$ hydrogen atoms were found and freely refined. Water hydrogen atoms were also found and refined, except for O6, where they were positioned geometrically (O–H = 0.870 Å) and refined using a riding model (AFIX 6) with $U_{iso}(H) = 1.5U_{eq}(O)$.

Crystal data and details on the structure refinement are given in Table 5. Graphics were drawn with DIAMOND [68]. Analyses on the supramolecular C–H···O, C–H···π and π-π-stacking interactions were done with *PLATON* for Windows [69]. CCDC 1450889 and 1450890 contain the supplementary crystallographic data for this paper. These data can be obtained free of charge via http://www.ccdc.cam.ac.uk/conts/retrieving.html.

Table 5. Crystal data and refinement details.

	1	2
Chemical formula	$C_{26}H_{21}O_{10}P_2 \cdot H_4N$	$C_{13}H_8O_5P \cdot CoH_{18}N_6 \cdot 4(H_2O)$
Mr	573.41	508.36
Crystal system, space group	Triclinic, P$\bar{1}$	Monoclinic, P2$_1$/n
Temperature (K)	150	173
a (Å)	5.9358(5)	7.0193(5)
b (Å)	7.5309(5)	35.454(3)
c (Å)	27.781(2)	9.2797(7)
α (°)	95.413(4)	90
β (°)	90.768(5)	111.921(4)
γ (°)	92.816(5)	90
V (Å^3)	1234.65(16)	2142.4 (3)
Z	2	4
μ (mm^{-1})	0.24	0.93
Crystal size (mm)	0.20 × 0.15 × 0.01	0.33 × 0.3 × 0.15
Absorption correction	Multi-scan, wR2(int) was 0.0937 before and 0.0571 after correction. The Ratio of minimum to maximum transmission is 0.9165. The l/2 correction factor is 0.0000.	Multi-scan, wR2(int) was 0.1520 before and 0.0844 after correction. The Ratio of minimum to maximum transmission is 0.6784. The l/2 correction factor is 0.0015.
T$_{min}$, T$_{max}$	0.683, 0.746	0.507, 0.748
No. of measured, independent and observed reflections	20157, 4937, 3104 [I > 2σ(I)]	89006, 4214, 4150 [I > 2σ(I)]
R$_{int}$	0.066	0.073
(sin θ/λ)max (Å^{-1})	0.617	0.617
R[F^2 > 2σ (F^2)], wR(F^2), S	0.049, 0.121, 1.02	0.038, 0.100, 1.04
No. of reflections	4937	4214
No. of parameters	379	374
Δρ$_{max}$, Δρ$_{min}$ (e·Å^{-3})	0.30, –0.34	1.10, –0.85

4. Conclusions

We investigated the hydrogen-bonding potential of the new organo-phosphonate linker, H$_3$BPPA, in its (partially) deprotonated forms H$_2$BPPA$^-$ and BPPA^{3-}. As expected, the protonated phosphonic acid and deprotonated phosphonate group enters into H-bonds with all of its P–O–H donors and P–O$^-$ acceptors. Remarkably and unexpectedly, an almost symmetric resonance-assisted hydrogen bond (RAHB) bond was found between the formal H$_2$BPPA$^-$ anion and the formally neutral H$_3$BPPA molecule in **1** to give the overall anion [H$_2$BPPA$\cdots$H$\cdots$H$_2$BPPA]$^-$.

Supplementary Materials: Supplementary Materials: The following are available online at http://www.mdpi.com/2073-4352/6/3/22/s001. Figure S1: Photographs of crystals of (a) $NH_4(HO_3P\text{-}(C_6H_4)_2\text{-}COOH)(H_2O_3P\text{-}(C_6H_4)_2\text{-}COOH)$, **1** and (b) $[Co(NH_3)_6](O_3P\text{-}(C_6H_4)_2\text{-}COO)\cdot 4H_2O$, **2** taken with a light microscope; Figure S2: FT-IR (ATR) spectrum of $NH_4(HO_3P\text{-}(C_6H_4)_2\text{-}COOH)(H_2O_3P\text{-}(C_6H_4)_2\text{-}COOH)$, **1**; Figure S3: FT-IR (ATR) spectrum of $[Co(NH_3)_6](O_3P\text{-}(C_6H_4)_2\text{-}COO)\cdot 4H_2O$, **2**. Figure S4: Comparison of the experimental PXRD pattern of **1** (black) with the simulated pattern from the X-ray data (red). Packing analyses.

Acknowledgments: Acknowledgments: The work was supported by the German Science Foundation (DFG) through grant Ja466/25-1.

Author Contributions: Author Contributions: Christian Heering designed the experiments, synthesized the ligand and compound **2**. Bahareh Nateghi carried out the reaction leading to **1**. Data analysis and measurements were performed by Christian Heering, while Christoph Janiak and Christian Heering wrote the manuscript.

Conflicts of Interest: Conflicts of Interest: The authors declare no conflict of interest. The founding sponsors had no role in the design of the study; in the collection, analyses, or interpretation of data; in the writing of the manuscript, and in the decision to publish the results.

References

1. Clearfield, A.; Demadis, K. *Metal Phosphonate Chemistry: From Synthesis to Applications*; Royal Society of Chemistry: Oxford, UK, 2012; pp. 45–128.
2. Deng, M.; Liu, X.; Zheng, Q.; Chen, Z.; Fang, C.; Yue, B.; He, H. Controllable preparation and structures of two zinc phosphonocarboxylate frameworks with MER and RHO zeolitic topologies. *CrystEngComm* **2013**, *15*, 7056–7061.
3. Gagnon, K.J.; Perry, H.P.; Clearfield, A. Conventional and unconventional metal-organic frameworks based on phosphonate ligands: MOFs and UMOFs. *Chem. Rev.* **2012**, *112*, 1034–1054.
4. Janiak, C. Engineering coordination polymers towards applications. *Dalton Trans.* **2003**, 2781–2804.
5. Taylor, J.M.; Vaidhyanathan, R.; Iremonger, S.S.; Shimizu, G.K.H. Enhancing water stability of metal-organic frameworks via phosphonate monoester linkers. *J. Am. Chem. Soc.* **2012**, *134*, 14338–14340.
6. Patterson, A.R.; Schmitt, W.; Evans, R.C. Lighting-Up Two-dimensional lanthanide phosphonates: Tunable structure-property relationships towards visible and near-infrared emitters. *J. Phys. Chem. C* **2014**, *118*, 10291–10301.
7. Jiménez-García, L.; Kaltbeitzel, A.; Pisula, W.; Gutmann, J.S.; Klapper, M.; Müllen, K. Phosphonated hexaphenylbenzene: A crystalline proton conductor. *Angew. Chem. Int. Ed.* **2009**, *48*, 9951–9953.
8. Corma, A.; García, H.; Llabrés i Xamena, F.X. Engineering metal organic frameworks for heterogeneous catalysis. *Chem. Rev.* **2010**, *110*, 4606–4655.
9. Shimizu, G.K.H.; Vaidhyanathan, R.; Taylor, J.M. Phosphonate and sulfonate metal organic frameworks. *Chem. Soc. Rev.* **2009**, *38*, 1430–1449.

10. Rojo, T.; Mesa, J.L.; Lago, J.; Bazan, B.; Pizarro, J.L.; Arriortua, M.I. Organically templated open-framework phosphite. *J. Mater. Chem.* **2009**, *19*, 3793–3818.
11. Hou, S.-Z.; Cao, D.-K.; Liu, X.-G.; Li, Y.-Z.; Zheng, L.-M. Metal phosphonates based on (4-carboxypiperidyl)-*N*-methylene-phosphonate: In situ ligand cleavage and metamagnetism in $Co_3(O_3PCH_2\text{-}NHC_5H_9\text{-}COO)_2(O_3PCH_2\text{-}NC_5H_{10})(H_2O)$. *Dalton Trans.* **2009**, *15*, 2746–2750.
12. Habib, H.A.; Gil-Hernández, B.; Abu-Shandi, K.; Sanchiz, J.; Janiak, C. Iron, copper and zinc ammonium-1-hydroxyalkylidene-diphosphonates with zero-, one- and two-dimensional covalent metal-ligand structures extended into three-dimensional supramolecular networks by charge-assisted hydrogen-bonding. *Polyhedron* **2010**, *29*, 2537–2545.
13. Abu-Shandi, K.; Winkler, H.; Janiak, C. Structure and mössbauer study of the first mixed-valence iron diphosphonate. *Z. Anorg. Allg. Chem.* **2006**, *632*, 629–633.
14. Zhao, X.; Bell, J.G.; Tang, S.-F.; Li, L.; Thomas, K.M. Kinetic molecular sieving, thermodynamic and structural aspects of gas/vapor sorption on metal organic framework $[Ni_{1.5}(4,4'\text{-bipyridine})_{1.5}(H_3L)\text{-}(H_2O)_3][H_2O]_7$ where H_6L = 2,4,6-trimethylbenzene-1,3,5-triyl tris(methylene)triphosphonic acid. *J. Mater. Chem. A* **2016**, *4*, 1353–1365.
15. Zhai, F.; Zheng, Q.; Chen, Z.; Ling, Y.; Liu, X.; Weng, L.; Zhou, Y. Crystal transformation synthesis of a highly stable phosphonate MOF for selective adsorption of CO_2. *CrystEngComm* **2013**, *15*, 2040–2043.
16. Kinnibrugh, T.L.; Ayi, A.A.; Bakhmutov, V.I.; Zon, J.; Clearfield, A. Reversible dehydration behavior reveals coordinatively unsaturated metal sites in microporous aluminum phosphonates. *Cryst. Growth Des.* **2013**, *13*, 2973–2981.
17. Menelaou, M.; Dakanali, M.; Raptopoulou, C.P.; Drouza, C.; Lalioti, N.; Salifoglou, A. pH-Specific synthetic chemistry, and spectroscopic, structural, electrochemical and magnetic susceptibility studies in binary Ni(II)-(carboxy)phosphonate systems. *Polyhedron* **2009**, *28*, 3331–3339.
18. Ling, Y.; Deng, M.; Chen, Z.; Xia, B.; Liu, X.; Yang, Y.; Zhou, Y.; Weng, L. Enhancing CO_2 adsorption of a Zn-phosphonocarboxylate framework by pore space partitions. *Chem. Commun.* **2013**, *49*, 78–80.
19. Breeze, B.A.; Shanmugam, M.; Tuna, F.; Winpenny, R.E.P. A series of nickel phosphonate-carboxylate cages. *Chem. Commun.* **2007**, *48*, 5185–5187.
20. Rueff, J.-M.; Perez, O.; Leclaire, A.; Couthon-Gourvès, H.; Jaffrès, P.-A. Lead(II) Hybrid Materials from 3- or 4-Phosphonobenzoic Acid. *Eur. J. Inorg. Chem.* **2009**, 4870–4876.
21. Liao, T.-B.; Ling, Y.; Chen, Z.-X.; Zhou, Y.-M.; Weng, L.-H. A rutile-type porous zinc(II)-phosphonocarboxylate framework: Local proton transfer and size-selected catalysis. *Chem. Commun.* **2010**, *46*, 1100–1102.
22. Svoboda, J.; Zima, V.; Beneš, L.; Melánová, K.; Trchová, M.; Vlček, M. New barium 4-carboxyphenylphosphonates: Synthesis, characterization and interconversions. *Solid State Sci.* **2008**, *10*, 1533–1542.

23. Pütz, A.-M.; Carrella, L.M.; Rentschler, E. A distorted honeycomb motive in divalent transition metal compounds based on 4-Phosphonbenzoic acid and exchange coupled Co(II) and Cu(II): Synthesis, structural description and magnetic properties. *Dalton Trans.* **2013**, *42*, 16194–16199.
24. Rueff, J.-M.; Barrier, N.; Boudin, S.; Dorcet, V.; Caignaert, V.; Boullay, P.; Hix, G.B.; Jaffrès, P.-A. Remarkable thermal stability of Eu(4-phosphonobenzoate): Structure investigations and luminescence properties. *Dalton Trans.* **2009**, *47*, 10614–10620.
25. Li, J.-T.; Guo, L.-R.; Shen, Y.; Zheng, L.-M. LiF-assisted crystallization of zinc 4-carboxyphenylphosphonates with pillared layered structures. *CrystEngComm* **2009**, *11*, 1674–1678.
26. Rueff, J.-M.; Perez, O.; Caignaert, V.; Hix, G.; Berchel, M.; Quentel, F.; Jaffrès, P.-A. Silver-based hybrid materials from meta- or para-phosphonobenzoic acid: Influence of the topology on silver release in water. *Inorg. Chem.* **2015**, *54*, 2152–2159.
27. Zima, V.; Svoboda, J.; Beneš, L.; Melánová, K.; Trchová, M.; Dybal, J. Synthesis and characterization of new strontium 4-carboxyphenylphosphonates. *J. Solid State Chem.* **2007**, *180*, 929–939.
28. Adelani, P.O.; Albrecht-Schmitt, T.E. Comparison of thorium (IV) and uranium (VI) carboxyphosphonates. *Inorg. Chem.* **2010**, *49*, 5701–5705.
29. Melánová, K.; Klevcov, J.; Beneš, L.; Svoboda, J.; Zima, V. New layered functionalized titanium (IV) phenylphosphonates. *J. Phys. Chem. Solids* **2012**, *73*, 1452–1455.
30. Li, J.-T.; Cao, D.-K.; Akutagawa, T.; Zheng, L.M. $Zn_3(4\text{-}OOCC_6H_4PO_3)_2$: A polar metal phosphonate with pillared layered structure showing SHG-activity and large dielectric anisotropy. *Dalton Trans.* **2010**, *39*, 8606–8608.
31. Chen, Z.; Zhou, Y.; Weng, L.; Zhao, D. Mixed-solvothermal syntheses and structures of six new zinc phosphonocarboxylates with zeolite-type and pillar-layered frameworks. *Cryst. Growth Des.* **2008**, *8*, 4045–4053.
32. Merkushev, E.B.; Shvartzberg, M.S. *Organoiodine Compounds and Syntheses Based on Them*; Tomsk Gos. Pedagogicheskii Institut: Tomsk, Soviet Union, 1982; pp. 2598–2601.
33. Etter, M.C. Encoding and decoding hydrogen-bond patterns of organic compounds. *Acc. Chem. Res.* **1990**, *23*, 120–126.
34. Janiak, C. A critical account on π–π stacking in metal complexes with aromatic nitrogen-containing ligands. *J. Chem. Soc. Dalton Trans.* **2000**, *21*, 3885–3896.
35. Nishio, M. CH/π hydrogen bonds in crystals. *CrystEngComm* **2004**, *6*, 130–158.
36. Lin, Z.; Lei, X.-Q.; Bai, S.-D.; Ng, S.W. Ammonium benzenephosphonate. *Acta Crystallogr. Sect. E.—Struct Rep. Online* **2008**, *64*, o1607.
37. Gilli, G.; Gilli, P. *On Noncovalent Interactions in Crystals: Supramolecular Chemistry: From Molecules to Nanomaterials*; Steed, J., Gale, P.A., Eds.; Wiley: Chichester, UK, 2012; Volume 6, pp. 2829–2868.
38. Góra, R.W.; Maj, M.; Grabowski, S.J. Resonance-assisted hydrogen bonds revisited. Resonance stabilization *vs.* charge delocalization. *Phys. Chem. Chem. Phys.* **2013**, *15*, 2514–2522.

39. Sanz, P.; Mó, O.; Yáñez, M.; Elguero, J. Resonance-assisted hydrogen bonds: A critical examination. structure and stability of the enols of β-diketones and β-enaminones. *J. Phys. Chem. A* **2007**, *111*, 3585–3591.
40. Gilli, P.; Bertolasi, V.; Pretto, L.; Ferretti, V.; Gilli, G. Covalent *versus* electrostatic nature of the strong hydrogen bond: Discrimination among single, double, and asymmetric single-well hydrogen bonds by variable-temperature X-ray crystallographic methods in β-diketone enol RAHB systems. *J. Am. Chem. Soc.* **2004**, *126*, 3845–3855.
41. Gilli, P.; Bertolasi, V.; Ferretti, V.; Gilli, G. Evidence for intramolecular N–H···O resonance-assisted hydrogen bonding in enaminones and related heterodienes. A combined crystal-structural, IR and NMR spectroscopic, and quantum-mechanical investigation. *J. Am. Chem. Soc.* **2000**, *122*, 10405–10417.
42. Feng, C.; Liang, M.; Jiang, J.; Huang, J.; Liu, H. Synergistic effect of a Novel triazine Charring Agent and ammonium polyphosphate on the flame retardant properties of Halogen-Free Flame Retardant Polypropylene composites. *Thermochim. Acta* **2016**.
43. Wilk, M.; Janczak, J.; Videnova-Adrabinska, V. Hexaaquacobalt(II) bis[hydrogen bis(4-carboxyphenyl-phosphonate)] dihydrate. *Acta Crystallogr. Sect. C—Cryst. Struct. Commun.* **2011**, *67*, 9–12.
44. Wang, X.-Y.; Justice, R.; Sevov, S.C. Hydrogen-bonded metal-complex sulfonate (MCS) inclusion compounds: Effect of the guest molecule on the host framework. *Inorg. Chem.* **2007**, *46*, 4626–4631.
45. Morral, F.R. Alfred werner and cobalt complexes. *Adv. Chem.* **2009**, *6*, 70–77.
46. Reddy, D.S.; Duncan, S.; Shimizu, G.K.H. A Family of supramolecular inclusion solids based upon second-sphere interactions. *Angew. Chem. Int. Ed.* **2003**, *42*, 1360–1364.
47. Dorn, T.; Chamayou, A.-C.; Janiak, C. Hydrophilic interior between hydrophobic regions in inverse bilayer structures of cation–1,1'-binaphthalene- 2,2'-diyl phosphate salts. *New J. Chem.* **2006**, *30*, 156–167.
48. Maclaren, J.K.; Sanchiz, J.; Gili, P.; Janiak, C. Hydrophobic-exterior layer structures and magnetic properties of trinuclear copper complexes with chiral amino alcoholate ligands. *New J. Chem.* **2012**, *36*, 1596–1609.
49. Enamullah, M.; Vasylyeva, V.; Janiak, C. Chirality at metal and helical ligand folding in optical isomers of chiral bis(naphthaldiminato)nickel(II) complexes. *Inorg. Chim. Acta* **2013**, *408*, 109–119.
50. Ward, M.D. Design of crystalline molecular networks with charge-assisted hydrogen bonds. *Chem. Commun.* **2005**, *47*, 5838–5842.
51. Gil-Hernández, B.; Maclaren, J.K.; Höppe, H.A.; Pasan, J.; Sanchiz, J.; Janiak, C. Homochiral lanthanoid(III) mesoxalate metal-organic frameworks: Synthesis, crystal growth, chirality, magnetic and luminescent properties. *CrystEngComm* **2012**, *14*, 2635–2644.
52. Maclaren, J.K.; Janiak, C. Amino-acid based coordination polymers. *Inorg. Chim. Acta* **2012**, *389*, 183–190.

53. Chamayou, A.-C.; Neelakantan, M.A.; Thalamuthu, S.; Janiak, C. The first vitamin B6 zinc complex, pyridoxinato-zinc acetate: A 1D coordination polymer with polar packing through strong inter-chain hydrogen bonding. *Inorg. Chim. Acta* **2011**, *365*, 447–450.
54. Drašković, B.M.; Bogdanović, G.A.; Neelakantan, M.A.; Chamayou, A.-C.; Thalamuthu, S.; Avadhut, Y.S.; Schmedt auf der Günne, J.; Banerjee, S.; Janiak, C. N-o-Vanillylidene-L-histidine: Experimental charge density analysis of a double zwitterionic amino acid Schiff-base compound. *Cryst. Growth Des.* **2010**, *10*, 1665–1676.
55. Collins, L.W.; Wendlandt, W.W.; Gibson, E.K. The thermal dissociation of the [Co(NH$_3$)$_5$Cl]Cl$_2$ and [Co(NH$_3$)$_5$Br]Br$_2$ complexes in vacuo. *Thermochim. Acta* **1974**, *8*, 303–306.
56. Saito, A. Thermal decomposition of the complexes [Co(NH$_3$)$_6$][Nd(SO$_4$)$_3$]·nH$_2$O. *Thermochim. Acta* **1986**, *102*, 373–386.
57. Dines, M.B.; DiGiacomo, P.M. Derivatized lamellar phosphates and phosphonates of M(IV) ions. *Inorg. Chem.* **1981**, *20*, 92–97.
58. Mittemeijer, E.J.; Welzel, U. *Modern Diffraction Methods*; Wiley: Weinheim, Germany, 2013.
59. Pecharsky, V.; Zavalij, P. *Fundamentals of Powder Diffraction and Structural Characterization of Materials*, 2nd ed.; Springer: New York, NY, USA, 2008.
60. Dollase, W.A. Correction of intensities for preferred orientation in powder diffractometry: Application of the March model. *J. Appl. Cryst.* **1986**, *19*, 267–272.
61. Macrae, C.F.; Edgington, P.R.; McCabe, P.; Pidcock, E.; Shields, G.P.; Taylor, R.; Towler, M.; van de Streek, J. Mercury: Visualization and analysis of crystal structures. *J. Appl. Cryst.* **2006**, *39*, 453–457.
62. Heering, C.; Francis, B.; Nateghi, B.; Makhloufi, G.; Janiak, C. Syntheses, structures and properties of group 12 element (Zn, Cd, Hg) coordination polymers with a phosphonate-biphenyl-carboxylate linker. *CrystEngComm* **2016**. (to be submitted).
63. APEX2. *SAINT, Data Reduction and Frame Integration Program for the CCD Area-Detector System, Bruker Analytical X-ray Systems*; Data Collection program for the CCD Area-Detector System: Madison, WI, USA, 1997–2006.
64. Sheldrick, G. *SADABS: Area-Detector Absorption Correction*; University of Göttingen: Göttingen, Germany, 1996.
65. Hübschle, C.B.; Sheldrick, G.M.; Dittrich, B. ShelXle: A graphical user interface for SHELXL. *J. Appl. Cryst.* **2011**, *44*, 1281–1284.
66. Bourhis, L.J.; Dolomanov, O.V.; Gildea, R.J.; Howard, J.A.K.; Puschmann, H. The anatomy of a comprehensive constrained, restrained refinement program for the modern computing environment—Olex2 dissected. *Acta Crystallogr. Sect. A* **2015**, *71*, 59–75.
67. Dolomanov, O.V.; Bourhis, L.J.; Gildea, R.J.; Howard, J.A.K.; Puschmann, H. OLEX2: A complete structure solution, refinement and analysis program. *J. Appl. Cryst.* **2009**, *42*, 339–341.
68. Brandenburg, K. *DIAMOND*; version 3.2; Crystal and Molecular Structure Visualization; Crystal Impact—K. Brandenburg & H. Putz Gbr: Bonn, Germany, 2009.
69. Spek, A.L. Structure validation in chemical crystallography. *Acta Crystallogr. Sect. D—Biol. Crystallogr.* **2009**, *65*, 148–155.

Structural Elucidation of α-Cyclodextrin-Succinic Acid *Pseudo* Dodecahydrate: Expanding the Packing Types of α-Cyclodextrin Inclusion Complexes

Sofiane Saouane and Francesca P. A. Fabbiani

Abstract: This paper reports a new packing type of α-cyclodextrin inclusion complexes, obtained here with succinic acid under low-temperature crystallization conditions. The structure of the 1:1 complex is characterized by heavy disorder of the guest, the solvent, and part of the host. The crystal packing belongs to the known channel-type structure; the basic structural unit is composed of cyclodextrin trimers, as opposed to the known isolated molecular or dimeric constructs, packed along the *c*-axis. Each trimer is made of crystallographically independent molecules assembled in a stacked vase-like cluster. A multi-temperature single-crystal X-ray diffraction analysis reveals the presence of dynamic disorder.

Reprinted from *Crystals*. Cite as: Saouane, S.; Fabbiani, F.P.A. Structural Elucidation of α-Cyclodextrin-Succinic Acid *Pseudo* Dodecahydrate: Expanding the Packing Types of α-Cyclodextrin Inclusion Complexes. *Crystals* **2016**, *6*, 2.

1. Introduction

Native cyclodextrins (CDs) are cyclic oligosaccharides with low chemical toxicity [1,2]. They are formed by several α-D-glucopyranose units, linked by α(1→4) *O*-glycosidic bonds. A CD molecule has a hollow truncated cone-like shape where the primary hydroxy groups sit on the narrow side and the secondary ones on the wide side of the truncated cone. The architecture of CD molecules favors the partition of the structure into an outer surface and an inner cavity, which are hydrophilic and hydrophobic in character, respectively [3]. This molecular feature boosted the interest of using CDs as "molecular cages" [4] in both the solid and solution states, particularly in the pharmaceutical field where CDs are known to improve the aqueous solubility, and thus bioavailability, of poorly soluble active pharmaceutical ingredients [3]. The formation of CD inclusion complexes in the solid state, particularly with organic molecules, has been pioneered by Saenger [5–7] and Harata [8]. Three packing types have been described for both CD hydrates and inclusion complexes: two of these belong to the cage type and are known as herringbone- and brick-type packing, and the third one is the channel type [7]. The packing preference of CD molecules for one packing type is closely related to the size and shape of the guest molecule [7].

The work presented here is part of ongoing investigations of cyclodextrin inclusion complex and hydrate formation at both ambient- and high-pressure crystallization conditions using water as crystallization medium [9]. α-CD (Figure 1), the smallest natural cyclic oligosaccharide with six sugar units, was chosen for studying inclusion complex formation with small molecules (here defined as molecules with a molecular weight <500 Daltons), particularly to compare and contrast the effects of non-ambient (low-temperature and high-pressure) crystallization conditions on complex formation. A Cambridge Structural Database [10] (CSD) search (the CSD, V 5.36 including updates to November 2014 was searched for structures with 3-D coordinates) indicates that α-CD crystallizes in the channel-packing type in 52% of the total 96 hits, followed by the brick-type in 26% and the herringbone-type in 22% of the structures. In the channel-packing type favored by α-CD, guest molecules are inserted lengthwise into the host's cavity, forming infinite columns.

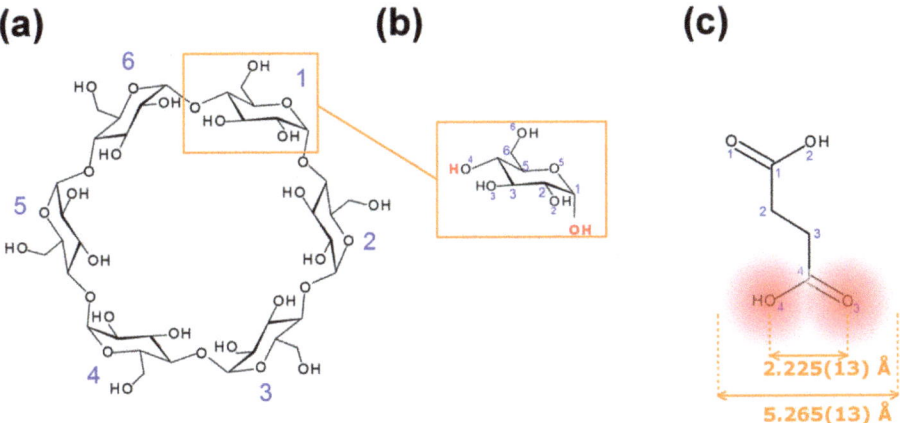

Figure 1. Diagrams and numbering schemes of: (a) α-cyclodextrins; (b) α-D-glucopyranose; (c) succinic acid (SA): the mean distance O(3)–O(4) was computed from 72 structures of SA in the Cambridge Structural Database (CSD), the width of SA is the (mean distance O(3)–O(4) plus twice the van der Waals radius of an oxygen atom.

Succinic acid (SA, Figure 1), an aliphatic dicarboxylic acid, is essential in aerobic cellular metabolism by intervening in the citric acid cycle, a metabolic pathway for the regeneration of adenosine triphosphate (ATP), which is the main energy source of most cellular functions [11]. SA is a FDA-Generally Recognized As SAFE (GRAS) substance also used in the pharmaceutical industry for the preparation of succinate ester derivatives of active pharmaceutical ingredients. The structure of a β-CD·SA inclusion complex (CSD refcode KIJSEC) has been previously obtained while investigating the enhancement of succinic anhydride's reactivity using β-CD as molecular cages in aqueous solutions [12]. We found that this complex can easily be obtained with SA instead of succinic anhydride; the large cavity size of β-CD, which is 6.0–6.5 Å in diameter [13], can easily accommodate SA and the crystal structure of the complex shows several intermolecular interactions between host, guest and solvent molecules [12]. We hypothesize that with a width (5.265(13) Å, see Figure 1) commensurable with the cavity diameter of α-CD (4.7–5.3 Å) [13], SA could in principle form a crystalline complex with α-CD. The literature shows that similar linear compounds form inclusion complexes with α-CD (see for example CSD refcodes BUPDEV [14], CDKABA [15] and XIGBOE [16]).

2. Results and Discussion

Low-temperature crystallization of an equimolar mixture of α-CD with SA in water led to the formation of hexagonal prism shaped crystals. A first polarized microscopy analysis of the crystals through the hexagonal face indicated an absence of light extinction characteristic of uniaxial crystals. This microscopic analysis endorsed the choice of the unit cell from X-ray diffraction, with successful indexing of the reflections using a rhombohedral unit cell. The reflections could also be indexed using a lower symmetry monoclinic unit cell, through the transformation matrix $(-1/3\ -2/3\ -2/3,\ 1\ 0\ 0,\ -1/3\ -2/3\ 1/3)$. The choice of a trigonal crystal system dictates that both host and guest molecules sit on a 3-fold rotation axis going through α-CD cavities. Heavy disorder of the guest, evident from the electron density maps, could be modelled in the higher symmetry space group and this was finally chosen for refinement.

The α-CD·SA 1:1 inclusion complex crystallizes in space group $R32$ with $3 \times 1/3$ α-CD molecules in the asymmetric unit, here named A, B and C (Figure 2). Each α-CD sits on the 3-fold rotation axis and encloses one SA molecule disordered over six positions; crystallographically, this is modelled by having two SA molecules per cavity each with $1/6$ site occupancy. α-CD C is further disordered over two positions with $2/3$ (C) and $1/3$ (C') site occupancies. In addition the unit cell contains *ca.* 12 water molecules making the complex a *pseudo* dodecahydrate.

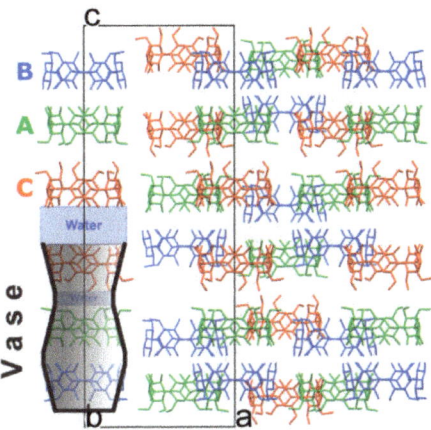

Figure 2. Crystal packing of α-CD molecules viewed along the *b*-axis. H atoms, disorder of α-CD C, SA and water molecules have been omitted for clarity. Symmetry-equivalent molecules are color coded.

In the solid state α-CD molecules have always been known, to the best of our knowledge, to pack as either distinct entities or dimers [6,8]. In this work, a new building block is observed, namely α-CD trimers packed along the *c*-axis. Each trimer is made of crystallographically independent molecules assembled in a stacked vase-like cluster (Figure 2). The stacked trimer motif is not unknown for γ-CD molecules (see for example CSD refcodes FEJFIJ, FEJFOP, NUNRIX, SIBJAO, SIBJES) [17–19]; however the concept of vase-like packing has not been previously reported. It has also been reported that β-CD molecules crystallize as trimers (CSD refcodes RIPKIL, OCIGAK) [20,21] or tetramers [22,23]. The structure of the title compound appears to be very similar to that of α-CD·hexa-ethylene glycol reported by Harada *et al.* (CSD refcode LOJTUZ, no 3D coordinates deposited in the CSD) [24]; however, this publication neither describe the distinctive trimer arrangement in detail, nor identifies a new packing type. In contrast, in a conference abstract, Caira *et al.* have recently reported that the α-CD-lipoic acid system "crystallizes in the trigonal system, space group *R*32, with three independent CD molecules in the asymmetric unit and is not isostructural with any known CD complex" [25]. These observations make the reported packing type rare but not unknown.

The vase-like cluster in the α-CD·SA inclusion complex is formed by two sub-dimers: (1) a head-to-head dimer, which is stabilized by H-bonds between secondary hydroxy groups of α-CDs A and B; and (2) a tail-to-tail dimer between α-CDs A and C interconnected through a cluster of water molecules. Analysis of planes formed by glycosidic O-atoms shows that the interplanar distance in the head-to-head dimer is 7.022 Å, compared to a mean distance of 9.060 Å for the tail-to-tail dimer (Figure 3). Successive vase-like clusters, stacked via two-fold rotation symmetry, are separated by a layer of water molecules forming a complex H-bonded network that holds two vase-like structures together.

The three α-CD molecules forming the vase-like cluster exhibit an almost ideal cylindrical shape and form infinite linear channels extending along the *c*-axis of the unit cell in a honeycomb-type arrangement. A honeycomb or *quasi*-hexagonal pattern ensures effective close packing [19,26], and for CDs it was reported for the first time by Saenger in 1980, while describing the general arrangement of CDs in crystals, as *"hexagonal packing of stacks"* [5]. A careful analysis of the reported channel-type structures of α-CD inclusion complexes, including hydrates, in the CSD shows that honeycomb packing is observed in 27 out of 50 structures (CSD refcodes of these structures are reported in the Supplementary Information), though this particular structural feature has been rarely reported explicitly. Instead, authors usually report the type of α-CD dimer arrangement, head-to-head or head-to-tail, in channel-type structures.

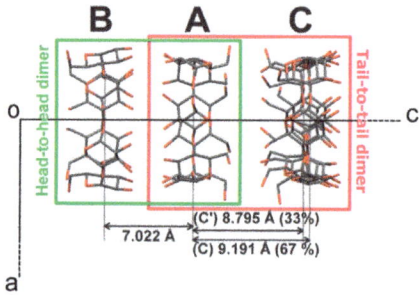

Figure 3. Projection of α-CD·SA structure along the *b*-axis. The crystallographically-independent CD molecules are named A, B and C, see main text for details. H atoms, SA and water molecules have been omitted for clarity. The distances refer to gap in Å between the least-square planes formed by all O(4) atoms involved in the glycosidic bonds.

One of the features of CDs is the flexibility of the primary-hydroxy groups. The conformation of the hydroxy group is defined by the value of the O(5)–C(5)–C(6)–O(6) torsion angle (Figure 1): a preferred (−)-gauche conformation in which the hydroxy group is facing the exterior of the cavity, and a less preferred (+)-gauche conformation where the hydroxy group is facing the inner cavity [27]. All but one primary hydroxy groups in α-CD exhibit the (−)-gauche conformation and are directed away from the cavity. The rotation of the primary hydroxy group to the less favored (+)-gauche conformation in molecule C can be explained by the formation of the short H-bonds [2.52(4) and 2.79(1) Å] with a disordered water molecule (named O(10)_91 in the structure). A CSD search (V 5.36) shows that (−)-gauche conformers are not unknown for α-CD inclusion complexes and indicate the absence of direct

H-bonds between the guest and the host molecules, which could explain the high degree of rotational disorder of the guest observed here. In contrast, in the structure of β-CD·SA which contains one full guest and host molecules in the asymmetric unit, SA is H-bonded to two primary CD hydroxy groups, which are facing the inner cavity, through one water molecule each. Interestingly, the SA molecule in the β-CD·SA inclusion complex does not exhibit disorder.

A search in the CSD (V 5.36 including updates to November 2014) based on the C(1)–C(2)–C(3)–C(4) torsion angle of the guest (Figure 1) shows that SA molecules exhibit two conformations in the solid state, trans and gauche, with an incidence of 90 and 10%, respectively, out of a total of 165 structures. Lisnyak *et al.* reported that the trans conformer is energetically more favorable than the gauche one with an energy difference of 31.4 kJ·mol^{-1} [12]. In β-CD·SA (CSD refcode KIJSEC [12]), SA molecules lie almost equatorially in the β-CD cavity and exhibit the gauche conformer. In contrast, the electron density maps of α-CD·SA point to the trans conformation with SA molecules extended along the c-axis. The difference in conformation may be directly related to the cavity size differences between α- and β-CD, with an axial inclusion mode being more commensurable with the size of the α-CD cavity. Although as a result of this one might expect a tighter fit of SA inside α-CD, the guest is actually found be disordered. This is in line with the almost ideal cylindrical shape of the host, yielding a more evenly distributed cavity space, and the absence of direct H-bonds between the guest and host molecules. In α-CD·SA the axial inclusion mode is described by the almost orthogonal angular differences between the planes formed by the glycosidic O-atoms of the ordered CD molecules A and B and to the planes formed by the SA molecules inside the respective cavities (SA planes were calculated using C(1)–C(2)–C(3)–C(4)). For α-CD C, similar calculations show that the same planes are 13 to 18° off from being orthogonal. The different inclusion modes of SA inside α- and β-CD are also associated with different crystal packing: the trans SA molecules in α-CD·SA form a H-bonded linear chain (Figure 4) that is associated with the formation of a channel-type packing. On the other hand, in the β-CD·SA complex gauche SA molecules are "buried" in the cavities and the herringbone-type packing observed in the hydrated β-CD and small-guest molecules structures is conserved. This comparative analysis based on size, shape and functionality of the guest molecule supports the work by Saenger and Steiner [7].

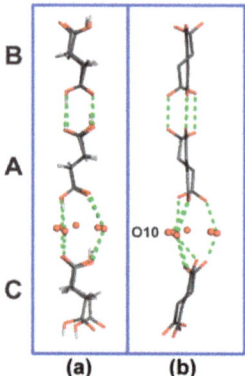

Figure 4. H-bonded motif formed by SA molecules in the asymmetric unit viewed (**a**) along the *a*-axis and (**b**) along the *b*-axis. O–O contacts are represented by dashed green lines. H-atoms (in b), α-CD and water molecules have been omitted for clarity. SA molecules are contained in the cavities of the crystallographically-independent CD molecules named A, B and C, see main text for details.

2.1. Insight into Disorder

In an attempt to gain a better insight into the disorder of α-CD·SA inclusion complex and to investigate whether a temperature-dependent phase transition takes place, we extended our initial work to a multi-temperature investigation covering the 100–270 K temperature range, in an ascending temperature ramp.

The crystallographic data of all structures, summarized in Table S1, show small differences in the lattice constants across the temperature range. The lattice parameters were normalized to the unit cell values of the 100 K crystal structure and are visually represented in Figure 5. Approximately uniform thermal expansion up to 180 K is observed for both the *c*- and *a*-axes. For the unit cell volume, the rate of thermal expansion is approximately linear up to this temperature. From 180 to 240 K, the complex overall expands but the *c*-axis shows negative thermal expansion. The negative thermal expansion for this axis is marked between 240 and 270 K so that a slight volume contraction is concomitantly observed.

The variations in the unit-cell parameters can be related to the following structural features: (1) lengthening of the *a*-axis ($a = b$) as a function of increasing temperature is associated with lengthening of the intermolecular contacts between adjacent α-CD molecules along the same axis; (2) the negative thermal expansion along the *c*-axis is associated with a compression of the vase-like cluster (see Figure S1), mainly arising from compression of the thin water layer that stabilizes the α-CD A–C tail-to-tail dimer (Figure 2). This thin water layer is formed by one disordered water molecule (named O(10)_91 in the structure), and its symmetry equivalents, H-bonded to the (+)-gauche primary hydroxy group of α-CD molecule C (Figure 4).

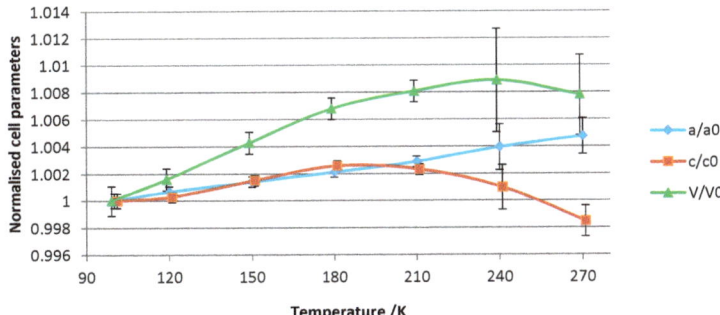

Figure 5. The evolution of the normalized unit-cell parameters as function of temperature (errors at 3σ level) taken from the structures in Supplementary Table S1, see Supplementary Information for details.

An apparent ordering of α-CD C is observed at 270 K although it should be noted that the structure has large temperature factors. In fact, the low-temperature study points to the presence of dynamic disorder, whereby low temperatures freeze the movement of α-CD C into two distinct positions. Ordering at 270 K is associated with a rearrangement of hydroxy groups: α-CD C has two (+)-gauche primary hydroxy conformers, instead of one observed at low temperature. The new (+)-gauche conformer, pointing towards the inner cavity, is H-bonded to the disordered water molecule O(10)_91 with a distance of 2.52(9) Å. This indicates that there is interplay between movement of α-CD and water molecules towards stabilization of the structure. The multi-temperature experiment did not provide information on the nature of the disorder of the guest molecules, for which it is likely that a combination of dynamic and static disorder is present.

The 100 K structural model was refined based on the atomic coordinates of the 90 K model, and was then used as a starting model for all other temperature points. Modeling of disordered water molecules in a highly hydrated structure is subject to user bias. Temperature factors and site occupancy parameters suffer from correlation and in the absence of independent information, e.g., on water content determined by thermal analysis, the user must make judicious use of refinement tools in order to obtain a reasonable model, which does not necessarily lead to the lowest R-factor. Comparing Fourier and difference Fourier maps can give an indication on the mean position and site occupancy of the atoms provided a consistent data reduction and refinement strategy for all data points is applied. To minimize model bias and double check the stability of the model, the *SHELXL* WIGL command was used as a cross-validation method during the refinement [28]. Two sets of results were then generated. In the first set, all water molecules were placed and refined according to the difference Fourier electron density maps (results reported in Table S1). In the second set, all water molecules were omitted during refinement in order to compare the evolution of the maps as a function of increasing temperature. The latter set of structures can be used to emphasize the change in the positions of water molecules and site occupancies by two methods:

(a) An analysis with *SHELXLE* [29] allows a qualitative analysis of the changes in the difference density as a function of temperature. Figure 6 clearly illustrates the loss of this signal from the solvent region as temperature is increased.

(b) Void analysis performed with *MERCURY* [30] using a probe radius of 1.2 Å and a grid spacing of 0.2 Å, calculated using the contact surface algorithm, reveals that the volume occupied by the solvent does not change significantly as temperature is increased (Figure S2).

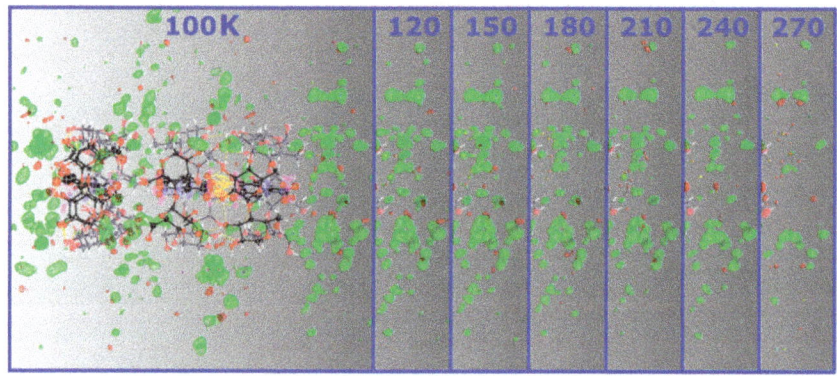

Figure 6. *SHELXLE* F_{obs}-F_{calc} maps (at 0.40 e$^-$/Å^3) of α-CD·SA inclusion complex at 100 K and the respective solvent region at the bottom (right-hand side in the picture) of the vase-like cluster at the different temperatures showing the decrease of the signal as a function of increasing temperature. A similar decrease, not shown here, is observed for the other side.

An analysis, based on the fully refined structures, of the anisotropic displacement parameters (ADP) as function of temperature indicates larger thermal motion at higher temperatures, as expected (see Figure S3). Analysis of the refined U_{ij} values of O(1)→O(6) water molecules, which are located in the space between adjacent α-CD columns (Figure 2), shows comparable expansion of the three diagonal elements U_{11}, U_{22}, U_{33} of the ADP tensor. These elements are parallel to the reciprocal unit-cell axes a^*, b^* and c^*, respectively, and in the case of a rhombohedral crystal system, U_{33} is parallel to the c-axis. For α-CD atoms, the thermal expansion of U_{33} is much more pronounced compared to the two other directions. Hence, while water molecules increase their vibrations in all directions as temperature increases, α-CD molecules have more freedom along the *c*-direction.

To summarize, the loss of diffraction quality at higher temperatures can be directly ascribed to structural disorder, primarily of the solvent region. It is not unlikely that at even higher temperatures, water molecules vibrate excessively leading to disruption of the crystal packing and structural instability as testified by complete loss of diffraction at 298 K. Given the channel-type packing adopted by the structure, it is conceivable that water molecules find their way out of the structure. Moreover, the multi-temperature study also shows that the guest molecules inside α-CD cavities are freely rotating with no apparent ordering induced by lowering the temperature. This is due to the lack of H-bond interactions between the guest and the host molecules, which, in theory, should help to stabilize the position of the guest molecules. A CSD search reveals that the lack of direct intermolecular contacts between host and guest molecules is not unusual for CD inclusion complexes, and that the guest is indirectly H-bonded to the host *via* water molecules.

2.2. High-Pressure Crystallization Results

We have investigated complex formation at high pressure up to 0.7 GPa and in the 277–293 K temperature range. Two avenues were explored: first, a diamond-anvil cell (DAC) was loaded with a saturated 1:1 aqueous solution of α-CD and SA and pressure was increased to induce precipitation. No crystals were obtained in this manner. In a second set of experiments, a single crystal of the α-CD·SA complex crystallized at ambient-pressure conditions was loaded in a DAC together with either its mother liquor or water. Pressure was increased and gradual dissolution of the crystal was observed, similar to what was reported for pure α-CD hydrate form I [9]. Subsequent to dissolution, no crystallization event was recorded. Upon releasing pressure, there was no microscopic evidence for crystallization of either SA or α-CD alone. While our experiments were not comprehensive, the lack of complex formation at high pressure may be ascribed to concentration, the formation of a particularly stable complex in solution stabilized at high pressure and kinetic effects; it is also conceivable that structural disorder is essential for stability of the α-CD·SA complex (the entropic contributions of structural disorder in molecular compounds have for instance been discussed in [31–34]) and that high pressure, favoring more dense, and in general more ordered states, hinders the *in situ* formation of a complex.

3. Experimental Section

3.1. Ambient-Pressure Crystallization

Both α-CD and SA were bought from Fluka (Munich, Germany) and SIGMA (Munich, Germany), respectively, and used without further treatment. An undersaturated solution with a 1:1 molar ratio mixture of α-CD (194.57 mg) and succinic acid (23.6 mg) was prepared in approximately 2 mL of demineralized water. Fast evaporation of the solution yielded a glass at ambient conditions. A combination of cold temperatures (*ca.* 277 K) and very slow evaporation over the course of six months led to the crystallization of hexagonal prism-shaped crystals. When the experiment was repeated using a more concentrated solution, crystallization occurred within a week at the same low-temperature conditions.

3.2. High-Pressure Crystallization

The high-pressure crystallization experiment was carried out using an in-house modified Merrill-Bassett DAC [35] with a half opening angle of 45°. The DAC was equipped with 800 μm culet diamonds of low fluorescence grade and Inconel gaskets with a starting diameter hole of 350 μm. Pressure was monitored using the ruby fluorescence method described by Piermarini *et al.* [36].

3.3. Data Collection and Reduction

Several crystals were tested on the diffractometer. When left out of the mother liquor, crystals were found to be unstable at room-temperature conditions, yielding poor diffraction quality even when measured at low temperature. Maintaining low temperature from the moment the crystal was extracted from the mother liquor until final mounting on a glass fiber on the goniometer's head was essential to ensure good diffraction quality: in order to achieve this a drop containing mother liquor and crystals was pipetted onto a microscope slide that had previously been cooled using an ice bath beneath the slide. Subsequently, the crystal was moved from the mother liquor to a cold drop of mounting oil and then to the goniometer's head under a cold and dry-nitrogen stream [37].

Data collection was undertaken using a Bruker-AXS APEX II diffractometer (Bruker-AXS, Karlsruhe, Germany) equipped with graphite-monochromatic Mo-Kα radiation and an Oxford Cryosystems low-temperature device (Oxoford Cryosystems Ltd., Oxford, UK). X-ray data were first collected on a single crystal specimen at 90 K. A subsequent experiment was performed at low-temperature conditions on a second single-crystal specimen. Data sets were collected on the same single crystal using the same data collection strategy at 100, 120, 150, 180, 210, 240 and 270 K. The purpose of this experiment was to monitor structural changes, e.g., phase transitions, and possible crystal decay as function of temperature.

Data integration and global-cell refinement were performed with the program *SAINT* (Bruker-AXS, Madison, WI, USA) [38]. Absorption correction was performed with *SADABS* (Bruker-AXS, Madison, WI, USA) [39]. Structure solution of this medium-sized structure was based on Patterson-seeded dual-space recycling in the *SHELXD* program (Göttingen, Germany) [40]. Structures were refined by full-matrix least squares against F^2 using *SHELXL-2014/7* (Göttingen, Germany) [41] through the *SHELXLE GUI* (Göttingen, Germany) [29]. Due to the limited resolution and structural complexity of the model, soft restraints on bond lengths and angles of the host and guest molecules were applied using the *GRADE* webserver (Globar Phasing Ltd., Cambridge, UK) [42]. A *GRADE* dictionary for *SHELXL* contains target values and standard deviations for 1,2-distances (DFIX) and 1,3-distances (DANG), as well as restraints for planar groups (FLAT). Following a CSD search on structures containing SA-SA H-bonded dimer, SA molecules were restrained to be planar. Anisotropic displacement parameters were refined using the new rigid bond restraint (RIGU) implemented in the *SHELXL* program [43]; SIMU restraints were also applied where necessary. ISOR restrains were used on a handful of atoms to cure ill-defined ADPs. All H-atoms of α-CD and SA were placed geometrically and allowed to ride on the parent atoms. H-atoms belonging to ordered water molecules were clearly visible in difference Fourier maps and their positions were refined subject to distance restraints. H-atoms belonging to disordered water molecules were not placed during refinement but were taken into account for the calculation of F_{000} and derived properties. $U_{iso}(H)$ values were assigned in the range 1.2–1.5 times U_{eq} of the parent atom. Details on the treatment of disorder are given in the discussion below and crystallographic details can be found in Table S1. CCDC 1437063-1437070 contain the supplementary crystallographic data for this paper. The data can be obtained free of charge from The Cambridge Crystallographic Data Center via www.ccdc.cam.ac.uk/getstructures.

3.4. Disorder Modeling

The structural refinement of α-CD·SA was particularly challenging due to the heavy disorder of the guest and solvent molecules as well as one of the CD molecules. The disordered α-CD C molecule was modelled using a free variable which for the 90 K structure refines to a final value of 0.668(5) for the major component. The effect of disorder is to elongate the vase-like cluster (Figure 2). At 90 K, a careful inspection of the difference Fourier maps permits the refinement of 33 positions of water molecules, which are distributed heterogeneously over the asymmetric unit and of which only two are fully occupied. For these two water molecules, H-atoms could be located in difference Fourier maps.

The channel-type structure of α-CD·SA favors the guest to be in the trans, planar conformation. The three-fold rotation axis, going through the center of the host, is incompatible with the guest molecular symmetry and imposes the presence of disorder, which was modelled as exemplified by α-CD molecules B (Figure 7) as follows: electron density maps inside the α-CD cavity show tetrahedral features, incompatible with a single SA molecule. The difference electron density peak Q(1), sitting on the 3-fold rotation axis, was assigned to the carbon atom C(1) of the carboxylic group. The coordinates of C(1) were constrained to make model building more straightforward. By using stereochemical restrains generated with the *GRADE* webserver [42] all other Q peaks were correctly assigned, *i.e.*, Q(4) to O(1) and O(2), Q3 to C(2), Q(2) to C(4) (Figure 7). Finally, all atomic coordinates were freely refined. Two partially occupied SA molecules fit inside the disordered electron density; the negative PART instruction was used in the program *SHELXL* to exclude the generation of special position constraints and dissociate bonding to symmetry-generated atoms within the same PART group.

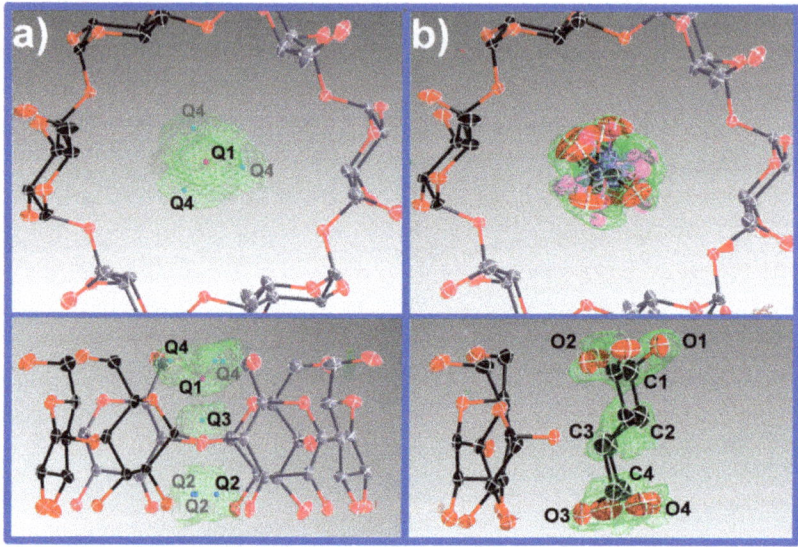

Figure 7. F_{obs}-F_{calc} maps in green showing the electron density before (**a**) and after (**b**) modeling SA inside α-CD B molecule. The peaks Q1 and 3 × (Q4) are forming a tetrahedron. H-atoms and the rest of the structure have been omitted for clarity. Displacement ellipsoids are drawn at the 50% probability level.

After several trials, the best refinement strategy consisted of gradually modeling the electron density of the host molecules and the reliably localized solvent peaks, followed by the guests, and finally the rest of the disordered solvent molecules. In order to avoid over-parametrization, site-occupancy factors of the disordered water molecules were fixed to the initially refined values. Those of the guest molecules were constrained to $1/6$ each.

4. Conclusions

The crystal structure of a 1:1 α-CD·SA pseudododecahydrate inclusion complex has been elucidated using single-crystal X-ray diffraction. The complex crystallizes in a high-symmetry space group, previously unreported for α-CD inclusion complexes. The structure exhibits a honeycomb arrangement of channels, in which the guest molecules are heavily disordered. The building block of the channels has been described as a vase-like cluster formed by a trimer of α-CD molecules.

Structural disorder has been analyzed by means of a multi-temperature X-ray diffraction experiment. Careful analysis of difference Fourier maps proved that the nature of disorder is, at least partially, dynamic. In the low-temperature regime disorder is attributed to thermal vibrations; above 210 K vibrations of water molecules accentuate the disorder and lead eventually to the decomposition of the crystal at ambient temperature.

Supplementary Materials: The following are available online at www.mdpi.com/2073-4352/6/1/2, Table S1: Crystallographic data of the multi-temperature structures of α-CD·SA inclusion complex, Figure S1: Evolution of the normalized distances between dimers in α-CD·SA. Distances were calculated based on the O(4) glycosidic planes using *MERCURY*, Figure S2: The solvent accessible void volume calculated using *MERCURY* (grid spacing = 0.2 Å; probe radius = 1.2 Å), Figure S3: U_{11}, U_{22} and U_{33} elements of the ADP tensor for selected atoms plotted as a function of temperature. The value of the ADPs were taken from the result file of the refinement of SHELX (res file). O(1) to O(6) are water molecules located in the interstices. The numbering suffix refers to the residue number in the structure.

Acknowledgments: The authors gratefully acknowledge funding from the German Science Foundation (DFG, Emmy Noether Grant to FPAF, FA 946/1-1) and technical support from Ulf Kahmann (Göttingen). The authors also wish to thank Werner F. Kuhs (Göttingen) for critical reading of the manuscript. The authors would like to thank anonymous Reviewers for helpful comments.

Author Contributions: Experimental investigations, data interpretation and manuscript writing were performed by Sofiane Saouane with the support of Francesca P. A. Fabbiani.

Conflicts of Interest: The authors declare no conflict of interest.

References

1. Irie, T.; Uekama, K. Pharmaceutical Applications of Cyclodextrins. III. Toxicological Issues and Safety Evaluation. *J. Pharm. Sci.* **1997**, *86*, 147–162.
2. Uekama, K.; Hirayama, F.; Irie, T. Cyclodextrin Drug Carrier Systems. *Chem. Rev.* **1998**, *98*, 2045–2076.
3. Loftsson, T.; Brewster, M.E. Pharmaceutical Applications of Cyclodextrins. 1. Drug Solubilization and Stabilization. *J. Pharm. Sci.* **1996**, *85*, 1017–1025.
4. Roux, M.; Perly, B.; Djedaïni-Pilard, F. Self-Assemblies of Amphiphilic Cyclodextrins. *Eur. Biophys. J.* **2007**, *36*, 861–867.
5. Saenger, W. Cyclodextrin Inclusion Compounds in Research and Industry. *Angew. Chemie Int. Ed. Engl.* **1980**, *19*, 344–362.
6. Saenger, W.; Jacob, J.; Gessler, K.; Steiner, T.; Hoffmann, D.; Sanbe, H.; Koizumi, K.; Smith, S.M.; Takaha, T. Structures of the Common Cyclodextrins and Their Larger Analogues-Beyond the Doughnut. *Chem. Rev.* **1998**, *98*, 1787–1802.

7. Saenger, W.; Steiner, T. Cyclodextrin Inclusion Complexes: Host-Guest Interactions and Hydrogen-Bonding Networks. *Acta Crystallogr. Sect. A* **1998**, *54*, 798–805.
8. Harata, K. Structural Aspects of Stereodifferentiation in the Solid State. *Chem. Rev.* **1998**, *98*, 1803–1827.
9. Granero-García, R.; Lahoz, F.J.; Paulmann, C.; Saouane, S.; Fabbiani, F.P.A. A Novel Hydrate of A-Cyclodextrin crystallized under High-Pressure Conditions. *CrystEngComm* **2012**, *14*, 8664–8670.
10. Allen, F.H. The Cambridge Structural Database: A Quarter of a Million Crystal Structures and Rising. *Acta Crystallogr. Sect. B Struct. Sci.* **2002**, *58*, 380–388.
11. Zeikus, J.G.; Jain, M.K.; Elankovan, P. Biotechnology of Succinic Acid Production and Markets for Derived Industrial Products. *Appl. Microbiol. Biotechnol.* **1999**, *51*, 545–552.
12. Lisnyak, Y.V.; Martynov, A.V.; Baumer, V.N.; Shishkin, O.V.; Gubskaya, A.V. Crystal and Molecular Structure of β-Cyclodextrin Inclusion Complex with Succinic Acid. *J. Incl. Phenom. Macrocycl. Chem.* **2007**, *58*, 367–375.
13. Del Valle, E.M.M. Cyclodextrins and Their Uses: A Review. *Process Biochem.* **2003**, *39*, 1033–1046.
14. Harata, K.; Uekama, K.; Otagiri, M.; Hirayama, F. Crystal Structures Pentahydrate of *a*-Cyclodextrin-3-Lodopropionic Acid (1:1) Complex and Hexakis(2, 3, 6-Tri-O-Methyl)-a-Cyclodextrin-Lodoacetic Acid (1:1) Complex Monohydrate. *Jpn. Chem. Soc. J.* **1983**, *2*, 173–180.
15. Tokuoka, R.; Abe, M.; Matsumoto, K.; Shirakawa, K.; Fujiwara, T.; Tomita, K. Structure of the α-Cyclodextrin (α-CD) Inclusion Complex with the Potassium Salt of γ-Aminobutyric Acid (GABA). *Acta Crystallogr. Sect. B Struct. Crystallogr. Cryst. Chem.* **1981**, *37*, 445–447.
16. Sicard-Roselli, C.; Perly, B.; le Bas, G. The Respective Benefits of X-ray Crystallography and NMR for the Structural Determination of the Inclusion Complex Between Butyl-Isothiocyanate and Alpha-Cyclodextrin. *J. Incl. Phenom. Macrocycl. Chem.* **2001**, *39*, 333–337.
17. Ding, J.; Steiner, T.; Saenger, W. Structure of the γ-cyclodextrin-1-propanol-17H$_2$O Inclusion Complex. *Acta Crystallogr. Sect. B Struct. Sci.* **1991**, *47*, 731–738.
18. Kamitori, S.; Hirotsu, K.; Higuchi, T. Crystal and Molecular Structures of Double Macrocyclic Inclusion Complexes Composed of Cyclodextrins, Crown Ethers, and Cations. *J. Am. Chem. Soc.* **1987**, *109*, 2409–2414.
19. Steiner, T.; Saenger, W. Channel-Type Crystal Packing in the Very Rare Space Group P4212 with Z' = 3/4: Crystal Structure of the Complex γ-Cyclodextrin-Methanol-n-Hydrate. *Acta Crystallogr. Sect. B Struct. Sci.* **1998**, *54*, 450–455.
20. Chatziefthimiou, S.; Yannakopoulou, K.; Mavridis, I.M. β-Cyclodextrin trimers enclosing an unusual organization of the guest: The inclusion complex of β-cyclodextrin/4-pyridinealdazine. *CrystEngComm* **2007**, *9*, 976–979.
21. Cheng, X.; Lu, Z.; Li, Y.; Wang, Q.; Lu, C.; Meng, Q. An interesting molecular-assembly of β-cyclodextrin pipelines with embedded hydrophilic nickel maleonitriledithiolate. *Dalton Trans.* **2011**, *40*, 11788–11794.

22. Caira, M.; Dodds, D.; Nassimbenni, L.R. Diverse modes of guest inclusion in a β-cyclodextrin: X-ray structural and thermal characterization of a 4:3 β-cyclodextrin-cyclizine complex. *Supramol. Chem.* **2001**, *13*, 61–70.
23. Liu, Y.; Zhong, R.Q.; Zhang, H.Y.; Song, H.B. A unique tetramer of 4:5 β-cyclodextrin-ferrocene in the solid state. *Chem. Commun.* **2005**, *17*, 2211–2213.
24. Harada, A.; Li, J.; Kamachi, M.; Kitagawa, Y.; Katsube, Y. Structures of Polyrotaxane Models. *Carbonhydr. Res.* **1998**, *305*, 127–129.
25. Caira, M.; Bourne, S.; Mzondo, B. Cyclodextrin inclusion complexes of the antioxidant α-lipoic acid. *Acta Crystallogr. Sect. A Found. Cryst.* **2014**, *70*, C992.
26. Saenger, W. Nature and Size of Included Guest Molecule Determines Architecture of Crystalline Cyclodextrin Host Matrix. *Isr. J. Chem.* **1985**, *25*, 43–50.
27. Chacko, K.K.; Saenger, W. Topography of Cyclodextrin Inclusion Complexes. 15. Crystal and Molecular Structure of the Cyclohexaamylose-7.57 Water Complex, Form III. Four- and Six-Membered Circular Hydrogen Bonds. *J. Am. Chem. Soc.* **1981**, *103*, 1708–1715.
28. Gruene, T.; Hahn, H.W.; Luebben, A.V.; Meilleur, F.; Sheldrick, G.M. Refinement of Macromolecular Structures against Neutron Data with SHELXL2013. *J. Appl. Crystallogr.* **2014**, *47*, 462–466.
29. Hübschle, C.B.; Sheldrick, G.M.; Dittrich, B. ShelXle: A Qt Graphical User Interface for SHELXL. *J. Appl. Crystallogr.* **2011**, *44*, 1281–1284.
30. Macrae, C.F.; Bruno, I.J.; Chisholm, J.A.; Edgington, P.R.; McCabe, P.; Pidcock, E.; Rodriguez-Monge, L.; Taylor, R.; van de Streek, J.; Wood, P.A. Mercury CSD 2.0—New Features for the Visualization and Investigation of Crystal Structures. *J. Appl. Crystallogr.* **2008**, *41*, 466–470.
31. Cruz-Cabeza, A.J.; Day, G.M. Structure prediction, disorder and dynamics in a DMSO solvate of carbamazepine. *Phys. Chem. Chem. Phys.* **2011**, *13*, 12808–12816.
32. Nyman, J.; Day, G.M. Static and lattice vibrational energy differences between polymorphs. *CrystEngComm* **2015**, *17*, 5154–5165.
33. Price, S.L. Predicting crystal structures of organic compounds. *Chem. Soc. Rev.* **2014**, *43*, 2098–2111.
34. Neumann, M.A.; van de Streek, J.; Fabbiani, F.P.A.; Hidber, P.; Grassmann, O. Combined Crystal Structure Prediction And High-Pressure Crystallization in Rational Pharmaceutical Polymorph Screening. *Nat. Commun.* **2015**, *6*, 7793.
35. Merrill, L.; Bassett, W.A. Miniature Diamond Anvil Pressure Cell for Single Crystal X-ray Diffraction Studies. *Rev. Sci. Instrum.* **1974**, *45*, 290–294.
36. Piermarini, G.J.; Block, S.; Barnett, J.D.; Forman, R.A. Calibration of the Pressure Dependence of the R1 Ruby Fluorescence Line to 195 Kbar. *J. Appl. Phys.* **1975**, *46*, 2774–2780.
37. Kottke, T.; Stalke, D. Crystal Handling at Low Temperatures. *J. Appl. Crystallogr.* **1993**, *26*, 615–619.
38. Bruker-AXS Inc. *SAINT Version 8.34A*; Bruker-AXS Inc: Madison, WI, USA, 2013.
39. Sheldrick, G.M. *SADABS Version 2012/1, 1*; Bruker-AXS Inc: Madison, WI, USA, 2012.
40. Sheldrick, G.M. A Short History of SHELX. *Acta Crystallogr. A* **2008**, *64*, 112–122.

41. Sheldrick, G.M. Crystal Structure Refinement with SHELXL. *Acta Crystallogr. Sect. C Struct. Chem.* **2015**, *71*, 3–8.
42. Grade Web Server, Global Phasing Ltd. Grade Web Server. Available online: http://grade.globalphasing.org/ (accessed on 15 January 2015).
43. Thorn, A.; Dittrich, B.; Sheldrick, G.M. Enhanced Rigid-Bond Restraints. *Acta Crystallogr. Sect. A Found. Crystallogr.* **2012**, *68*, 448–451.

The Hydrogen Bonded Structures of Two 5-Bromobarbituric Acids and Analysis of Unequal C5–X and C5–X' Bond Lengths (X = X' = F, Cl, Br or Me) in 5,5-Disubstituted Barbituric Acids

Thomas Gelbrich, Doris E. Braun, Stefan Oberparleiter, Herwig Schottenberger and Ulrich J. Griesser

Abstract: The crystal structure of the methanol hemisolvate of 5,5-dibromobarbituric acid (**1MH**) displays an H-bonded layer structure which is based on N–H⋯O=C, N–H⋯O(MeOH) and (MeOH)O–H⋯O interactions. The barbiturate molecules form an H-bonded substructure which has the **fes** topology. 5,5'-Methanediylbis (5-bromobarbituric acid) **2**, obtained from a solution of 5,5-dibromobarbituric acid in nitromethane, displays a N–H⋯O=C bonded framework of the **sxd** type. The conformation of the pyridmidine ring and the lengths of the ring substituent bonds C5–X and C5–X' in crystal forms of 5,5-dibromobarbituric acid and three closely related analogues (X = X' = Br, Cl, F, Me) have been investigated. In each case, a conformation close to a C5-endo envelope is correlated with a significant lengthening of the axial C5–X' in comparison to the equatorial C5–X bond. Isolated molecule geometry optimizations at different levels of theory confirm that the C5-endo envelope is the global conformational energy minimum of 5,5-dihalogenbarbituric acids. The relative lengthening of the axial bond is therefore interpreted as an inherent feature of the preferred envelope conformation of the pyrimidine ring, which minimizes repulsive interactions between the axial substituent and pyrimidine ring atoms.

Reprinted from *Crystals*. Cite as: Gelbrich, T.; Braun, D.E.; Oberparleiter, S.; Schottenberger, H.; Griesser, U.J. The Hydrogen Bonded Structures of Two 5-Bromobarbituric Acids and Analysis of Unequal C5–X and C5–X' Bond Lengths (X = X' = F, Cl, Br or Me) in 5,5-Disubstituted Barbituric Acids. *Crystals* **2016**, *6*, 47.

1. Introduction

Barbiturates are derivatives of barbituric acid which have the ability to act as nervous system depressants. A number of 5,5-disubstituted species have been used widely as sedative, hypnotic and anticonvulsant agents [1–4]. These barbiturates are known for their high propensity to crystallize in multiple solid forms, and they are a model polymorphic system in which a set of competing H-bonded structures (HBSs) occurs. In the course of our systematic study of this group of compounds [5–9], we have investigated polymorphs of 5,5-dibromobarbituric acid (**1**) and 5,5-dichlorobarbituric acid (**3**), some of which had been first described by Groth more than 100 years ago [10]. The polymorphs of **1** and **3** display four distinct 1D, 2D or 3D hydrogen-bond motifs, and three of these motifs are unique in that they have not been encountered in any of the more than 50 known crystal structures of analogous barbiturates bearing two organic groups as ring substituents R and R′ at atom C5 (Scheme 1) [11,12]. Few crystal structures of solvates and hydrates of 5,5-disubstituted barbiturates have been reported so far [7,13–16]. The formation of H-bond motifs which are distinct from those of the corresponding unsolvated species is expected in cases where additional H-bond donor and/or acceptor functions are present in the solvent molecule.

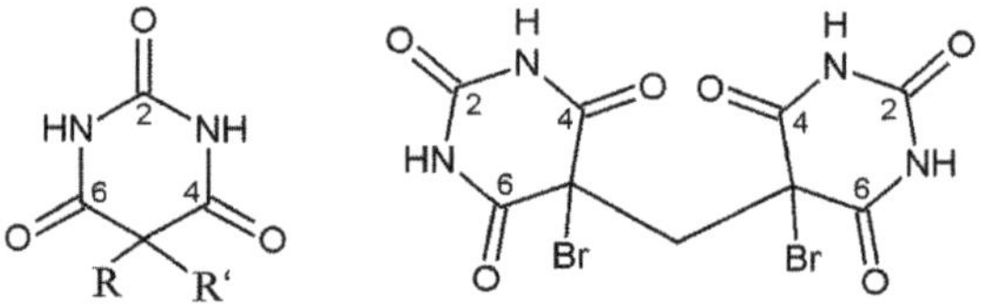

	R, R′	CAS
1	Br	[511-67-1]
3	Cl	[699-40-1]
4	F	[55052-01-2]
5	Me	[24448-94-0]

Scheme 1. Structure formulas of compounds discussed in this report.

Herein we report the crystal structure of the methanol hemisolvate of dibromobarbituric acid (**1MH**) and that of a new derivative, 5,5′-methanediylbis (5-bromobarbituric acid) (**2**) (Scheme 1), obtained from an unexpected reaction of **1** with nitromethane. The complex HBSs of these crystals will be discussed in detail and a specific feature of the 5,5-dibromobarbituric acid molecule in the crystal structure of **1MH**, namely the unequal lengths of the two C5–Br bonds, will be analyzed by means of comparison with crystal structures of structural analogues and their geometry optimized molecular structures. Additionally, a low-temperature redetermination of the crystal structure of 5,5-dimethylbarbituric acid (**5**) is reported.

2. Results and Discussion

*2.1. Crystal Structure of 5,5-Dibromobarbituric Acid Hemisolvate (**1MH**)*

The title structure has the monoclinic space group symmetry $P2_1/n$ (Table 1). Its asymmetric unit (Figure 1) consists of two dibromobarbituric acid molecules (labeled A and B) and one MeOH molecule. The pyrimidine rings adopt a C5-endo envelope conformation, characterized by these ring puckering parameters according to Cremer and Pople [17]: $Q = 0.201(4)$ Å, $θ = 50.7(11)°$, $φ = 242.9(14)°$ (molecule A) and $Q = 0.196(4)$ Å, $θ = 139.9(12)°$, $θ = 61.6(17)°$ (molecule B). The positions of the atoms C5 and C5′ of A and B, respectively, deviate by 0.437(4) and 0.450(4) Å from the mean plane defined by the other five atoms of the pyrimidine ring. In both molecules, the bond distance to the equatorial bromo substituent, C5–Br1, 1.907(4) Å, and C5′–Br1′, 1.912(4) Å, is significantly shorter than that to the axial substituent, C5–Br2, 1.963(4) Å, and C5′–Br2′, 1.967(3) Å.

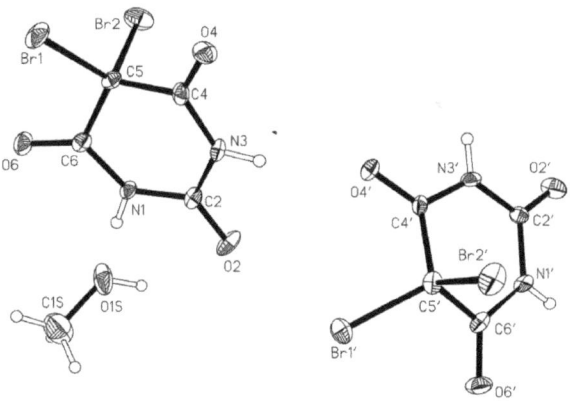

Figure 1. Asymmetric unit of **1MH**. Thermal ellipsoids are drawn at the 50% probability level. H-atoms are drawn as spheres of arbitrary size.

Table 1. Crystallographic details for **1MH**, **2** and **5**.

Compound	1MH	2	5
Moiety formula	2(C$_4$H$_2$Br$_2$N$_2$O$_3$)·CH$_4$O	C$_9$H$_6$Br$_2$N$_4$O$_6$	C$_6$H$_8$N$_2$O$_3$
Formula mass	603.83	426.00	156.14
Crystal system	monoclinic	orthorhombic	triclinic
Space group	$P2_1/n$	$P2_12_12_1$	$P\bar{1}$
Z	4	4	2
a/Å	14.6451(4)	6.8347(3)	5.6667(7)
b/Å	6.73660(16)	10.6206(5)	6.4172(7)
c/Å	17.1159(4)	17.0982(9)	10.5800(8)
α/°	90	90	84.429(8)
β/°	90.238(2)	90	81.341(8)
γ/°	90	90	64.916(12)
Unit cell volume/Å^3	1688.61(7)	1241.13(10)	344.24(7)
Temperature/K	173	173	173
No. of reflections measured	5387	9500	2223
No. of independent reflections	3388	2256	1329
R_{int}	0.0277	0.1130	0.0268
No. of parameters	238	202	117
Absolute structure parameter (Flack)	-	0.014(18)	-
Final R_1 value ($I > 2σ(I)$)	0.0310	0.0477	0.0331
Final $wR(F^2)$ value (all data)	0.0620	0.1220	0.0821

Molecules A and B are each H-bonded, via one-point connections, to four other molecules (Table 2). However, the two molecule types are distinct with respect to their H-bond connectivity. Via its NH H-bond donor groups, molecule A is linked to one B molecule, N3–H···O=C4', and to one MeOH molecule, N1–H···O=C(MeOH). Additionally, it is connected via its C2 and C4 carbonyl functions, to another two B-type molecules. By contrast, both NH groups of a B-type molecule link to A-molecules, N1'–H···O=C2^i and N3'–H···O=C4^{ii}. Moreover, its C4' carbonyl group links to a third A-molecule and the C2' group to a MeOH molecule. The latter serves as a bridge between an A and a B molecule in such a way that $R_3^3(10)$ rings [18,19] are generated (Figure 2a). The resulting HBS is a layer structure which lies parallel to $(10\bar{1})$ and also contains $R_4^4(16)$ rings which involve two A and two B molecules. The latter ring motif connecting four barbiturate molecules is not encountered in either the orthorhombic or the monoclinic polymorph of **1**, both of which display rings connecting six H-bonded molecules via N–H···O=C interactions. However, the $R_3^3(10)$ ring is reminiscent of the $R_2^2(8)$ motif via the C2 carbonyl function (Figure 2b), which is present in both previously described polymorphs of **1** (the orthorhombic polymorph **1a** is the desolvation product of **1MH**, see Figure S1 of the Supplementary Materials).

Table 2. Geometric parameters for hydrogen bonds in **1MH**.

Type	D–H···A	d(D–H)/Å	d(H···A)/Å	d(D···A)/Å	∠(DHA)/°
A→MeOH	N1–H1···O1S	0.855(10)	1.881(15)	2.721(4)	167(5)
A→B	N3–H3···O4'	0.862(10)	2.071(15)	2.910(4)	164(3)
B→A	N1'–H1'···O2 ^{i}	0.851(10)	1.974(13)	2.814(4)	169(4)
B→A	N3'–H3'···O4 ^{ii}	0.850(10)	2.162(19)	2.942(4)	152(3)
MeOH→B	O1S–H1S···O2' ^{iii}	0.838(10)	1.97(2)	2.757(4)	157(5)

Symmetry transformations: ^{i} $-x + 1/2, y + 1/2, -z + 1/2$; ^{ii} $-x + 1, -y + 1, -z + 1$; ^{iii} $-x + 1/2, y - 1/2, -z + 1/2$.

The diagram in Figure 3 is based on the topological net of the HBS of **1MH**, generated in the manner described by Baburin & Blatov [20]. The nodes represent molecules, and links between these nodes represent H-bond connections. As proposed by Hursthouse et al. [21], arrows have been added to indicate the type and direction of each H-bond interaction. Only this latter modification reveals the subtle differences between the two molecule types A and B (Figure 3a, Table 2) with respect to their H-bond connectivity. If the bridging MeOH molecules are not taken into account, the barbiturate molecules alone form a 2-periodic 3-connected uninodal subnet which has the 4.8^2-**fes** topology [22] (Figure 3b). The topology symbol for the complete H-bonded net composed of 4-connected A and B molecules and 2-connected MeOH (Figure 3a) is $(3.4.8^2.9^2)_2(3)$. The short symbol according to Hursthouse et al. [21] for the complete HBS is $L(4_4)_2.2_2[(3.4.8^2.9^2)_2(3)]$, and for the

A + B substructure it is L(4$_4$)$_2$[4.8^2-**fes**]. The stacking of multiple H-bonded layers in the crystal is shown in Figure S2 of the Supplementary Materials.

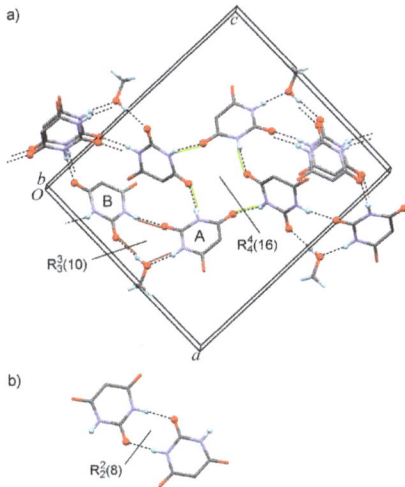

Figure 2. (**a**) N–H···O=C, N–H···O(MeOH) and (MeOH)O–H···O=C-bonded layer structure of **1MH**. O and H atoms involved in hydrogen bonding are drawn as spheres and hydrogen bonds are drawn as dashed lines. Br atoms are omitted for clarity; (**b**) R$_2^2$(8) ring motif involving the C2 carbonyl functions of two barbiturate molecules, found in the orthorhombic and monoclinic polymorphs of **1**.

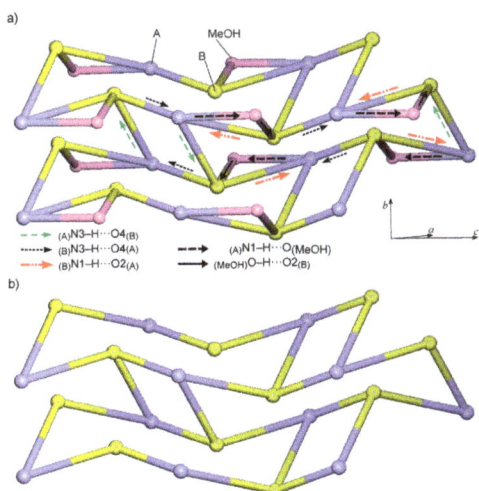

Figure 3. HBS of **1MH**, represented as a graph according to ref. [21]: (**a**) the complete L(4$_4$)$_2$.2$_2$[(3.4.8^2.9^2)$_2$(3)] structure and (**b**) the L(4$_4$)$_2$[4.8^2-**fes**] substructure formed by the H-bonded molecules of **1**.

2.2. Crystal Structure of 5,5′-Methanediylbis(5-bromobarbituric acid) (2)

This compound has the orthorhombic space group symmetry $P2_12_12_1$ (Table 1), and its asymmetric unit consists of a single molecule (Figure 4). The two six-membered rings adopt C5-endo envelope conformations of the same handedness. The Cremer-Pople puckering parameters are $Q = 0.240(9)$ Å, $\theta = 65(2)°$, $\phi = 224(2)°$ for the pyrimidinetrione unit A (N1, C2, N3, C4, C5, C6), and they are $Q = 0.156(10)$ Å, $\theta = 45(2)°$, $\phi = 235(5)°$ for unit B (N1′, C2′, N3′, C4′, C5′, C6′). The axial C5–Br1 and C5′–Br1′ bond distances of 1.973(8) Å and 1.982(8) are in good agreement with the corresponding axial values found in **1MH**. The angle between the mean planes of the two heterocyclic rings is 50.1(3)°.

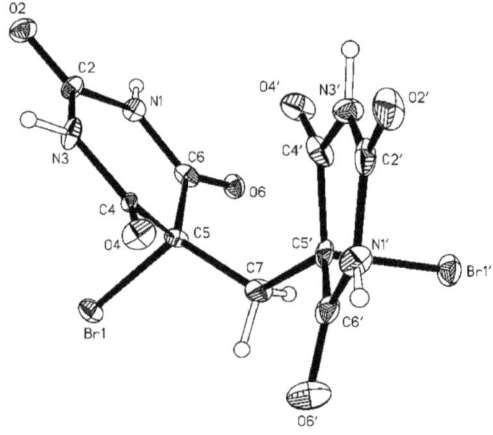

Figure 4. Molecular structure of **2**. Thermal ellipsoids are drawn at the 70% probability level and H-atoms drawn as spheres of arbitrary size.

Each molecule is linked to six neighboring molecules via four one-point and two two-point N–H···O=C connections. The two independent pyrimidinetrione units differ somewhat from one another in their H-bond connectivity (Table 3). Two-point connections between neighboring molecules involve the C6 and C2 carbonyl groups of the A and B rings, N1–H1···O2′[i] and N3′–H3′···O6[iv], resulting in an asymmetrical $R_2^2(8)$ ring. With the second H-bond donor function, rings A and B are connected to the C6 and C2 carbonyl groups, respectively, of A-type rings belonging to two different molecules, N3–H3···O6[ii] and N1′–H1′···O2[iii]. As a result, the C2 carbonyl functions of both rings and the C6 carbonyl group of ring A are engaged in one and two H-bonds, respectively, whereas the C6 carbonyl group of ring B is not involved in N–H···O=C bonding. The HBS resulting from all these interactions is an H-bonded framework (Figure 5) which has the topology of the $3^3.4^6.5^5.6$-**sxd** net [22]. Figure 6 shows a graphical representation of this HBS, whose short symbol according to ref. [21] is F64[$3^3.4^6.5^5.6$-**sxd**]. As in the structure of **1MH**, each NH group is engaged in exactly one N–H···O=C interaction, but the resulting H-bonded framework is completely different from the H-bonded layer structure of **1MH**.

Table 3. Geometric parameters for hydrogen bonds in **2**.

#	Type	D–H···A	d(D–H)/Å	d(H···A)/Å	d(D···A)/Å	∠(DHA)/°
a	A→B	N1–H1···O2' [i]	0.863(14)	2.34(7)	3.085(11)	144(10)
b	A→A	N3–H3···O6 [ii]	0.861(14)	2.13(5)	2.929(10)	153(10)
c	B→A	N1'–H1'···O2 [iii]	0.861(14)	2.12(5)	2.933(11)	158(10)
d	B→A	N3'–H3'···O6 [iv]	0.859(14)	2.05(4)	2.871(10)	161(11)

Symmetry transformations: [i] $x - 1/2, -y + 1/2, -z + 2$; [ii] $x + 1, y, z$; [iii] $-x + 3/2, -y + 1, z + 1/2$; [iv] $x + 1/2, -y + 1/2, -z + 2$.

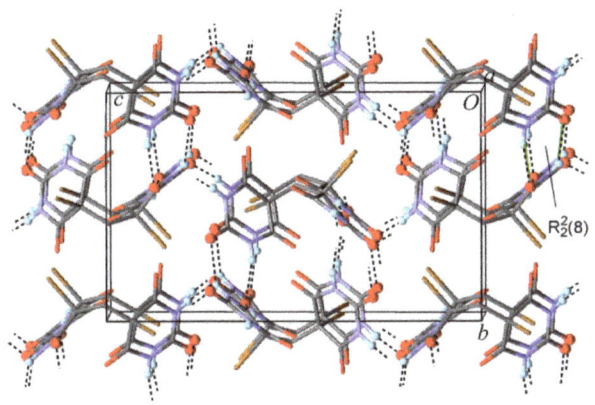

Figure 5. N–H···O=C-bonded framework of **2**. O and H atoms involved in hydrogen bonding are drawn as spheres and hydrogen bonds are drawn as dashed lines.

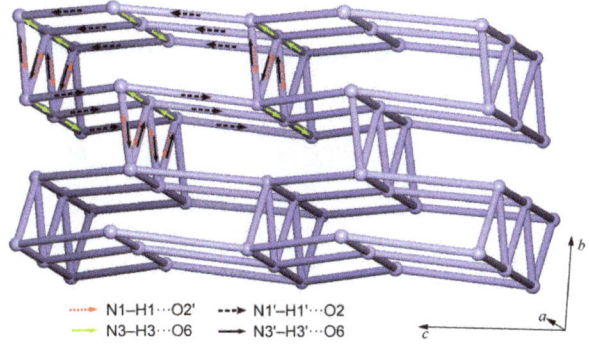

Figure 6. Graph according to ref. [21] for the HBS of **2** with the symbol F64[$3^3.4^6.5^5.6$-**sxd**].

The crystal of **2** contains two intermolecular contacts Br1···O2(1 − x, y + 0.5, 1.5 − z) and Br1'···O2'(x − 1, y, z) whose Br···O distances of 2.81 and 2.98 Å, respectively, are considerably smaller than the sum of the van der Waals radii of Br and O (3.37 Å [23]), and the corresponding C–Br···O angles are 165° and 174° (Supplementary Materials, Figure S3). We have used the semi-classical density sums (SCDS-PIXEL) [24–27] method to assess the importance of these contacts. This method enables to calculate total energy contributions (E_T) associated with individual molecule–molecule pairs, which are partitioned into contributions from Coulombic (E_C), polarization (E_P), dispersion (E_D) and repulsion (E_R) effects. Table S1 of the Supplementary Materials shows the results of these calculations for the seven most important pairs of symmetry-equivalent molecule–molecule interactions. The most stabilizing interaction pair (denoted as #1/1' in Table S1) accounts for more than 33% of the total PIXEL energy of the crystal, $E_{T,Cry}$, of −152.3 kJ·mol^{-1} and involves two-point N–H···O-bond connections between neighboring molecules (based on the interactions #a and #d in Table 3). This is followed by two pairs of symmetry-equivalent molecule–molecule interactions involving one-point N–H···O-bond connections (#b and #c in Table 3) which contribute 19% (#3/3') and 18% (#5/5') to $E_{T,Cry}$. Two pairs of molecule–molecule interactions (#7/7' and #9/9'), which are dominated by dispersion effects due to extensive van der Waals contacts, each account for 13% of $E_{T,Cry}$. By contrast, the stabilization effect of a single molecule–molecule interaction involving the aforementioned short Br1···O2(1 − x, y + 0.5, 1.5 − z) contact is very small (E_T = −1.5 kJ·mol^{-1}) as its significant Coulombic energy of −16.9 kJ·mol^{-1} is counterbalanced by strong repulsion (31.8 kJ·mol^{-1}). As a result, the corresponding symmetry-equivalent interaction pair (#13/13') affects $E_{T,Cry}$ by just 1%. Effects arising from the second short Br···O contact, Br1'···O2'(x − 1, y, z) in (#3'/3), are overlaid by those of the N3–H3···O6(x, y + 1, z) bond. However, a comparison with the other interaction pair (#5/5') involving a one-point N–H···O-bond connection indicates that any additional stabilization due to the short Br···O contact should be very small. Overall, the SCDS-PIXEL calculations indicate that the two short Br···O contacts contribute only little to the stabilization of the lattice in **2**.

2.3. Analysis of C5–X and C5–X' Bond Lengths (X = X' = F, Cl, Br, Me) in 5,5-Disubstituted Barbituric Acids

As reported above, the 5,5-dibromobarbituric acid molecule in **1MH** shows a significant disparity between its two C5–Br bond distances as well as a C5-endo envelope conformation. The elongated axial bond could be the result of the specific spatial characteristics of the axial substituent position. As such, it could be interpreted as an effort to avoid repulsive close interactions between the electron clouds of the axial substituent and pyrimidine ring atoms, specifically C4 and C6 (the corresponding Br···C distances for the axial and equatorial Br atom are 2.77–2.79 Å and 2.83–2.85 Å, respectively). In order to test this hypothesis we will first establish whether a similar correlation between equatorial and axial C5–X and C5–X' bond distances on the one hand and pyrimidine-ring puckering parameters on the other hand can be found for the previously reported [11] polymorphic forms **1a** and **1b**. The same kind of analysis will also be carried out for the crystal structures of the dichloro, difluoro and dimethyl analogues of **1** contained in the Cambridge Structural Database (CSD; version 5.37 [28]). Additional theoretical calculations of molecular structures in the gas phase were carried out in order to ascertain the possible influence of crystal packing effects (see next section).

Table 4 contains information on the pucker of pyrimidine rings (derived from Cremer-Pople parameters [17]) and the corresponding equatorial (d_{C5-X}) and axial ($d_{C5-X'}$) bond distances for the polymorphs of dibromo- (**1a, 1b** [11]) and dichlorobarbituric (**3a, 3b** [11], **3c** [12]) acid and for the crystal structures of difluoro- and dimethylbarbituric acid (**4** [29], **5** [30]). As the difference between d_{C5-X} and $d_{C5-X'}$ in the room-temperature structure of **5** is relatively small a redetermination at 173 K (Table 1) was carried out in an effort to obtain very accurate data.

Table 4. Conformation and puckering amplitudes Q [17] of pyrimidine rings, equatorial (d_{C5-X}) and axial ($d_{C5-X'}$) bond distances at ring atom C5 for the independent molecules (A, B ...) in the crystal structures of dibromobarbituric acid (**1**) and several close analogues (**3–5**; see Scheme 1).

Structure	CSD Refcode	Reference	X, X'	Mol.	Ring Conformation [a]	Q (Å)	d_{C5-X} (Å)	$d_{C5-X'}$ (Å)	Δd (Å) [b]
1MH	-	this work	Br	A	E	0.201(4)	1.907(4)	1.963(4)	0.056(5)
				B	E	0.196(4)	1.912(4)	1.967(3)	0.055(5)
1a	UXIZAD	[11]	Br	A	E	0.217(9)	1.916(8)	1.943(8)	0.027(11)
				B	E	0.091(9)	1.922(8)	1.957(8)	0.035(11)
				C	E→HC	0.232(9)	1.925(9)	1.967(9)	0.042(12)
1b	UXIZAD01	[11]	Br	A	E→C	0.184(6)	1.904(6)	1.970(5)	0.066(8)
				B	E→SB	0.246(6)	1.908(5)	1.965(5)	0.057(7)
3a	UXIYOQ	[11]	Cl	A	E→SB	0.146(4)	1.762(3)	1.782(3)	0.020(5)
				B	E→SB	0.154(4)	1.743(3)	1.787(3)	0.044(5)
				C	E→SB	0.143(4)	1.756(3)	1.782(3)	0.026(4)
				D	E→SB	0.156(4)	1.742(3)	1.788(3)	0.046(5)
3b	UXIYOQ01	[11]	Cl	A	E→HC	0.220(2)	1.742(2)	1.7849(19)	0.043(3)
				B	E	0.112(2)	1.7521(18)	1.7755(18)	0.023(3)
				C	E→HC	0.256(2)	1.743(2)	1.793(2)	0.050(3)
3c	UXIYOQ02	[12]	Cl	-	SB→E	0.136(3)	1.7472(18)	1.7868(18)	0.040(3)
4	HEKTIA	[29]	F	-	E	0.173	1.324(6)	1.359(7)	0.035(9)
5 (173 K)	-	this work	Me	-	E	0.201(2)	1.518(2) [c]	1.5586(18) [c]	0.041(3)
5 (293 K)	NUXTAC	[30]	Me	-	E	0.200	1.532(3) [c]	1.562(3) [c]	0.030(4)

[a] E = C5-endo envelope; HC = half chair; C = chair; SB = skew boat. [b] $\Delta d = d_{C5-X'} - d_{C5-X}$.
[c] Length of the C5–C or C5–C' bond.

All pyrimidine rings listed in Table 4 adopt either an envelope conformation with C5 at the flap or a conformation which is very near to it. The puckering amplitudes Q lie between 0.14 and 0.26 Å. The equatorial C5–X bond is always significantly shorter (taking into account standard uncertainties) than the axial C5–X' bond. The difference Δd between axial and equatorial distances are 0.027–0.066 Å for (X, X' = Br), 0.020 – 0.050 Å for (X, X' = Cl), 0.35 Å (for X, X' = F) and 0.041 Å for (X,

X' = Me). A survey of Br−C(sp^3)−Br structure fragments contained in the CSD [28] indicates that the systematic differences between C5−Br and C5−Br' bonds distances found in **1MH, 1a** and **1b** are unusual (see Figure S4 of the Supplementary Materials).

2.4. Geometry Optimization of 5,5-Disubstituted Barbituric Acid Analogues (X = X' = F, Cl, Br, Me)

The molecular structures of the four 5,5-disubstituted barbituric acid analogues (X = X' = Br, Cl, F, Me) were optimized in gas phase and the calculated parameters (Table 5) have been contrasted to the observed conformations. The C5-endo envelope was calculated to be the conformational energy minimum for all three halogen-substituted barbituric acid analogues, independent of the used level of theory. In each case, it is correlated with a lengthening of the axial relative to the equatorial bond. These results indicate that the non-planarity of the pyrimidine ring is not an effect of crystal packing, *i.e.* of intermolecular forces. Instead, it is associated with the energetically preferred molecular structure. The computed puckering amplitudes Q and the difference between axial and equatorial distances are in agreement with the experimental values for the halogen barbituric acid analogues listed in Table 4. Except for X = X' = Me, changing the basis set from 6-31G(d,p) to aug-cc-pVTz results in slightly smaller puckering amplitudes (Table 5).

Table 5. Calculated parameters, conformation and puckering amplitudes Q [17] of pyrimidine rings and the corresponding equatorial (d_{C5-X}) and axial ($d_{C5-X'}$) bond distances at ring atom C5, derived from gas phase *ab initio* models for four 5,5-disubstituted barbituric acid analogues (X = X' = F, Cl, Br, Me).

Compound (X, X')	Level of Theory	Ring Conformation [a]	Q (Å)	d_{C5-X} (Å)	$d_{C5-X'}$ (Å)	Δd (Å) [b]
1 (Br)	PBE0/6-31G(d,p)	E	0.252	1.902	1.967	0.064
	PBE0/aug-cc-pVTz	E	0.174	1.909	1.959	0.050
	MP2/6-31G(d,p)	E	0.288	1.919	1.976	0.057
3 (Cl)	PBE0/6-31G(d,p)	E	0.245	1.748	1.800	0.052
	PBE0/aug-cc-pVTz	E	0.203	1.744	1.794	0.050
	MP2/6-31G(d,p)	E	0.274	1.745	1.797	0.051
4 (F)	PBE0/6-31G(d,p)	E	0.251	1.331	1.363	0.032
	PBE0/aug-cc-pVTz	E	0.210	1.327	1.360	0.033
	MP2/6-31G(d,p)	E	0.233	1.344	1.378	0.033
5 (Me)	PBE0/6-31G(d,p)	nearly planar	-	1.535	1.541	0.006
	PBE0/aug-cc-pVTz	nearly planar	-	1.531	1.539	0.008
	MP2/6-31G(d,p)	E	0.197	1.522	1.543	0.021

[a] E = C5-endo envelope. [b] $\Delta d = d_{C5-X'} - d_{C5-X}$.

The MP2/6-31G(d,p) optimized molecular structure of 5,5-dimethylbarbituric acid is also a C5-endo envelope conformation, as observed in the experimental crystal structure, and is in agreement with previous calculations performed at the MP2/6-31G(d) level of theory [30]. However, using a DFT (PBE0) method with different basis sets, a nearly planar pyrimidine ring and Δd values of less than 0.01 Å were obtained. In a previous report, Roux and coworkers have shown that the planar conformation becomes more stable than the C5-endo envelope if another DFT method, B3LYP, and a different basis set or the combination MP2/6-31(3df,2p) are used [30]. The disagreement between the methods indicates that the steric effect of the Me group on the ring conformation is less pronounced than that of a halogen substituent. Moreover, a conformational change of the 5,5-dimethylbarbituric acid molecular structure from planar to C5-endo envelope geometry and vice versa requires less energy than the same change in each of the halogen analogues.

2.5. Correlation between Bond Parameters

The optimized molecular structure of **1**, which has the C5-endo envelope conformation, is depicted in Figure 7b. A corresponding molecular structure of **1**, obtained from an alternative optimization with a constrained planar ring geometry, is shown in Figure 7c. In the latter structure, the Br–C5–Br' bond angle is bisected by the trace of the plane defined by ring atoms C4, C5 and C6 so that all four bond angles of the type C4/C6–C5–Br/Br' are 106.7°. The bond geometry around C5 changes significantly in the C5-envelope conformation in that the plane defined by C5, Br and Br' is rotated by more than 6° against the plane defined by C5, C4, and C6 so that the axial C5–Br' bond moves towards the axis of the ring. As the orientation of the C5–Br and C5–Br' bonds relative to one another remains almost unchanged, the two axial C4/6–C5–Br' bond angles are decreased by 2.5°. The ensuing shortening effect on the axial 1,3-distances C···Br' is largely counterbalanced by the simultaneous elongation of the C5–Br' bond by 0.03 Å (envelope: C···Br' = 2.77 Å; planar: C···Br' = 2.79 Å). This suggests that the relative lengthening of the axial C5–Br' bond helps to prevent unfavorably close 1,3-contacts between the electron clouds of the axial substituent and the ring atoms C4 and C6. This relative lengthening would therefore be a steric effect of the energetically preferred C5-endo envelope conformation. Moreover, larger equatorial C4/6–C5–Br angles are accompanied by a shortening of the equatorial bond C5–Br by 0.02 Å in comparison to the molecule with a planar pyrimidine ring.

In order to establish the general correlation between the C–Br bond length and the corresponding Br–C–C bond angle (or the 1,3-distance C···Br), we have carried out a survey of the CSD [28] in which crystal structures containing the O=C(sp^2)–C(sp^3)–Br fragment have been considered. The C–Br bond lengths of 819 such structure fragments are plotted against the corresponding Br–C–C bond angles in Figure 7d. The distribution of data points in the relevant interval between 110° and 100° indicates that a decrease in the Br–C–C angle is generally correlated with a lengthening of the C–Br bond. All data points for the two optimized molecules of **1** shown in Figure 7b,c agree very well with this general trend.

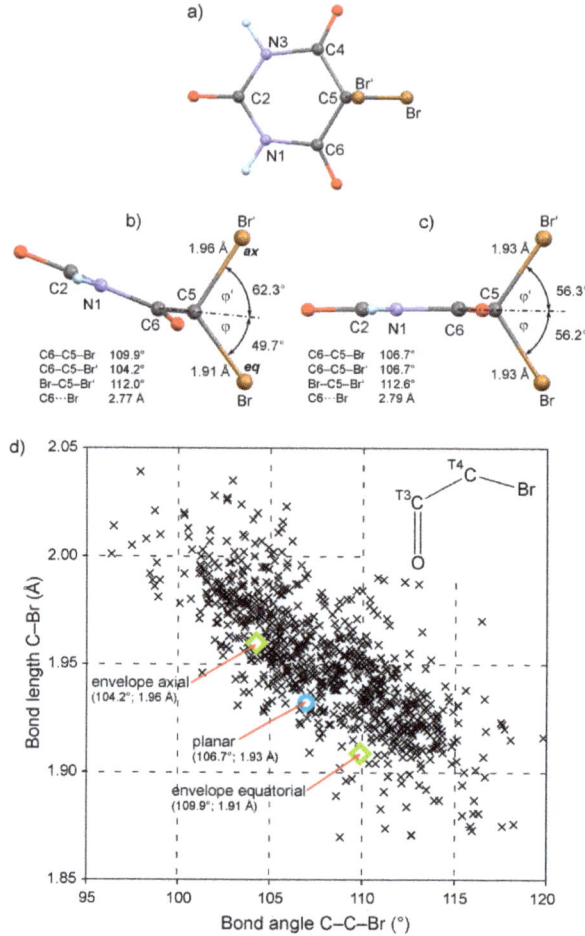

Figure 7. (a) Numbering scheme for **1** (gray = C, light blue = H, blue = N, red = O, brown = Br). (b) PBE0/aug-cc-pVTz optimized molecular structure of **1** showing a C5-endo envelope conformation of the pyrimidine ring [view parallel to the plane defined by (N1, C2, N3, C4, C6), with N3 and C4 superimposed by N1 and C6, respectively; the trace of the plane defined by (C4, C5, C6) is drawn as a dash-dot line; φ and φ' are defined as the angles formed between this plane and the equatorial C5–Br and the axial C5–Br' bond, respectively]; (c) analogous view of a PBE0/aug-cc-pVTz optimized molecular structure of **1** with constrained planar ring geometry. (d) The C–Br bond lengths in 819 O=C(sp^2)–C(sp^3)–Br fragments (inset in upper right-hand corner; from 541 crystal structures with $R < 0.075$, no disorder, no errors, not polymorphic, no ions, no powder structures, only organics) plotted against the corresponding C–C–Br bond angle. The bond parameters of the molecular structures shown in (b) and (c) are represented by the green rhombuses and the blue circle, respectively.

Similar surveys have also been carried out for the analogous $O=C(sp^2)–C(sp^3)–X$ fragments with X = Cl, F, Me. The resulting distance *vs.* angle plots (Supplementary Materials, Figure S5) show that smaller C–C–X bond angles are correlated with longer C–X bonds in the 110° to 100° range in each case. The diagrams in Figure 8 were obtained by superimposing these plots with the experimental data for barbiturates listed in Table 4. In each case, the bond parameters of the equatorial as well as axial ring substituents agree well with the general trend. This is also true for the axial C5–Br bonds of each of the two C5-endo envelope rings in the molecule of **2** (Figure 8a).

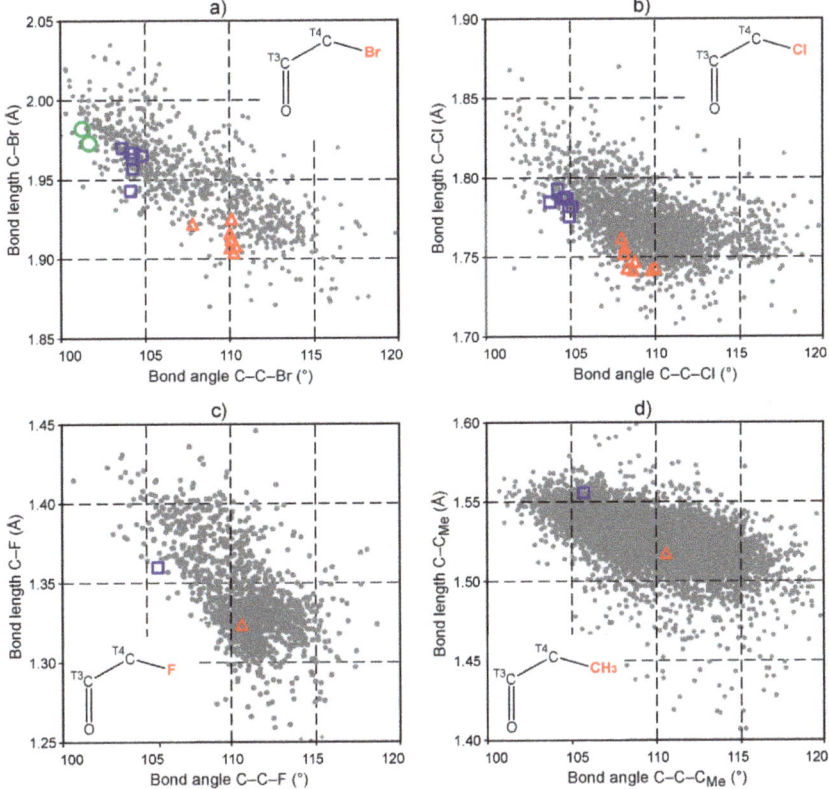

Figure 8. Bond lengths C–X plotted against bond angles C–C–X (gray dots) for structure fragments $O=C(sp^2)–C(sp^3)–X$ with X = Br (**a**); Cl (**b**); F (**c**) and CH_3 (**d**) identified in CSD surveys. Red triangles and blue squares represent the parameters of the equatorial C5–X and axial C5–X' bonds, respectively, listed in Table 4. Green circles in (**a**) represent data points for the axial C5–Br ring substituent bonds of the two C5-endo envelope rings in the molecule of **2**.

3. Experimental Section

3.1. Preparation of Crystal Forms

5,5-Dibromobarbituric acid (**1**) was purchased from Sigma-Aldrich, St. Louis, MO, USA (European affiliate, Steinheim, Germany). Crystals of **1MH** were obtained, at room temperature, from a solution of **1** in MeOH. Desolvation of **1MH** on air resulted in the orthorhombic polymorph **1a** (Supporting Information, Figure S1).

Single crystals of 5,5′-methanediylbis(5-bromobarbituric acid) (**2**) were obtained as products of an unexpected reaction of **1** in a nitromethane solution upon storage at room temperature for several weeks. An NMR tube with a perforated cap was used as a crystallization vessel in order to achieve a slow evaporation rate. The newly introduced methylene linker obviously originates from the nitromethane solvent. Considering the highly acidic character of nitromethane (including its tautomeric equilibrium between *aci*-form and *nitro*-form), it seems likely that, as a first step, a 5-bromo-5-nitromethylbarbituric acid intermediate was formed by nucleophilic replacement of one Br atom by a nitromethane anion. Regardless of the specific mode of dimerization and the nature of the organic nitroalkane reactant involved, the final methylene moiety of **2** would have to be cleaved from the nitro functionality. Such a conversion, e.g., with nitrite as a leaving group, would represent a kind of retro-Kornblum reaction that has been documented in numerous reports [31]. However, 5,5-disubstituted barbituric acid derivatives have also been described to exhibit an extraordinary rich chemistry, and in particular the first bromo group is known to be extremely active [32,33]. Therefore, the direct monodebromination of **1** by a nitromethane tautomer to form 5-bromobarbituric acid may also be considered the source of a potential participant in the formation of the dibromo dimer (Supplementary Materials, Scheme S1).

In summary, the nitromethane anion may be capable of bromobarbiturate alkylation or may serve as a debrominative reducing agent. In the absence of specific investigations, any mechanistic proposal remains however highly speculative. It is therefore unsurprising that our attempts to reproduce compound **2** according to any of the anticipated dimerization steps have been unsuccessful or inconclusive. Nevertheless, an independent synthetic route has been successfully established which starts from methylene bridged barbituric acid (CAS [27406-39-9]) [34] and involves direct bromination in glacial acetic acid of the preformed, already bridged system (Supplementary Materials, Scheme S2). The identity of the resulting precipitate with the investigated single crystal phase of **2** formed in nitromethane was confirmed by comparison of its X-ray powder pattern with the powder pattern calculated from the single crystal data of **2**. A full NMR-spectroscopic characterization of **2** was not possible due to its low solubility. The heating of **2** in solvents suitable for NMR measurements (caried out to achieve better solubility) resulted in a conversion to spiro[furo[2,3-*d*]pyrimidine-6(2*H*),5′(2′*H*)-pyrimidine]-2,2′

(CAS [1333529-88-6]) [35,36]. The identity of the elimination product was unequivocally confirmed by a single-crystal structure determination, which will be published in due course.

5,5-Dimethylbarbituric acid (**5**) was purchased from Sigma-Aldrich, St. Louis, MO, USA (European affiliate, Steinheim, Germany). Single crystals of **5** were obtained by sublimation on a hot bench at 230 °C.

3.2. Single-Crystal X-ray Structure Analyses

Intensity data were collected, using Mo radiation ($\lambda = 0.71073$ Å), on an Oxford Diffraction Gemini-R Ultra diffractometer operated by the CrysAlis software [37]. The data were corrected for absorption effects by means of comparison of equivalent reflections using the program *SADABS* [38]. The structures were solved using the direct methods procedure in *SHELXS97* [39] and refined by full-matrix least squares on F^2 using *SHELXL-2014* [40]. Non-hydrogen atoms were refined anisotropically. Hydrogen atoms were located in difference maps. All NH hydrogen atoms were refined with distance restraints of N–H = 0.86(2) Å and hydrogen atoms bonded to C atoms were refined using riding models. In the case of **1MH**, the O–H distance in the MeOH moiety was restrained to 0.86(2) Å, the U_{iso} parameters of NH hydrogen atoms were refined freely and those of H atoms in the solvent molecule were set to 1.2U_{eq} (OH group) or 1.5U_{eq} (CH$_3$ group) of the parent atom. In the structure of **2**, the U_{iso} parameters of all hydrogen atoms were set to 1.2U_{eq} of the parent N or C atom. The U_{iso} parameters of all hydrogen atoms in the crystal structure of **5** were refined freely.

CCDC 1441623–14416235 contains the supplementary crystallographic data for this paper. These data can be obtained free of charge via http://www.ccdc.cam.ac.uk/conts/retrieving.html (or from the CCDC, 12 Union Road, Cambridge CB2 1EZ, UK; Fax: +44 1223 336033; E-mail: deposit@ccdc.cam.ac.uk)

3.3. Analysis of Crystal Data

Puckering parameters for pyrimidine rings were calculated with *PLATON* [41]. The topology of hydrogen-bonded structures was determined and classified with the programs *ADS* and *IsoTest* of the *TOPOS* package [42] in the manner described by Baburin & Blatov [20].

3.4. Computational Modelling

Gas phase *ab initio* geometry optimizations for each of the four 5,5-dibsubstited barbituric acids (X = X′ = Br, Cl, F, Me) were performed at the PBE0/6-31G(d,p), PBE0/aug-cc-pVTz and MP2/6-31G(d,p) levels of theory using GAUSSIAN09 [43].

3.5. SCDS-PIXEL Calculation

Intermolecular interaction energies for **2** were calculated with the SCDS-PIXEL [24–27] method and the program *OPiX* [44]. The structure model of the CIF was used, and C–H and N–H distances were re-calculated to standard lengths within *OPiX*. No optimization of the molecular geometry was performed. An electron density map was calculated on a three-dimensional grid with a step size of 0.08 Å at the MP2/6-31G(d,p) level using GAUSSIAN09 [43]. A PIXEL condensation factor of 4 was applied, giving superpixels with dimensions 0.32 × 0.32 × 0.32 Å. The calculations yielded interaction energies partitioned into Coulombic, polarization, dispersion and repulsion terms with an expected accuracy of $1 - 2$ kJ·mol^{-1}.

4. Conclusions

The complex H-bonded structure of **1MH** is derived from the 4.8^2-**fes** net which is a well-known topology of two-dimensional MOFs [45], while the HBS of **2** is based on the $3^3.4^6.5^5.6$-**sxd** framework, which has been previously identified as a frequent topology type in organic crystals [20]. The $R_3^3(10)$ rings in the H-bonded structure of **1MH** are reminiscent of the $R_2^2(8)$ motif which has been found in the crystal structures of two polymorphs of 5,5-dibromobarbituric acid (**1a**, **1b**) and in those of many other barbiturates. It seems therefore possible that the desolvation of **1MH** and subsequent formation of the orthorhombic polymorph **1a** proceeds via a direct conversion of $R_3^3(10)$ into $R_2^2(8)$ rings. The minimum energy molecular conformation of 5,5-dihalogen substituted derivatives of barbituric acid is the C5-endo envelope geometry in which the two axial angles C4/C6–C5–X′ are significantly smaller than the corresponding equatorial C4/C6–C5–X angles. Simultaneously, the relatively long axial C5–X′ bond (in comparison with the equatorial C5–X bond) prevents unfavorably short 1,3-distances (C4···X′ and C6···X′) between the axial substituent and pyrimidine ring atoms. This interpretation of the axial C5–X′ and equatorial C5–X bond distances is consistent with general trends in the correlation between the C–C–X bond angles and C–X bond lengths of O=C(sp^2)–C(sp^3)–X structure fragments (X = Br, Cl, F, Me) contained in the CSD.

Supplementary Materials: The following are available online at http://www.mdpi.com/2073-4352/6/4/47/s1, a comparison of the PXRD characteristics of the desolvation product of **1MH** with those of polymorph **1a**, an additional diagram of the H-bonded structure of 1MH, details of the SCDS-PIXEL calculation for **2**, results of CSD surveys and information about the synthetic procedure for compound **2**.

Acknowledgments: We thank Volker Kahlenberg for access to the diffractometer used in this study. Doris E. Braun gratefully acknowledges funding by the Elise Richter programme of the Austrian Science Fund (FWF, project V436-N34).

Author Contributions: Thomas Gelbrich carried out the crystal structure determinations and analyzed crystal data. Doris E. Braun carried out energy calculations of molecular conformations. Synthetic work on the formation of compound **2** was carried out by Stefan

Oberparleiter. All authors were involved in the design of the study and the drafting of the manuscript. All authors have read and approved the final manuscript.

Conflicts of Interest: The authors declare no conflict of interest.

References

1. Brandstätter-Kuhnert, M.; Aepkers, M. Molecular compounds, crystalline solid solutions, and new cases of polymorphism in barbiturates. I. *Microchim. Acta* **1962**, *50*, 1041–1054.
2. Brandstätter-Kuhnert, M.; Aepkers, M. Molecular compounds, crystalline solid solutions, and new cases of polymorphism in barbiturates. II. *Microchim. Acta* **1962**, *50*, 1055–1074.
3. Brandstätter-Kuhnert, M.; Aepkers, M. Molecular compounds, crystalline solid solutions, and new cases of polymorphism in barbiturates. III. *Microchim. Acta* **1963**, *51*, 360–375.
4. Kuhnert-Brandstätter, M.; Vlachopoulos, A. Molecular compounds, crystalline solid solutions, and new cases of polymorphism in barbiturates. IV. *Microchim. Acta* **1967**, *55*, 201–217.
5. Zencirci, N.; Gelbrich, T.; Kahlenberg, V.; Griesser, U.J. Crystallization of metastable polymorphs of phenobarbital by isomorphic seeding. *Cryst. Growth Des.* **2009**, *9*, 3444–3456.
6. Rossi, D.; Gelbrich, T.; Kahlenberg, V.; Griesser, U.J. Supramolecular constructs and thermodynamic stability of four polymorphs and a co-crystal of pentobarbital (nembutal). *CrystEngComm* **2012**, *14*, 2494–2506.
7. Zencirci, N.; Griesser, U.J.; Gelbrich, T.; Kahlenberg, V.; Jetti, R.K.R.; Apperley, D.C.; Harris, R.K. New solvates of an old drug compound (phenobarbital): Structure and stability. *J. Phys. Chem. B* **2014**, *118*, 3267–3280.
8. Gelbrich, T.; Meischberger, I.; Griesser, U.J. Two polymorphs of 5-cyclohexyl-5-ethylbarbituric acid and their packing relationships with other barbiturates. *Acta Crystallogr. Sect. C-Struct. Chem.* **2015**, *71*, 204–210.
9. Gelbrich, T.; Braun, D.E.; Griesser, U.J. Specific energy contributions from competing hydrogen-bonded structures in six polymorphs of phenobarbital. *Chem. Cent. J.* **2016**, *10*.
10. Groth, P. *Chemische Krystallographie. Dritter Teil. Aliphatische und Hydroaromatische Kohlenstoffverbindungen*; Verlag von Wilhelm Engelmann: Leipzig, Germany, 1910; p. 579. (In German)
11. Gelbrich, T.; Rossi, D.; Häfele, C.A.; Griesser, U.J. Barbiturates with hydrogen-bonded layer and framework structures. *CrystEngComm* **2011**, *13*, 5502–5509.
12. Gelbrich, T.; Rossi, D.; Griesser, U.J. Tetragonal polymorph of 5,5-dichlorobarbituric acid. *Acta Crystallogr. Sect. E-Struct. Rep. Online* **2012**, *68*, o235–o236.
13. Williams, P.P. Polymorphism of phenobarbitone: The crystal structure of 5-ethyl-5-phenylbarbituric acid monohydrate. *Acta Crystallogr. Sect. B-Struct. Sci.* **1973**, *29*, 1572–1579.
14. Bhatt, P.M.; Desiraju, G.R. 5,5-Dibenzylbarbituric acid monohydrate. *Acta Crystallogr. Sect. E-Struct. Rep. Online* **2007**, *63*, o771–o772.

15. Gelbrich, T.; Rossi, D.; Griesser, U.J. Butallylonal 1,4-dioxane hemisolvate. *Acta Crystallogr. Sect. E-Struct. Rep. Online* **2010**, *66*, o2688.
16. Ravi Kiran, B.; Suchetan, P.A.; Amar, H.; Vijayakumar, G.R. Crystal structure of 5,5-bis(4-methylbenzyl) pyrimidine-2,4,6(1*H*,3*H*,5*H*)-trione monohydrate. *Acta Crystallogr. Sect. E-Struct. Commun.* **2015**, *71*, 19–21.
17. Cremer, D.; Pople, J.A. General definition of ring puckering coordinates. *J. Am. Chem. Soc.* **1975**, *97*, 1354–1358.
18. Etter, M.C.; MacDonald, J.C.; Bernstein, J. Graph-set analysis of hydrogen-bond patterns in organic crystals. *Acta Crystallogr. Sect. B-Struct. Sci.* **1990**, *46*, 256–262.
19. Bernstein, J.; Davis, R.E.; Shimoni, L.; Chang, N.-L. Patterns in hydrogen bonding: Functionality and graph set analysis in crystals. *Angew. Chem. Int. Ed.* **1995**, *34*, 1555–1573.
20. Baburin, I.A.; Blatov, V.A. Three-dimensional hydrogen-bonded frameworks in organic crystals: A topological study. *Acta Crystallogr. Sect. B-Struct. Sci.* **2007**, *63*, 791–802.
21. Hursthouse, M.B.; Hughes, D.S.; Gelbrich, T.; Threlfall, T.L. Describing hydrogen-bonded structures; topology graphs, nodal symbols and connectivity tables, exemplified by five polymorphs of each of sulfathiazole and sulfapyridine. *Chem. Cent. J.* **2015**, *9*.
22. O'Keeffe, M.; Peskov, M.A.; Ramsden, S.J.; Yaghi, O.M. The reticular chemistry structure resource (RCSR) database of, and symbols for, crystal nets. *Acc. Chem. Res.* **2008**, *41*, 1782–1789.
23. Bondi, A. Van der waals volumes and radii. *J. Phys. Chem.* **1964**, *68*, 441–451.
24. Dunitz, J.D.; Gavezzotti, A. Molecular recognition in organic crystals: Directed intermolecular bonds or nonlocalized bonding? *Angew. Chem. Int. Ed.* **2005**, *44*, 1766–1787.
25. Gavezzotti, A. *Molecular Aggregation: Structure Analysis and Molecular Simulation of Crystals and Liquids*; Oxford University Press: Oxford, UK, 2007.
26. Gavezzotti, A. Calculation of lattice energies of organic crystals: The PIXEL integration method in comparison with more traditional methods. *Z. Kristallogr.* **2005**, *220*, 499–510.
27. Gavezzotti, A. Quantitative ranking of crystal packing modes by systematic calculations on potential energies and vibrational amplitudes of molecular dimers. *J. Chem. Theory Comput.* **2005**, *1*, 834–840.
28. Groom, C.R.; Allen, F.H. The Cambridge Structural Database in retrospect and prospect. *Angew. Chem. Int. Ed.* **2014**, *53*, 662–671.
29. DesMarteau, D.D.; Pennington, W.T.; Resnati, G. Fluorinated barbituric acid derivatives. *Acta Crystallogr. Sect. C-Cryst. Struct. Commun.* **1994**, *50*, 1305–1308.
30. Roux, M.V.; Notario, R.; Foces-Foces, C.; Temprado, M.; Ros, F.; Emel'yanenko, V.N.; Verevkin, S.P. Experimental and computational thermochemical study and solid-phase structure of 5,5-dimethylbarbituric acid. *J. Phys. Chem. A* **2010**, *114*, 3583–3590.
31. Kornblum, N.; Smiley, R.A.; Blackwood, R.K.; Iffland, D.C. The mechanism of the reaction of silver nitrite with alkyl halides. The contrasting reactions of silver and alkali metal salts with alkyl halides. The alkylation of ambident anions. *J. Am. Chem. Soc.* **1955**, *77*, 6269–6280.

32. Brown, D.J.; Mason, S.F. *The Chemistry of Heterocyclic Compounds: The Pyrimidines*, 99th ed.; Wiley: New York, NY, USA; London, UK, 2009.
33. Grundke, G.; Keese, W.; Rimpler, M. 5,5-Dibrombarbitursäure, ein neues Reagenz zur Bromierung von gesättigten und α,β-ungesättigten Carbonylverbindungen. *Chem. Berichte* **1985**, *118*, 4288–4291.
34. Gysling, H.; Schwarzenbach, G. Metallindikatoren II. Beziehungen zwischen Struktur und Komplexbildungsvermögen bei Verwandten des Murexids. *Helv. Chim. Acta* **1949**, *32*, 1484–1504.
35. Jalilzadeh, M.; Pesyan, N.N.; Rezaee, F.; Rastgar, S.; Hosseini, Y.; Sahin, E. New one-pot synthesis of spiro[furo[2,3-d]pyrimidine-6,5'-pyrimidine]pentaones and their sulfur analogues. *Mol. Divers.* **2011**, *15*, 721–731.
36. Ara, T.; Khan, K.Z. Synthesis of some derivatives of dimedone, γ-pyrone and barbituric acid. *J. Pharm. Res. (Mohali, India)* **2014**, *8*, 786–790.
37. Oxford Diffraction Ltd. *Crysalis CCD and CrysAlis RED*; Oxford Diffraction Ltd.: Abingdon, UK, 2003.
38. Sheldrick, G.M. *SADABS. Version 2007/7*; Bruker AXS Inc.: Madison, WI, USA, 2007.
39. Sheldrick, G.M. A short history of SHELX. *Acta Crystallogr. Sect. A-Fundam. Crystallogr.* **2008**, *64*, 112–122.
40. Sheldrick, G. Crystal structure refinement with SHELXL. *Acta Crystallogr. Sect. C-Cryst. Struct. Chem.* **2015**, *71*, 3–8.
41. Spek, A. Structure validation in chemical crystallography. *Acta Crystallogr. Sect. D-Biol. Crystallogr.* **2009**, *65*, 148–155.
42. Blatov, V.A. Multipurpose crystallochemical analysis with the program package TOPOS. *IUCr Compcomm Newsl.* **2006**, *7*, 4–38.
43. Frisch, M.J.; Trucks, G.W.; Schlegel, H.B.; Scuseria, G.E.; Robb, M.A.; Cheeseman, J.R.; Scalmani, G.; Barone, V.; Mennucci, B.; Petersson, G.A.; et al. *Gaussian 09*; Gaussian Inc.: Wallingford, CT, USA, 2009.
44. Gavezzotti, A. *OPiX: A Computer Program Package for the Calculation of Intermolecular Interactions and Crystal Energies*; University of Milan: Milano, Italy, 2007.
45. Khamitova, D.R.; Blatov, V.A.; Carlucci, L.; Ciani, G.; Proserpio, D.M. Local and global topology of two-dimensional structural groups. *Acta Crystallogr. Sect. A-Fundam. Crystallogr.* **2009**, *65*, s306.

Isomorphous Crystals from Diynes and Bromodiynes Involved in Hydrogen and Halogen Bonds

Pierre Baillargeon, Édouard Caron-Duval, Émilie Pellerin, Simon Gagné and Yves L. Dory

Abstract: Isomorphous crystals of two diacetylene derivatives with carbamate functionality (BocNH-CH$_2$-diyne-X, where X = H or Br) have been obtained. The main feature of these structures is the original 2D arrangement (as supramolecular sheets or walls) in which the H bond and halogen bond have a prominent effect on the whole architecture. The two diacetylene compounds harbor neighboring carbamate (Boc protected amine) and conjugated alkyne functionalities. They differ only by the nature of the atom located at the penultimate position of the diyne moiety, either a hydrogen atom or a bromine atom. Both of them adopt very similar 2D wall organizations with antiparallel carbamates (as in antiparallel beta pleated sheets). Additional weak interactions inside the same walls between molecular bricks are H bond interactions (diyne-H···O=C) or halogen bond interactions (diyne-Br···O=C), respectively. Based on crystallographic atom coordinates, DFT (B3LYP/6-31++G(d,p)) and DFT (M06-2X/6-31++G(d,p)) calculations were performed on these isostructural crystals to gain insight into the intermolecular interactions.

Reprinted from *Crystals*. Cite as: Baillargeon, P.; Caron-Duval, É.; Pellerin, É.; Gagné, S.; Dory, Y.L. Isomorphous Crystals from Diynes and Bromodiynes Involved in Hydrogen and Halogen Bonds. *Crystals* **2016**, *6*, 37.

1. Introduction

One of the important scientific issues today involves the development of materials with controlled hierarchical structures organized in different sizes, especially at the nanoscopic scale [1]. Scientists exploit supramolecular chemistry (bottom-up strategy) to achieve this goal and control the organization of molecules into diverse 1D [2–4], 2D [5–9] or 3D shapes [10–12]. Owing to their directional potential, hydrogen bond [13–23] or halogen bond [24–36] are intermolecular forces that are often used to stick molecular building blocks together. Recently, in our laboratories, we exploited mainly H bonds to build supramolecular walls based on lactams [37] or proline derivatives [38].

Since isostructurality (isomorphism) is a successful approach adopted in the construction of organic assemblies [39–42], we decided to investigate if hydrogen bonds and halogen bonds could be used with similar results. We now present

2D layered isostructural crystals of two diacetylene derivatives (Figure 1) that are stabilized with hydrogen or halogen bonds. One of the main interests of these crystals is that they differ only by a diyne C–H···O=C or C–Br···O=C interaction. For this reason, it became very easy to isolate and compare these geometric arrangements since their environments are identical. The hydrogen bonds could be compared to the isosteric halogen bonds using DFT calculations based on X-ray atomic coordinates. Better knowledge of the behavior of these interactions could bring additional tools in crystal engineering and in material design, as well as for efficient and fruitful drug design. Indeed, terminal alkynes [43,44] and haloalkynes [45–47] are biologically important classes of molecules.

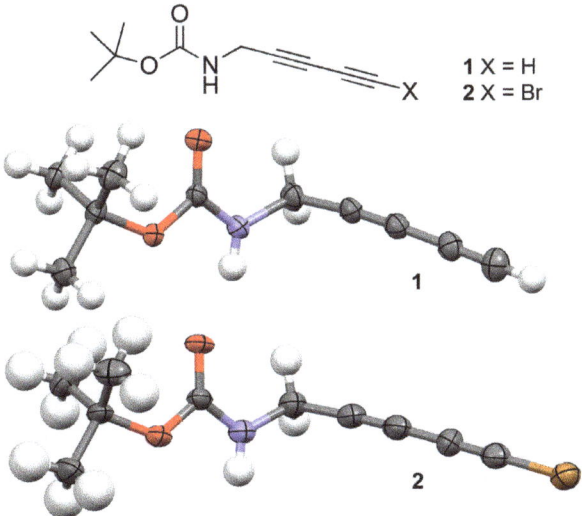

Figure 1. Chemical and crystal (ORTEP) structures of the diacetylene derivatives **1** (BocNHCH$_2$ CCCCH) and **2** (BocNHCH$_2$CCCCBr).

2. Results and Discussion

2.1. Synthesis

Both isosteric compounds **1** and **2** were obtained in a straightforward and efficient manner from commercially available N-Boc-propargylamine **3** and ethynyltrimethyl silane **4** (Scheme 1). These two reactants were coupled together with Hay catalyst [48] to yield the diyne **5** with a yield of 55%, easily separated from two symmetrical reaction byproducts **6** and **7**. The trimethyl silyl group of **5** was either cleaved with potassium carbonate to give the first target diyne **1** (60% yield) or it was replaced by a bromine atom to afford the other desired compound **2** with a yield of 30%. This was achieved with *N*-bromosuccinimide (NBS) in the presence of silver nitrate. In the process, some diyne **1** was also obtained (14%).

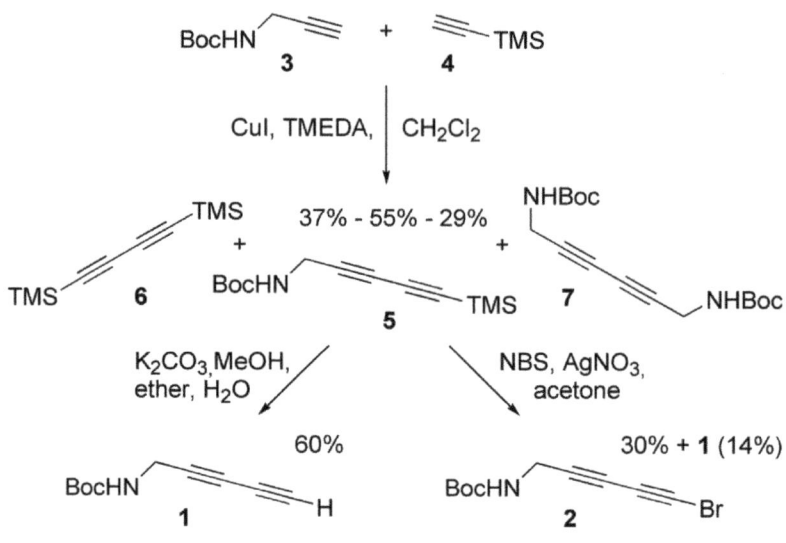

Scheme 1. Synthesis of diynes **1** and **2**.

2.2. Crystallographic Studies

Crystals of both diynes **1** and **2** are isostructural. Their unit cells are very similar, being slightly larger for bromide **2** (Table 1). The replacement of a hydrogen atom in **1** by a bromine atom in **2** leads to an increase in density. The *ab* planes of the crystal structures (Figure 2) are constituted of molecular walls whose bricks are maintained by weak forces like hydrogen bonds (alkyne **1**, Figure 2a) or a combination of hydrogen bonds and halogen bonds (bromoalkyne **2**, Figure 2b). For both crystals, the carbamates stack on top of each other along the *b*-axis and in an antiparallel way through NH···OC hydrogen bonds (the NH···OC hydrogen bond distances are very similar, being 2.054 Å and 2.088 Å for **1** and **2**, respectively). Consequently, the rigid diyne arms stick out from each side of the 1D carbamate tapes in an alternate manner. Their constitutive atoms also lay almost in the same *ab* plane. As an additional proof of isostericity, the *b* side of both unit cells displays almost identical length (b = 9.3135 Å for **1** and b = 9.2090 Å for **2**).

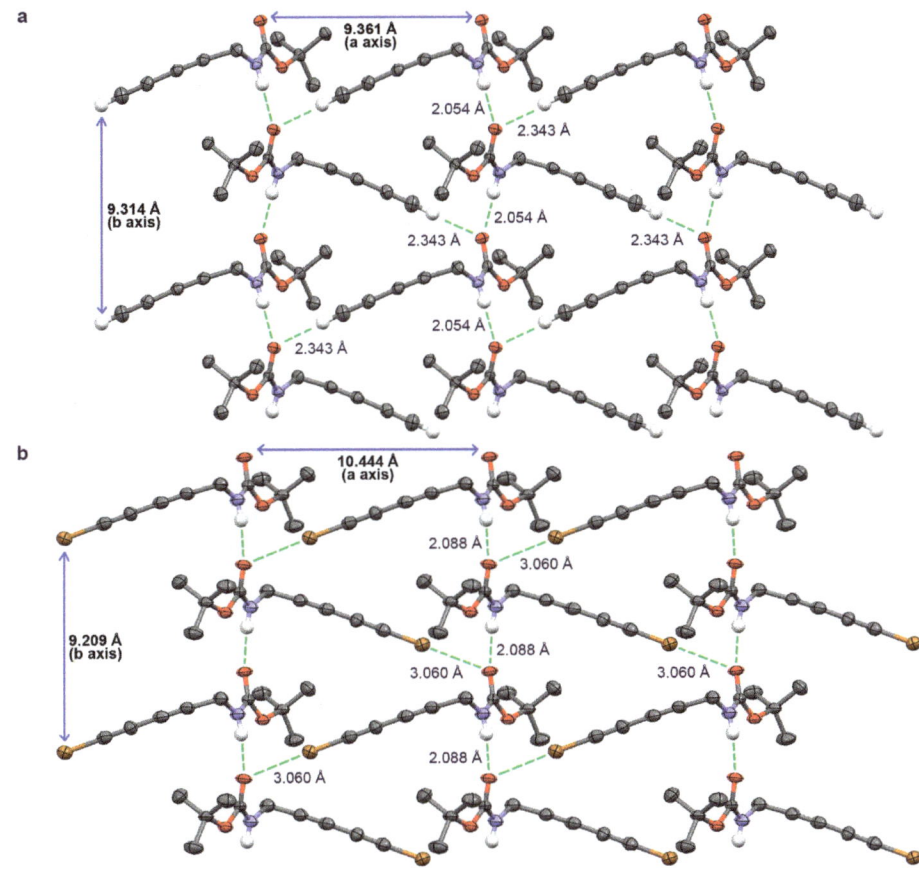

Figure 2. Halogen and/or hydrogen bonds inside the supramolecular walls of (**a**) diyne **1** and (**b**) bromodiyne **2**. The non polar hydrogen atoms have been removed for clarity.

Looking along the b-axis (Figure 3, top views of the supramolecular walls), there is no significant difference either. The repetitive unit along the c-axis is just a little bit longer in **2** ($c = 12.274$ Å) by comparison with **1** ($c = 11.898$ Å). All walls have opposite carbonyl orientation. The main difference between crystals **1** and **2** may be seen along the a-axis. It is a direct consequence of the replacement of an H atom by a Br atom. The C–Br bond being longer (1.800 Å) than the C–H bond (0.951 Å) leads to longer a side for the unit cell in **2** ($a = 10.4435$ Å) by comparison with **1** ($a = 9.3613$ Å). The elongation effect arises also from major differences existing between the non-conventional C–H···O=C hydrogen bond [49] length (2.343 Å) and the C–Br···O=C halogen bond length (3.060 Å). In fact, this latter distance is 9% shorter than the sum of the van der Waals radii of Br and O [50–52], strongly suggesting that there is more at stake than a simple van der Waals contact between the two heteroatoms. This is also supported by the alignment of the four atoms C–Br⋯O=C matching that of the C–H⋯O=C atoms involved in a hydrogen bond [53]. Few examples of such halogen bonds are reported in the Cambridge Structural Database [26,54–56].

Table 1. Crystallographic data for diynes **1** and **2**.

	Diyne 1	**Diyne 2**
formula	$C_{10}H_{13}NO_2$	$C_{10}H_{12}BrNO_2$
MW/g·mol^{-1}	179.21	258.12
crystal system	monoclinic	monoclinic
space group	P 21/c	P 21/c
a/Å	9.3613(15)	10.4435(16)
b/Å	9.3135(14)	9.2090(15)
c/Å	11.8981(19)	12.2744(19)
β/deg	102.497(5)	102.599(4)
V/Å^3	1012.8(3)	1152.1(3)
Z	4	4
ρ_{calc}/g·cm^{-3}	1.175	1.488
meas. reflns	5204	16523
ind. reflns	1848	2175
R_{int}	0.0403	0.0656
R_1 [$I > 2\sigma(I)$]	0.0382	0.0644
wR_2 [$I > 2\sigma(I)$]	0.0840	0.1662
GoF	1.027	1.133

Finally, for both crystals, slow degradation occurs over time at room temperature and in ambient atmosphere. At this moment, it this difficult to conclude what the "degradation" products might be. Although single-crystal-to-single-crystal dimerization has been observed in other bromodiacetylene derivatives, this process is very unlikely in the case of **1** and **2** [57]. Indeed, their geometric arrangements are inappropriate for this type of reaction. The diyne systems are parallel as requested; nevertheless, the angle of the reactive carbons and mostly their distance separations are not favorable, being distorted and much too long, respectively (Figure 4). Topochemical polymerization of diacetylene is also prevented because there is no long range parallel stacking of diynes [58–67].

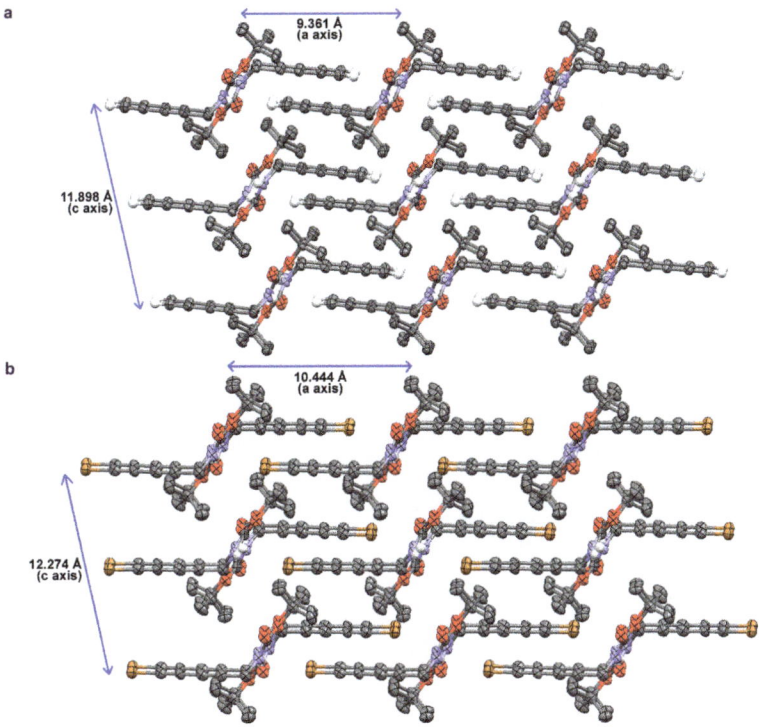

Figure 3. Top views of the supramolecular walls, seen along the b-axis of (**a**) diyne **1** and (**b**) bromodiyne **2**. The non polar hydrogen atoms have been removed for clarity.

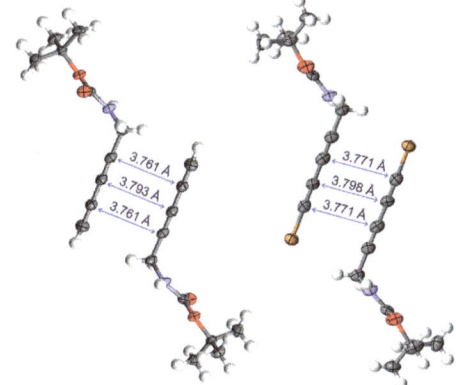

Figure 4. Geometrical characteristics of the closest parallel diyne rods in crystals of **1** and **2**.

2.3. Computational Studies

In order to gain more insight into the energetics at stake in the crystals of **1** and **2**, a DFT study was carried out on simplified models **8-8-9** and **8-8-10** of these crystals (Figure 5) [68–71].

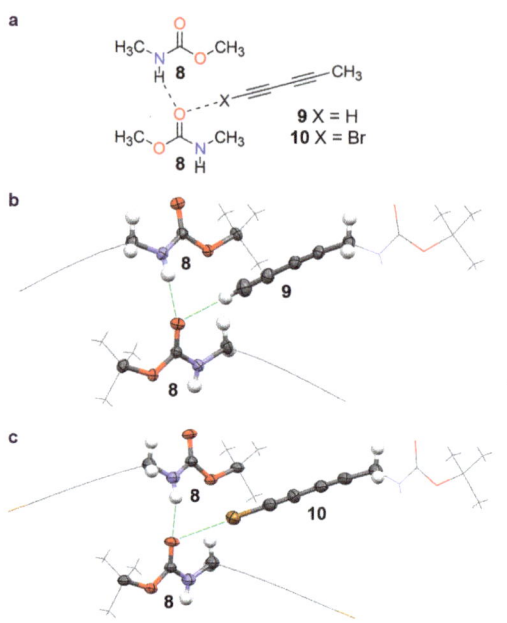

Figure 5. Summary of DFT calculations. (**a**) structures of systems **8-8-9** and **8-8-10**, in which **8**, **9** and **10** are *N,O*-dimethylcarbamate, penta-1,3-diyne and 1-bromopenta-1,3-diyne. Initial geometries for systems (**b**) **8-8-9** extracted from the crystal of **1** and (**c**) **8-8-10** corresponding to the crystal of **2**.

Thus, both starting geometries are constituted of two *N,O*-dimethylcarbamate molecules **8**, interacting with each other through a hydrogen bond. In each case, the carbamate carbonyl oxygen already involved in the hydrogen bond is also bound, via a bifurcated non-covalent bond [72], to the terminal atom of a simple diyne, either penta-1,3-diyne **9** (extracted from crystal of **1**) or 1-bromopenta-1,3-diyne **10** (extracted from crystal of **2**). This apparent oversimplification is justified by our desire to "isolate" the energy stored in the hydrogen and halogen bonds present in these crystals. Therefore, all atoms purely involved in these weak interactions were retained while the others were discarded. Without the crystal lattice constraint, calculations could obviously yield extremely different final structures, especially in the current cases. Indeed, the potential energy surface (PES) involving weak interactions is known to be shallow [73]. This means that important geometrical variations of distances and angles between H-bond and halogen bond partners may result in minute energy changes (e.g., <0.2 kcal·mol^{-1}). Nevertheless, it is still possible to gauge if the systems **8-8-9** and **8-8-10** are near a minimum on the PES by not being too stringent on the first derivative (maximum energy gradient) requirements during calculations.

Two DFT methods were applied to study these systems, namely the widely used B3LYP functional as well as the M06-2X method known to be more suitable to treat weak interactions and peptides [74,75]. The B3LYP minimizations led to RMSDs as small as 0.16 Å and 0.34 Å for the **8-8-9** and **8-8-10** systems, respectively. The RMSD figures were similar for the M06-2X method: 0.39 Å and 0.37 Å for the same systems (Figure 6). The distances between weak interaction partners (excluding van der Waals) remain close before and after calculations (distance C<u>O</u>×××<u>HN</u> of 2.93 Å, 3.03–3.02 Å, 2.98–2.95 Å in crystals **1** and **2**, B3LYP, M06-2X DFT calculations, respectively; distance C<u>O</u>×××H<u>C</u>CCC of 3.29 Å, 3.34 Å, 3.22 Å in crystal **1**, B3LYP, M06-2X DFT calculations, respectively; distance C<u>O</u>×××<u>Br</u>CCCC of 3.06 Å, 3.10 Å, 2.95 Å in crystal **2**, B3LYP, M06-2X DFT calculations, respectively). With these good matches between initial and equilibrium structures [76], it was then possible to convincingly evaluate, or at least rank, the strength of the weak hydrogen and halogen bonding interactions, partially responsible for the orientations of the monomers **1** and **2** in their respective crystals (Table 2).

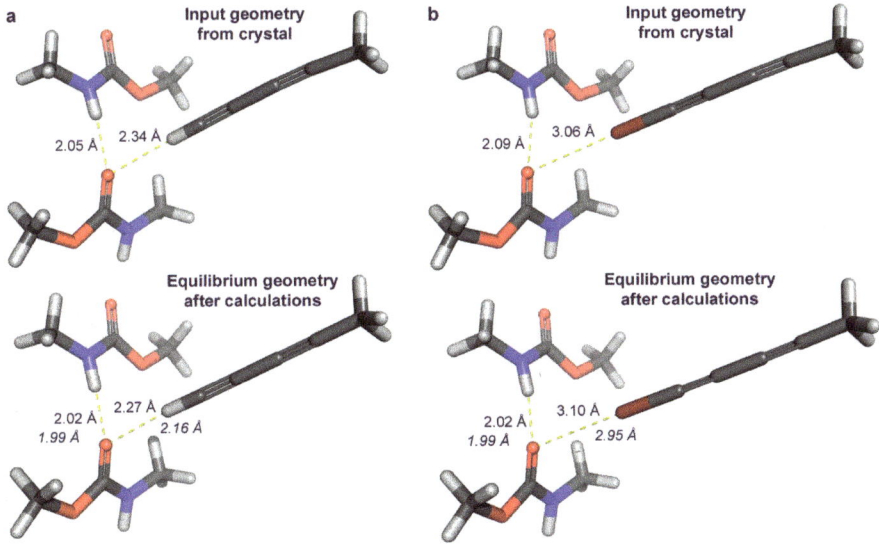

Figure 6. Starting (crystal) and equilibrium (DFT minimization) geometries for (**a**) system **8-8-9** and for (**b**) system **8-8-10**. Plain figures in the equilibrium geometries are for B3LYP calculations and italic figures for M06-2X calculations.

Table 2. DFT energies for the minimized geometries.

System	Energy (E_h) [1]	
	B3LYP	M06-2X
8	−323.59792	−323.62775
9	−192.70191	−192.73108
10	−2765.83708	−2766.01159
8-8	−647.20438	−647.26691
8-8-9	−839.90921	−840.00619
8-8-10	−3413.04394	−3413.28685

[1] E_h means Hartree (1 E_h = 627.5 kcal·mol^{-1}).

Examination of the energy figures obtained after DFT minimization shows that hydrogen or halogen bonds are consistently much weaker following the B3LYP method, in comparison with the M06-2X protocol. For example, B3LYP DFT minimizations yield an energy of −5.36 kcal·mol^{-1} for the hydrogen bond between carbamate NH (donor) and carbamate CO (acceptor) as shown in Figure 6 (E_8-8 − 2 × E_8, meaning Energy of system **8-8** minus twice the Energy of **8** as reported in Table 2). At the M06-2X level, the same hydrogen bond is even stronger with an energy of −7.16 kcal·mol^{-1}. The corresponding input coordinates of **8-8** (before its minimization) had been obtained from the minimized full systems **8-8-9** and **8-8-10**. It was finally possible to estimate the strength of the hydrogen bond between the alkyne hydrogen and the carbamate carbonyl by deducting the energies of system **8-8** and free diyne **9** to the energy of **8-8-9** (E_8-8-9 − E_8-8 − E_9). The strength of the halogen bond between the carbamate CO of **8** and the terminal bromine atom from bromodiyne **10** was evaluated in the same way (E_8-8-10 − E_8-8 − E_10).

Amazingly, both weak interactions involving alkyne hydrogen and alkyne bromine with a carbamate CO are very much comparable (same energy range), being weaker than a conventional hydrogen bond between carbamates. These hydrogen and halogen bonds have respective energies of −1.83 kcal·mol^{-1} and −1.56 kcal·mol^{-1} at the B3LYP DFT level and −5.15 kcal·mol^{-1} and −5.24 kcal·mol^{-1} following the M06-2X DFT method.

3. Experimental Section

3.1. Synthesis

Diynes **5-6-7**. To a solution of *N*-Boc-propargylamine **3** (1.0 g, 6.45 mmol) and ethynyltri methylsilane **4** (1.78 mL, 12.5 mmol) in DCM (50 mL) was added Hay catalyst [freshly prepared by stirring CuI (1.3 g, 6.8 mmol) and TMEDA (1.92 mL, 12.9 mmol) in DCM (10 mL) under argon]. The reaction mixture was stirred under oxygen atmosphere (balloon) for 110 min. The resulting deep brown residue was purified by flash chromatography (DCM, then ether:hexane 50:50, then ether), yielding the diyne **5** as a brown-orange oil (887 mg, 55%), the diyne **6** as a white solid (458 mg, 37%) and the diyne **7** as a white solid (290 mg, 29%).

For diyne **5**: Rf = 0.42 (hexane:EtOAc, 75:25). ^{1}H NMR (400 MHz, CDCl$_3$, δ ppm): 4.70 (br s, 1H), 4.00 (br s, 2H), 1.44 (s, 9H), 0.18 (s, 9H). HRMS (*m/z*) calcd for C$_{13}$H$_{21}$NO$_2$SiNa [MNa$^+$]: 274.1234, found: 274.1240.

For diyne **6**: ^{1}H NMR (400 MHz, CDCl$_3$, δ ppm): 0.00 (s, 18H).

For diyne **7**: ^{1}H NMR (400 MHz, CDCl$_3$, δ ppm): 4.91 (br, 1H), 3.96 (br, 2H), 1.42 (s, 18H).

Diyne **1**. A mixture of diyne **5** (208 mg, 0.83 mmol) and K$_2$CO$_3$ (508 mg, 3.68 mmol) in MeOH (5 mL) and Et$_2$O (5 mL) with a drop of water was stirred at room temperature for 70 min. The solvent was removed under reduced pressure and the residue was purified by chromatography (hexane: Et$_2$O, 70:30) to yield the diyne **1** as white crystals (89 mg, 60%). Rf = 0.40 (hexane:EtOAc, 75:25). ^{1}H NMR (300 MHz, CDCl$_3$ δ ppm): 4.72 (br s, 1H), 3.99 (br s, 2H), 2.10 (s, 1H), 1.45 (s, 9H). ^{13}C NMR (100 MHz, CDCl$_3$, δ ppm): 155.32, 80.58, 73.25, 67.81, 67.55, 67.33, 31.07, 28.53. HRMS (*m/z*) calcd for C$_{10}$H$_{13}$NO$_2$Na [MNa$^+$]: 202.0839, found: 202.0839.

Diyne **2**. AgNO$_3$ (36 mg, 0.21 mmol) and NBS (228 mg, 1.28 mmol) were added to a solution of diyne **5** (270 mg, 1.07 mmol) in acetone (10 mL) at room temperature. The resulting mixture was stirred for 18 h under N$_2$ in the absence of light. Purification of the crude product by flash chromatography on silica gel eluting with mixtures of EtOAc and hexane (gradient from 10:90 to 25: 75) provided the title compound **2** as an orange solid (82 mg, 30%) and the diyne **1** as a white solid (26 mg, 14%). Rf = 0.52 (hexane:EtOAc, 70:30). ^{1}H NMR (400MHz, CDCl$_3$, δ ppm): 4.72 (br s,1H), 3.98 (br s, 2H), 1.44 (s, 9H). ^{13}C NMR (100MHz, CDCl$_3$, δ ppm): 155.31, 80.58, 72.03, 68.26, 65.15, 40.49, 31.11, 28.53. HRMS (*m/z*) calcd for C$_{10}$H$_{12}$BrNO$_2$Na [MNa$^+$]: 279.9944, found: 279.9951

3.2. Crystallizations

Diyne **1** crystallized from a solution of acetone and CDCl$_3$ that was left to stand in a small vial at room temperature for several days. Colorless crystals of BocNHCH$_2$CCCCH, **1**, suitable for X-ray analysis were obtained. A Needle-like specimen of C$_{10}$H$_{13}$NO$_2$ (0.05 mm × 0.10 mm × 0.28 mm), was used for the X-ray crystallographic analysis.

The same technique was used to obtain colorless crystals of BocNHCH$_2$CCCCBr, **2**, from a solution of ether, CDCl$_3$ and hexane. A prism-like specimen of C$_{10}$H$_{12}$BrNO$_2$ (0.08 mm × 0.15 mm × 0.68 mm), was cropped for the X-ray crystallographic analysis.

3.3. X-Ray Crystallography

The X-ray intensity data were measured on a Bruker Apex DUO system equipped with a Cu Kα ImuS micro-focus source with MX optics (Bruker, Madison,

WI, USA) (λ = 1.54178 Å). The frames were integrated with the Bruker SAINT software package (Bruker, Madison, WI, USA) using a wide-frame algorithm. Data were corrected for absorption effects using the multi-scan method (SADABS). The structure was solved and refined using the Bruker SHELXTL Software Package (Bruker, Madison, WI, USA), using the space group P 21/c, with Z = 4 for the formula unit. Full details of the crystallographic data and refinement are presented in Table 1 and in the supporting information file (PDF). CCDC 1451745-1451746 contain the supplementary crystallographic data for this paper. These data can be obtained free of charge via http://www.ccdc.cam.ac.uk/conts/retrieving.html (or from the CCDC, 12 Union Road, Cambridge CB2 1EZ, UK; Fax: +44 1223 336033; E-mail: deposit@ccdc.cam.ac.uk).

3.4. Computational Details

All calculations were performed with the GAMESS program package (Iowa State University, Ames, IA, USA) [77] using the B3LYP/6-31++G(dp) and the M06-2X/6-31++G(d,p) density functional basis sets [78,79]. No zero point corrections were applied to the calculated raw energies following minimizations. The initial geometries were extracted from the single-crystal X-ray diffraction data (Figure 5). All atoms not directly involved in the hydrogen and halogen bonds were removed and hydrogen atoms were finally added to fill up the valence requirements. For the minimizations, the requested maximum energy gradient was 0.0005 $E_h \cdot a_0^{-1}$ (OPTTOL = 0.0005).

4. Conclusions

Despite their obvious differences in nature, both hydrogen and halogen bonds [80], between terminal diyne hydrogen or bromine atoms, respectively, with carbonyl oxygen as a partner, behave in the same way geometrically as well as energetically. Consequently, the data gathered in this work suggest that these non-covalent bonds can be used to produce identical patterns in crystals, the only noticeable difference arising from the van der Waals radii of hydrogen and bromine atoms.

Supplementary Materials: The following are available online at http://www.mdpi.com/2073-4352/6/4/37/s1. Spectral characterization data for compounds **1, 2, 5, 6** and **7** (PDF). RMSD parameters of DFT *x,y,*and *z* coordinates for systems **8-8-9** and **8-8-10**. Crystallographic information files for compounds **1** and **2** (CIF).

Acknowledgments: Financial support by NSERC Canada, the Fonds de recherche du Québec-Nature et technologies (FRQNT, Grant No. 2016-CO-194882), the Programme de Collaboration Universités-Collèges (Grant No. PCUC-9004, Université de Sherbrooke) and the Centre d'étude et de recherche transdisciplinaire étudiants-enseignants (CERTEE, Cégep de Sherbrooke) are gratefully acknowledged, as well as Calcul Québec for computational resources. Many thanks to Tarik Rahem and Victor Fan for their valuable help with the synthesis of diynes **1** and **2**.

Author Contributions: Pierre Baillargeon conceived and designed the experiments; Édouard Caron-Duval, Émilie Pellerin and Simon Gagné performed the experiments; Yves L. Dory carried out the calculations; Pierre Baillargeon and Yves L. Dory analyzed the data; Pierre Baillargeon and Yves L. Dory contributed reagents/materials/analysis tools; Pierre Baillargeon and Yves L. Dory wrote the paper.

Conflicts of Interest: The authors declare no conflict of interest.

Abbreviations

The following abbreviations are used in this manuscript:

NBS	*N*-BromoSuccinimide
DFT	Density-Functional Theory
DCM	DiChloroMethane
TMEDA	TetraMethylEthyleneDiAmine
Rf	Retardation Factor
IR	InfraRed
NMR	Nuclear Magnetic Resonance
HRMS	High-Resolution Mass Spectrometry
GAMESS	General Atomic and Molecular Electronic Structure System

References

1. Rao, C.N.R.; Cheetham, A.K. Science and technology of nanomaterials: Current status and future prospects. *J. Mater. Chem.* **2001**, *11*, 2887–2894.
2. Gauthier, D.; Baillargeon, P.; Drouin, M.; Dory, Y.L. Self-Assembly of Cyclic Peptides into Nanotubes and Then into Highly Anisotropic Crystalline Materials. *Angew. Chem. Int. Ed.* **2001**, *40*, 4635–4638.
3. Pasini, D.; Ricci, M. Macrocycles as Precursors for Organic Nanotubes. *Curr. Org. Synth.* **2007**, *4*, 59–80.
4. Baillargeon, P.; Bernard, S.; Gauthier, D.; Skouta, R.; Dory, Y.L. Efficient Synthesis and Astonishing Supramolecular Architectures of Several Symmetric Macrolactams. *Chem. Eur. J.* **2007**, *13*, 9223–9235.
5. Glaser, R. Polar Order by Rational Design: Crystal Engineering with Parallel Beloamphiphile Monolayers. *Acc. Chem. Res.* **2007**, *40*, 9–17.
6. George, S.; Nangia, A.; Bagieu-Beucher, M.; Masse, R.; Nicoud, J.-F. Crystal engineering of two-dimensional polar layer structures: Hydrogen bond networks in some *N*-meta-phenylpyrimidinones. *New J. Chem.* **2003**, *27*, 568–576.
7. Stone, A.J.; Tsuzuki, S. Intermolecular Interactions in Strongly Polar Crystals with Layer Structures. *J. Phys. Chem. B* **1997**, *101*, 10178–10183.
8. Govindaraju, T.; Avinash, M.B. Two-dimensional nanoarchitectonics: Organic and hybrid materials. *Nanoscale* **2012**, *4*, 6102–6117.

9. Palmore, G.T.R.; McBride, M.T. Engineering layers in molecular solids with the cyclic dipeptide of (S)-aspartic acid. *Chem. Commun.* **1998**, 145–146.
10. Hosseini, M.W. Molecular Tectonics: From Simple Tectons to Complex Molecular Networks. *Acc. Chem. Res.* **2005**, *38*, 313–323.
11. Wuest, J.D. Engineering crystals by the strategy of molecular tectonics. *Chem. Commun.* **2005**, 5830–5837.
12. Baillargeon, P.; Fortin, D.; Dory, Y.L. Hierarchical Self-Assembly of Lactams into Supramolecular CO-Spiked "Sea Urchins" and Then into a Channeled Crystal. *Cryst. Growth Des.* **2010**, *10*, 4357–4362.
13. Gordon, M.S.; Jensen, J.H. Understanding the Hydrogen Bond Using Quantum Chemistry. *Acc. Chem. Res.* **1996**, *29*, 536–543.
14. Steiner, T. The Hydrogen Bond in the Solid State. *Angew. Chem. Int. Ed.* **2002**, *41*, 48–76.
15. Taylor, R.; Kennard, O. Hydrogen-Bond Geometry in Organic Crystals. *Acc. Chem. Res.* **1984**, *17*, 320–326.
16. Desiraju, G.R. The C-H···O Hydrogen Bond: Structural Implications and Supramolecular Design. *Acc. Chem. Res.* **1996**, *29*, 411–449.
17. Bella, J.; Humphries, M.J. Cα-H···O=C hydrogen bonds contribute to the specificity of RGD cell-adhesion interactions. *BMC Struct. Biol.* **2005**, *5*.
18. Pierce, A.C.; Sandretto, K.L.; Bemis, G.W. Kinase Inhibitors and the Case for CH···O Hydrogen Bonds in Protein-Ligand Binding. *Proteins* **2002**, *49*, 567–576.
19. Weiss, M.S.; Brandl, M.; Sühnel, J.; Pal, D.; Hilgenfeld, R. More hydrogen bonds for the (structural) biologist. *Trends Biochem. Sci.* **2001**, *26*, 521–523.
20. Steiner, T.; Desiraju, G.R. Distinction between the weak hydrogen bond and the van der Waals interaction. *Chem. Commun.* **1998**, 891–892.
21. Desiraju, G.R. Hydrogen Bridges in Crystal Engineering: Interactions without Borders. *Acc. Chem. Res.* **2002**, *35*, 565–573.
22. Perlstein, J.; Steppe, K.; Vaday, S.; Ndip, E.M.N. Molecular Self-Assemblies. 5. Analysis of the Vector Properties of Hydrogen Bonding in Crystal Engineering. *J. Am. Chem. Soc.* **1996**, *118*, 8433–8443.
23. Ji, W.; Liu, G.; Li, Z.; Feng, C. Influence of C–H···O Hydrogen Bonds on Macroscopic Properties of Supramolecular Assembly. *ACS Appl. Mater. Interfaces* **2016**, *8*, 5188–5195.
24. Metrangolo, P.; Neukirch, H.; Pilati, T.; Resnati, G. Halogen Bonding Based Recognition Processes: A World Parallel to Hydrogen Bonding. *Acc. Chem. Res.* **2005**, *38*, 386–395.
25. Legon, A.C. Prereactive Complexes of Dihalogens XY with Lewis Bases B in the Gas Phase: A Systematic Case for the Halogen Analogue B···XY of the Hydrogen Bond B···HX. *Angew. Chem. Int. Ed.* **1999**, *38*, 2686–2714.
26. Auffinger, P.; Hays, F.A.; Westhof, E.; Ho, P.S. Halogen bonds in biological molecules. *Proc. Natl. Acad. Sci. USA* **2004**, *101*, 16789–16794.
27. Awwadi, F.F.; Willett, R.D.; Peterson, K.A.; Twamley, B. The Nature of Halogen···Halogen Synthons: Crystallographic and Theoretical Studies. *Chem. Eur. J.* **2006**, *12*, 8952–8960.
28. Metrangolo, P.; Pilati, T.; Resnati, G. Halogen bonding and other noncovalent interactions involving halogens: A terminology issue. *CrystEngComm* **2006**, *8*, 946–947.

29. Dey, A.; Jetti, R.K.R.; Boese, R.; Desiraju, G.R. Supramolecular equivalence of halogen, ethynyl and hydroxy groups. A comparison of the crystal structures of some 4-substituted anilines. *CrystEngComm* **2003**, *5*, 248–252.
30. Metrangolo, P.; Resnati, G.; Pilati, T.; Liantonio, R.; Meyer, F. Engineering Functional Materials by Halogen Bonding. *J. Polym. Sci. A Polym. Chem.* **2007**, *45*, 1–15.
31. Gilday, L.C.; Robinson, S.W.; Barendt, T.A.; Langton, M.J.; Mullaney, B.R.; Beer, P.D. Halogen Bonding in Supramolecular Chemistry. *Chem. Rev.* **2015**, *115*, 7118–7195.
32. Mukherjee, A.; Tothadi, S.; Desiraju, G.R. Halogen Bonds in Crystal Engineering: Like Hydrogen Bonds yet Different. *Acc. Chem. Res.* **2014**, *47*, 2514–2524.
33. Politzer, P.; Murray, J.S.; Janjić, G.V.; Zarić, S.D. σ-Hole Interactions of Covalently-Bonded Nitrogen, Phosphorus and Arsenic: A Survey of Crystal Structures. *Crystals* **2014**, *4*, 12–31.
34. Aakeröy, C.B.; Baldrighi, M.; Desper, J.; Metrangolo, P.; Resnati, G. Supramolecular Hierarchy among Halogen-Bond Donors. *Chem. Eur. J.* **2013**, *19*, 16240–16247.
35. Lieffrig, J.; Jeannin, O.; Frackowiak, A.; Olejniczak, I.; Swietlik, R.; Dahaoui, S.; Aubert, E.; Espinosa, E.; Auban-Senzier, P.; Fourmigué, M. Charge-Assisted Halogen Bonding: Donor-Acceptor Complexes with Variable Ionicity. *Chem. Eur. J.* **2013**, *19*, 14804–14813.
36. Cavallo, G.; Metrangolo, P.; Milani, R.; Pilati, T.; Priimagi, A.; Resnati, G.; Terraneo, G. The Halogen Bond. *Chem. Rev.* **2016**, *116*, 2478–2601.
37. Baillargeon, P.; Dory, Y.L. Supramolecular Walls from Cyclic Peptides: Modulating Nature and Strength of Weak Interactions. *Cryst. Growth Des.* **2009**, *9*, 3638–3645.
38. Baillargeon, P.; Lussier, T.; Dory, Y.L. Hydrogen Bonds between Acidic Protons from Alkynes (C–H···O) and Amides (N–H···O) and Carbonyl Oxygen Atoms as Acceptor Partners. *J. Crystallogr.* **2014**, *2014*, 1–5.
39. SeethaLekshmi, S.; Varughese, S.; Girl, L.; Pedireddi, V.R. Molecular Complexes of 4-Halophenylboronic Acids: A Systematic Exploration of Isostructurality and Structural Landscape. *Cryst. Growth Des.* **2014**, *14*, 4143–4154.
40. Dechambenoit, P.; Ferlay, S.; Kyritsakas, N.; Hosseini, M.W. Playing with isostructurality: From tectons to molecular alloys and composite crystals. *Chem. Commun.* **2009**, 1559–1561.
41. Aakeröy, C.B.; Schultheiss, N.C.; Rajbanshi, A.; Desper, J.; Moore, C. Supramolecular Synthesis Based on a Combination of Hydrogen and Halogen Bonds. *Cryst. Growth Des.* **2009**, *9*, 432–441.
42. Ebenezer, S.; Muthiah, P.T.; Butcher, R.J. Design of a Series of Isostructural Co-Crystals with Aminopyrimidines: Isostructurality through Chloro/Methyl Exchange and Studies on Supramolecular Architectures. *Cryst. Growth Des.* **2011**, *11*, 3579–3592.
43. Ekkebus, R.; van Kasteren, S.I.; Kulathu, Y.; Scholten, A.; Berlin, I.; Geurink, P.P.; de Jong, A.; Goerdayal, S.; Neefjes, J.; Heck, A.J.R.; *et al.* On Terminal Alkynes that Can React with Active-Site Cysteine Nucleophiles in Proteases. *J. Am. Chem. Soc.* **2013**, *135*, 2867–2870.

44. Sommer, S.; Weikart, N.D.; Linne, U.; Mootz, H.D. Covalent inhibition of SUMO and ubiquitin-specific cysteine proteases by an *in situ* thiol-alkyne addition. *Bioorg. Med. Chem.* **2013**, *21*, 2511–2517.
45. Wu, W.; Jiang, H. Haloalkynes: A Powerful and Versatile Building Block in Organic Synthesis. *Acc. Chem. Res.* **2014**, *47*, 2483–2504.
46. Mevers, E.; Liu, W.-T.; Engene, N.; Mohimani, H.; Byrum, T.; Pevzner, P.A.; Dorrestein, P.C.; Spadafora, C.; Gerwick, W.H. Cytotoxic Veraguamides, Alkynyl Bromide-Containing Cyclic Depsipeptides from the Marine Cyanobacterium *cf.* Oscillatoria margaritifera. *J. Nat. Prod.* **2011**, *74*, 928–936.
47. Salvador, L.A.; Biggs, J.S.; Paul, V.J.; Luesch, H. Veraguamides A-G, Cyclic Hexadepsipeptides from a Dolastatin 16-Producing Cyanobacterium *Symploca cf. hydnoides* from Guam. *J. Nat. Prod.* **2011**, *74*, 917–927.
48. Hay, A.S. Oxidative Coupling of Acetylenes II. *J. Org. Chem.* **1962**, *27*, 3320–3321.
49. Alkorta, I.; Rozas, I.; Elguero, J. Non-conventional hydrogen bonds. *Chem. Soc. Rev.* **1998**, *27*, 163–170.
50. Hassel, O. Structural Aspects of Interatomic Charge-Transfer Bonding. *Science* **1970**, *170*, 497–502.
51. Bolton, O.; Lee, K.; Kim, H.-J.; Lin, K.Y.; Kim, J. Activating efficient phosphorescence from purely organic materials by crystal design. *Nat. Chem.* **2011**, *3*, 205–210.
52. Metrangolo, P.; Resnati, G. Halogen Bonding: A Paradigm in Supramolecular Chemistry. *Chem. Eur. J.* **2001**, *7*, 2511–2519.
53. Priimagi, A.; Cavallo, G.; Metrangolo, P.; Restani, R. The Halogen Bond in the Design of Functional Supramolecular Materials: Recent Advances. *Acc. Chem. Res.* **2013**, 2686–2695.
54. Cody, V.; Murray-Rust, P. Iodine⋯X(O, N, S) intermolecular contacts: Models of thyroid hormone protein binding interactions using information from the cambridge crystallographic data files. *J. Mol. Struct.* **1984**, *112*, 189–199.
55. Ouvrard, C.; le Questel, J.-Y.; Berthelot, M.; Laurence, C. Halogen-bond geometry: A crystallographic data-base investigation of dihalogen complexes. *Acta Crystallogr. B* **2003**, *59*, 512–526.
56. Lommerse, J.P.M.; Stone, A.J.; Taylor, R.; Allen, F.H. The Nature and Geometry of Intermolecular Interactions between Halogens and Oxygen or Nitrogen. *J. Am. Chem. Soc.* **1996**, *118*, 3108–3116.
57. Hoheisel, T.N.; Schrettl, S.; Marty, R.; Todorova, T.K.; Corminboeuf, C.; Sienkiewicz, A. A multistep single-crystal-to-single-crystal bromodiacetylene dimerization. *Nat. Chem.* **2013**, *5*, 327–334.
58. Curtis, S.M.; Le, N.; Nguyen, T.; Ouyang, X.; Tran, T.; Fowler, F.W.; Lauher, J.W. What have We Learned about Topochemical Diacetylene Polymerizations? *Supramol. Chem.* **2005**, *17*, 31–36.
59. Okuno, T.; Yamane, K.; Sandman, D.J. Solid State Polymerization of Diacetylenes with Amide Groups. *Mol. Cryst. Liq. Cryst.* **2006**, *456*, 45–53.

60. Luo, L.; Wilhelm, C.; Sun, A.; Grey, C.P.; Lauher, J.W.; Goroff, N.S. Poly(diiododiacetylene): Preparation, isolation, and full characterization of a very simple poly(diacetylene). *J. Am. Chem. Soc.* **2008**, *130*, 7702–7709.
61. Li, Z.; Fowler, F.W.; Lauher, J.W. Weak Interactions Dominating the Supramolecular Self-Assembly in a Salt: A Designed Single-Crystal-to-Single-Crystal Topochemical Polymerization of a Terminal Aryldiacetylene. *J. Am. Chem. Soc.* **2009**, *131*, 634–643.
62. Jin, H.; Plonka, A.M.; Parise, J.B.; Goroff, N.S. Pressure induced topochemical polymerization of diiodobutadiyne: A single-crystal-to-single-crystal transformation. *CrystEngComm* **2013**, *15*, 3106–3110.
63. Haridas, V.; Sadanandan, S.; Collart-Dutilleul, P.-Y.; Gronthos, S.; Voelcker, N.H. Lysine-Appended Polydiacetylene Scaffolds for Human Mesenchymal Stem Cells. *Biomacromolecules* **2014**, *15*, 582–590.
64. Xu, W.L.; Smith, M.D.; Krause, J.A.; Greytak, A.B.; Ma, S.; Read, C.M.; Shimizu, L.S. Single Crystal to Single Crystal Polymerization of a Self-Assembled Diacetylene Macrocycle Affords Columnar Polydiacetylenes. *Cryst. Growth Des.* **2014**, *14*, 993–1002.
65. Wang, S.; Li, Y.; Liu, H.; Li, J.; Li, T.; Wu, Y.; Okada, S.; Nakanishi, H. Topochemical polymerization of unsymmetrical aryldiacetylene supramolecules with nitrophenyl substituents utilizing C–H···π interactions. *Org. Biomol. Chem.* **2015**, *13*, 5467–5474.
66. Lauher, J.W.; Fowler, F.W. Single-Crystal-to-Single-Crystal Topochemical Polymerizations by Design. *Acc. Chem. Res.* **2008**, *41*, 1215–1229.
67. Jelinek, R.; Ritenberg, M. Polydiacetylenes—Recent molecular advances and applications. *RSC Adv.* **2013**, *3*, 21192–21201.
68. Kohn, W.; Sham, L. Self-Consistent Equations Including Exchange and Correlation Effects. *J. Phys. Rev.* **1965**, *140*, A1133–A1138.
69. Becke, A.D. Perspective: Fifty years of density-functional theory in chemical physics. *J. Phys. Chem.* **2014**, *140*, 18A301.
70. Rezác, J.; Hobza, P. Benchmark Calculations of Interaction Energies in Noncovalent Complexes and Their Applications. *Chem. Rev.* **2016**.
71. Kolář, M.H.; Hobza, P. Computer Modeling of Halogen Bonds and Other σ-Hole Interactions. *Chem. Rev.* **2016**.
72. Feldblum, E.S.; Arkin, I.T. Strength of a bifurcated H bond. *Proc. Natl. Acad. Sci.* **2014**, *111*, 4085–4090.
73. Adalsteinsson, H.; Maulitz, A.H.; Bruice, T.C. Calculation of the Potential Energy Surface for Intermolecular Amide Hydrogen Bonds Using Semiempirical and Ab Initio Methods. *J. Am. Chem. Soc.* **1996**, *118*, 7689–7693.
74. Li, A.; Muddana, H.S.; Gilson, M.K. Quantum Mechanical Calculation of Noncovalent Interactions: A Large-Scale Evaluation of PMx, DFT, and SAPT Approaches. *J. Chem. Theory Comput.* **2014**, *10*, 1563–1575.
75. Zhao, Y.; Truhlar, D.G. Density Functionals with Broad Applicability in Chemistry. *Acc. Chem. Res.* **2008**, *41*, 157–167.
76. Ivanova, B.; Spiteller, M. Binding affinity of terrestrial and aquatic humics toward organic xenobiotics. *J. Environ. Chem. Eng.* **2016**, *4*, 498–510.

77. Schmidt, M.W.; Baldridge, K.K.; Boatz, J.A.; Elbert, S.T.; Gordon, M.S.; Jensen, J.H.; Koseki, S.; Matsunaga, N.; Nguyen, K.A.; Su, S.; *et al.* General Atomic and Molecular Electronic Structure System. *J. Comput. Chem.* **1993**, *14*, 1347–1363.
78. Becke, A.D. Density-functional thermochemistry. III. The role of exact exchange. *J. Chem. Phys.* **1993**, *98*, 5648–5652.
79. Zhao, Y.; Truhlar, D.G. The M06 suite of density functionals for main group thermochemistry, thermochemical kinetics, noncovalent interactions, excited states, and transition elements: Two new functionals and systematic testing of four M06-class functionals and 12 other function. *Theor. Chem. Acc.* **2008**, *120*, 215–241.
80. Shirman, T.; Boterashvili, M.; Orbach, M.; Freeman, D.; Shimon, L.J.W.; Lahav, M.; van der Boom, M.E. Finding the Perfect Match: Halogen *vs.* Hydrogen Bonding. *Cryst. Growth Des.* **2015**, *15*, 4756–4759.

Constructor Graphs as Useful Tools for the Classification of Hydrogen Bonded Solids: The Case Study of the Cationic (Dimethylphosphoryl)methanaminium (*dpma*H$^+$) Tecton

Guido J. Reiss

Abstract: The structural chemistry of a series of *dpma*H (*dpma*H = (dimethylphosphoryl) methanaminium) salts has been investigated using constructor graph representations to visualize structural dependencies, covering the majority of known *dpma*H salts. It is shown that the structurally related α-aminomethylphosphinic acid can be integrated in the systematology of the *dpma*H salts. Those *dpma*H salts with counter anions that are weak hydrogen bond acceptors (ClO_4^-, $SnCl_6^{2-}$, $IrCl_6^{2-}$, I^-) tend to form head-to-tail hydrogen bonded moieties purely consisting of *dpma*H$^+$ cations as the primarily structural motif. In structures with weak to very weak hydrogen bonds between the *dpma*H$^+$ cations and the counter anions, the anions fill the gaps in the structures. In salts with medium to strong hydrogen bond acceptor counter ions (Cl^-, NO_3^-, $PdCl_4^{2-}$), the predominant structural motif is a double head-to-tail hydrogen bonded (*dpma*H$^+$)$_2$ dimer. These dimeric units form further NH$\cdots$X hydrogen bonds to neighboring counter anions X, which results in one-dimensional and two-dimensional architectures.

> Reprinted from *Crystals*. Cite as: Reiss, G.J. Constructor Graphs as Useful Tools for the Classification of Hydrogen Bonded Solids: The Case Study of the Cationic (Dimethylphosphoryl)methanaminium (*dpma*H$^+$) Tecton. *Crystals* **2016**, *6*, 6.

1. Introduction

Neutral (dimethylphosphoryl)methanamine (*dpma*) has been synthesized decades ago [1,2], and its crystal structure is well-known [3]. Furthermore, it has been shown that *dpma* is a potent bidentate ligand stabilizing mononuclear [3–5] and oligonuclear [6] complexes as well as coordination polymers [7]. In recent years, we have shown that the N-protonated *dpma*H$^+$ cation is an excellent monodentate ligand. Based on its conformational flexibility—via rotation about the central P–C bond—(Scheme 1) the *dpma*H$^+$ ligand is able to fit with the needs of coordination, packing and hydrogen bonding in complexes such as [Mn(H$_2$O)$_2$Cl$_3$(*dpma*H)], [MCl$_2$(*dpma*H)$_4$][MCl$_4$]$_2$, (M = Co(II), Cu(II)) [8].

Above all, the *dpma*H⁺ cation is an excellent tecton to construct hydrogen bonded architectures. In terms of the concepts of Crystal Engineering, the *dpma*H⁺ is a building block with a threefold hydrogen donor functionality at one end (the NH$_3$ group) and a hydrogen-bond acceptor functionality at the oxygen atom of the phosphoryl moiety at the other end (Scheme 1, left part). *dpma*H⁺ can thus be abstracted by a block-shaped informative icon (Scheme 1, right part) which reduces the functionality of the tecton to the striking symbols of a blue plus sign (indicating the NH$_3$ group) and a red dot (indicating the oxygen atom), which enables a general und uniform description of simple and especially complicated structures e.g., for *dpma*H⁺ salts.

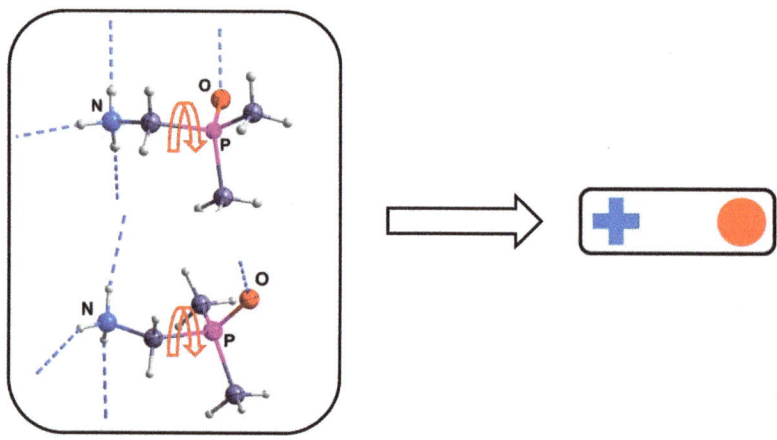

Scheme 1. Hydrogen bonding functionality of the *dpma*H⁺ tecton (**left part**) and the corresponding icon useable in constructor graphs (**right part**).

Coding of individual hydrogen bonding motifs by graph sets classification and notation has been widely used in the literature [9–11]. The main disadvantage of classical graph sets in the context of a broader comparison of chemically varied compounds is that similar connectivities may not be displayed in the corresponding graph set descriptors. A system less frequently used which is more favorable to classify larger sections of hydrogen bonded structures and to compare related structures is based on so-called constructor graphs [10,11]. In this methodology, each molecule is substituted by a dot and hydrogen bonds are symbolized by arrows pointing from the donor to the acceptor group. Infrequent use of constructor graphs may be related to the fact that the advantage of the abstraction of multifunctional molecules to dots is not intuitively comprehensible whereas the individual graph sets coding similar hydrogen-bond motifs are very handy for the description of one isolated structure or structure type. Further developments in this area concerns automated analysis and interpretation of hydrogen bonded networks [12]. The present contribution is meant to exemplarily show the ability of constructor graphs to classify extended hydrogen bonded structures. Furthermore, the usage of informative icons instead of simple dots in these constructor graphs is suggested, which should be more intuitive. Modified constructor graphs illustrated by such informative icons will be used throughout this contribution. Hopefully, a broader use of constructor graphs may support activities to systematize structurally and hierarchically related hydrogen bonded solids.

2. Results and Discussion

The hydrogen-bond functionality of *dpma*H$^+$ predestines this cation for the formation of head-to-tail connected dimers, chains or higher dimensional polymers. Most *dpma*H$^+$ salts containing weakly hydrogen bond accepting counter anions have the tendency to form (*dpma*H$^+$)$_n$ polymers. Salts with stronger hydrogen bond accepting counter anions mainly comprise (*dpma*H$^+$)$_2$ of dimers which are further connected by the counter anions.

2.1. Salt Structures Based on Head-to-Tail Connected Hydrogen Bonded Polymers: (dpmaH$^+$)$_n$

Hexahalogenometallate dianions of tetravalent metals like tin [13], iridium [14] and platinum [15] and some other perhalogenometallates are well known to be weak hydrogen bond acceptors, showing a tendency to bifurcated hydrogen bonds [16]. The synthesis of the (*dpma*H)$_2$[MCl$_6$] with M = Sn, Ir gave structures that show a highly symmetrical arrangement of the [MCl$_6$]$^{2-}$ anions in a monoclinic unit cell (Figure 1). The substructure formed by the anions seems to be a common feature of many hexahalogenidometallate salts, ranging from K$_3$[IrCl$_6$] [17] to analogous alkylammonium salts [13–15,18]. Hydrogen bonded (*dpma*H$^+$)$_n$ polymers form the one-dimensional cationic substructure.

This polymeric chain is formed by a head-to-tail connection of *dpma*H$^+$ units, which is an ubiquitous motif throughout structural chemistry. A prominent example would be dicarboxylic acids [19], which are well documented neutral compounds forming such a hydrogen bonding scheme. However, a few cationic chains are also documented in combination with weakly coordinating anions [20–22]. Starting from linear zig-zag arrangements, it is obvious that the formation of a ring motif should be possible.

In the structure of *dpma*HI exactly such a four-membered ring is present (Crystal data of *dpma*HI (C$_3$H$_{11}$INOP): M_r = 235.00, T = 292 K, crystal size: 0.15 × 0.13 × 0.07 mm^3, monoclinic, $P2_1/n$, a = 12.0399(5), b = 8.7857(4), c = 15.4854(5) Å, β = 90.689(4)°, MoKα-radiation (λ = 0.71073 Å), 11651 measured reflections (R_{int} = 0.0363), completeness > 99% up to 2theta = 51°, 3014 unique reflections, 2664 observed reflections ($I > 2\sigma(I)$), solution and refinement were carried out with the SHELX system [23], 136 refined parameters, $R^1(I > 2\,(I)) = 0.0461$, wR^2(all unique reflections) = 0.0906. CCDC 1417987 contains the supplementary crystallographic data for this crystal structure. The data can be obtained free of charge from The Cambridge Crystallographic Data Centre via www.ccdc.cam.ac.uk/getstructures). Each four-membered centrosymmetric (*dpma*H$^+$)$_4$ moiety (Figure 2a, graph set descriptor $R^4_4(20)$) is furthermore connected to neighboring iodide counter anions to form a two-dimensional hydrogen bonded polymer.

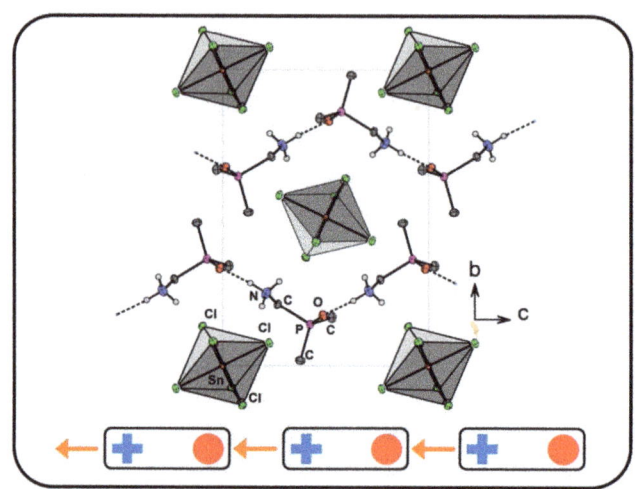

Figure 1. (**Upper part**): Structure of (*dpma*H)$_2$[SnCl$_6$] with view along [100]; (**Lower part**): modified constructor graph of the polymeric, cationic (*dpma*H$^+$)$_n$ substructure.

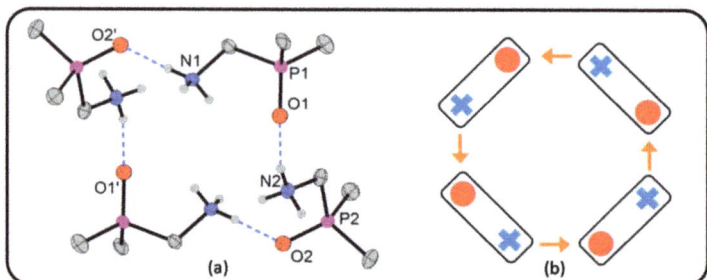

Figure 2. (**a**) Showing the primary structural motif of a four-membered, hydrogen bonded ring (N–H···I hydrogen bonds to the adjacent iodide counter anions are omitted); (**b**) modified constructor graph of the tetrameric (*dpma*H$^+$)$_4$ ring unit.

The next, more complex motif can be obtained by connecting two adjacent hydrogen bonded parallel chains. Figure 3a shows the corresponding strand (cationic substructure) of the structure of the *dpma*H[ClO$_4$] salt [24]. To realize an optimal interaction between the neighboring chains, they are shifted against each other by one half of the repetition unit. Consequently, hydrogen bonded ring motifs with the graph set descriptor $R^2_3(9)$ characterize the connections between parallel chains.

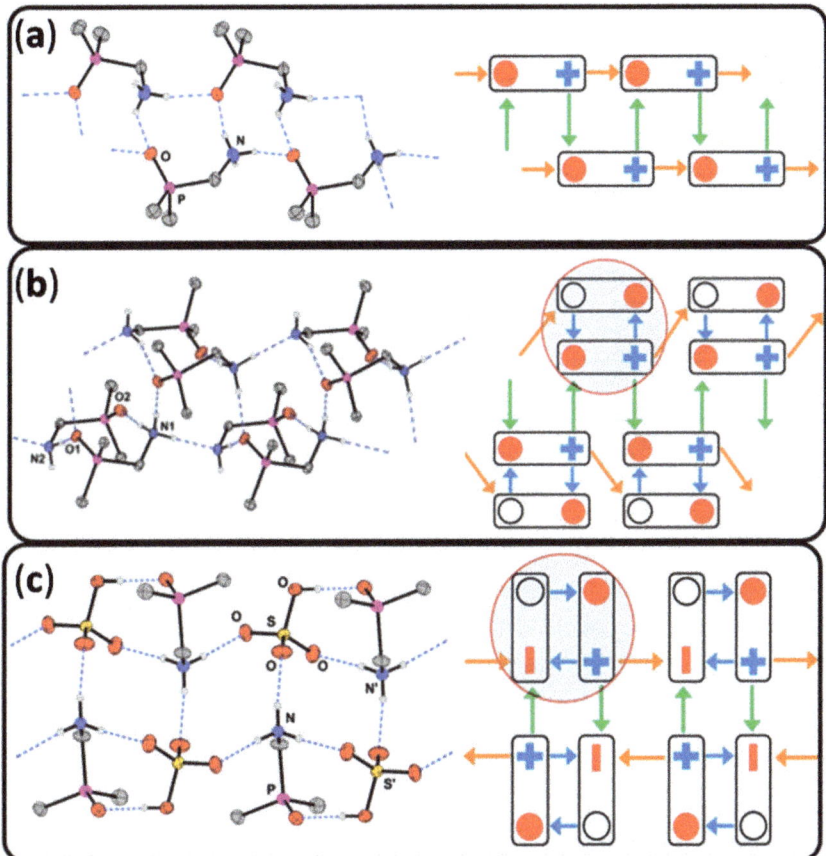

Figure 3. (a) part of the hydrogen bonded chain structure and modified constructor graph of dpmaH[ClO$_4$]; (b) hydrogen bonded strands in the structure of *dpmaHX·dpma* (X = [ClO$_4$]$^-$ [24], I$^-$ [25]) and its modified constructor graph (the *dpma* molecule is shown as brick-shaped icon composed of a black circle and a red dot; red colored circle indicates a composite tecton); (c) hydrogen bonded strands constructed from head-to-tail connected {*dpmaH*}$^+$[HSO$_4$]$^-$ ion pairs [26] and the corresponding modified constructor graph (brick-shaped icon composed of a black circle and a red minus sign is shown for the HSO$_4^-$ anion; red colored circle indicates a composite tecton).

A more complex cationic, one-dimensional strand-type motif with the general formula *dpmaHX·dpma* (X = [ClO$_4$]$^-$ [24], I$^-$ [25]) is closely related to the double chain structure discussed before (Figure 3b). In this more complex case, the *dpma*H⁺ cation (icon as introduced before) and one neutral *dpma* molecule (brick-shaped icon composed of a black circle and a red dot) can formally combined to give a new tecton which is indicated by a red colored circle in Figure 3b. This new assembly generally has the same functionality as the "pure" *dpma*H⁺ tecton alone. Each *dpma*HX·*dpma* unit is head-to-tail connected to two adjacent ones. In addition, this primary chain structure is connected to another parallel chain in the same way as in the prototype structure of *dpma*H[ClO$_4$] (Figure 3a). The structural similarity between them can be easily summarized by the two constructor graphs shown in Figure 3a and b, but not by the associated individual graph set descriptors of the hydrogen bonded rings ($R^2_2(10)$; $R^3_4(11)$; $R^4_6(15)$) in the structure of *dpma*HX·*dpma* and those in the structure of *dpma*H[ClO$_4$].

A neutral chain-type structure similar to that present in the structures of the *dpma*HX·*dpma* salts can be obtained by combination of the *dpma*H⁺ cation with an HSO$_4^-$ anion. The HSO$_4^-$ anion is a typical tecton which is predestined to do a head-to-tail connection. In this case, the implementation of a brick-shaped icon composed of a black circle and a red minus sign for the HSO$_4^-$ anion is useful (Figure 3c). In the structure of *dpma*H[HSO$_4$] [26], we find the desired head-to-tail connected ion pairs (Figure 3c; highlighted by a red colored circle) which are connected to the neighboring ones in the same manner as in the structure of *dpma*HX·*dpma*. This similarity can easily be visualized by the three corresponding constructor graphs in Figure 3. Within the {*dpma*H}⁺[HSO$_4$]⁻ ion pair, the HSO$_4^-$ anion is a medium-strong hydrogen bond donor and at the same time a hydrogen bond acceptor. Therefore, the HSO$_4^-$ anion in this structure forms a hydrogen bonding scheme which is identical to that of neutral *dpma* molecule in *dpma*HX·*dpma*. For this structure, the individual graph set descriptors for the ring motifs ($R^4_4(12)$ and $R^2_2(9)$) again do not fit with those of the structurally related structures of *dpma*HX·*dpma* and *dpma*H[ClO$_4$].

As an intermediate conclusion, two things should be mentioned:

- A comparison of the chain-type and strand-type structures discussed before has been simplified by the use of the modified constructor graphs.
- The intuitive dissection of these complex hydrogen bonded structures into subunits (tectons) that look like as they were taken from a "chemical toolbox", unquestionable supports the understanding of the individual structures and should support the prediction of structural features.

The next systematic step towards a higher dimensional framework would be the construction of layered structure consisting of parallel chains. For steric reasons, this motif is not possible for the *dpma*H⁺ tecton. Furthermore, it would be very unusual for a phosphoryl group to accept more than two hydrogen bonds. A systematic search for a related tecton shows that the zwitterionic α-aminomethylphosphinic acid ($H_3N–CH_2PO_2Me$) [27] exactly forms the predicted layered structure. This alternative tecton can be easily generated from the *dpma*H$^+$ tecton by the exchange of a methyl group with an oxido group. This modification of the basic tecton generates one more hydrogen-bond acceptor atom at the formal negatively charged part of the zwitterionic α-aminomethylphosphinic acid molecule. An informative icon with a plus and a minus sign is showing the zwitterionic character of the α-aminomethylphosphinic acid (Figure 4).

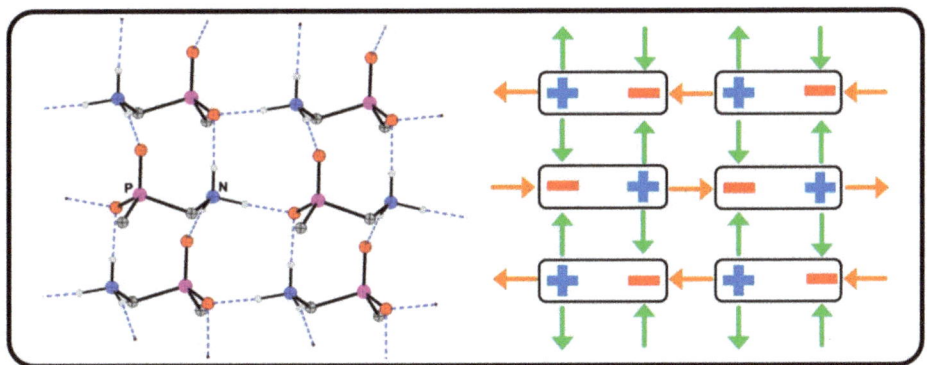

Figure 4. Hydrogen bonding pattern of the zwitterionic α-aminomethylphosphinic acid [27] and its modified constructor graph.

In the structure of α-aminomethylphosphinic acid, each of the three hydrogen atoms of the aminium group forms an unbifurcated hydrogen bond to a symmetry related molecule. The formation of a three-dimensional hydrogen bonded connection of α-aminomethylphosphinic acid tectons or of closely related molecules seems to be possible, in general, but to the best of the author's knowledge, such a structure has not been determined to date.

2.2. Salts Based on Hydrogen Bonded (dpmaH$^+$)$_2$ Dimers

For salts containing *dpma*H$^+$ and counter anions, which are good or excellent hydrogen-bond acceptors, a very different structural chemistry can be observed. The predominant motif in these structures are head-to-tail connected (*dpma*H$^+$)$_2$ units (Figure 5). Two out of six hydrogen atoms of the two aminium groups are used for the formation of the dimer. The four other hydrogen atoms are generally available for the formation of hydrogen bonds to the counter anions.

In the structure of *dpma*HCl [28], such (*dpma*H$^+$)$_2$ dicationic units are present. Each dicationic dimer donates four NH···Cl hydrogen bonds, and each chloride anion accepts two of them to form a chain structure (Figure 5). The constructor graph for this structure illustrates the structural similarity to *dpma*H[ClO$_4$]·*dpma* (Figure 3b), which also consist of head-to-tail connected basic units. In the structures of *dpma*H[ClO$_4$]·*dpma*, the oxygen atom of the phosphoryl group accepts two hydrogen bonds to realize chain propagation. In the structure of *dpma*HCl, there is no need to form such a bifurcated hydrogen bond at the phosphoryl group because of the presence of chloride anion, an excellent hydrogen bond acceptor.

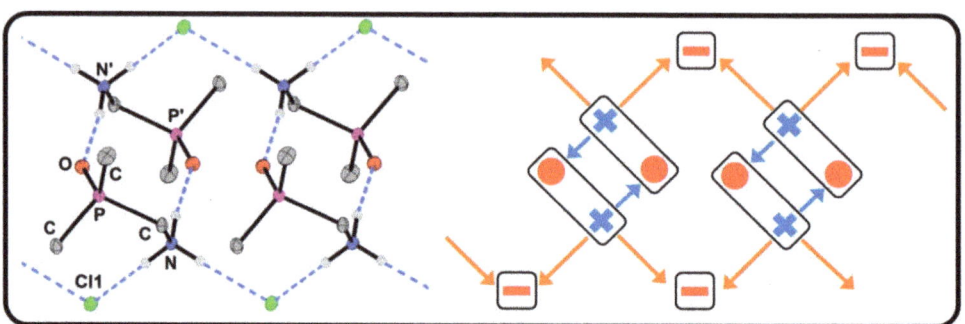

Figure 5. Hydrogen bonded structure of *dpma*HCl containing (*dpma*H$^+$)$_2$ dimers and chloride counter anions; a modified constructor graph for a section of the chain is shown.

An alternative arrangement for a *dpma*HX salt with a cation to anion ratio of 1:1 has been reported for *dpma*H[NO$_3$] [29]. This two-dimensionally connected structure can be understood as the next stage of aggregation (Figure 6a). Again, each dimeric ((*dpma*H$^+$)$_2$ unit is connected to four anions and each nitrate anion is connected to two dicationic moieties. Two of the three oxygen atoms of each nitrate anion form hydrogen bonds. The reason for additional structural complexity must be associated with the nitrate counter anion, which obviously causes a larger distance of the (*dpma*H$^+$)$_2$ dimers. Similar hydrogen bonded networks are well known for (H$_2$NR$_2$)$_2$[MCl$_6$] (R = alkyl, M = Sn, Ir, Pt [13–15,18]) and ((CH$_3$)$_2$NH$_2$)$_4$[MCl$_6$]Cl (M = Ru, W [30,31]) which can be discussed as an inverse structure type, as the dialkylaminium cations form two hydrogen bonds whereas there are four for each halogenometallate anions.

Finally, the (*dpma*H)$_2$[PdCl$_4$] salt [32] should be discussed. In this structure, each (*dpma*H$^+$)$_2$ dimer and each [PdCl$_4$]$^{2-}$ ions form two hydrogen bonded connections to adjacent counter ions (Figure 6b). The chain structure generated by this connection can be understood as a structural motif already known from *dpma*H[NO$_3$] (Figure 6a; blue-shaded area). This is a nice proof of the potential of this system in terms of crystal engineering. The "toolbox" for salts with strong hydrogen-bond acceptors contains dimeric (*dpma*H$^+$)$_2$ units that tends to donate four hydrogen bonds, whereas the counter anions are only twofold connected.

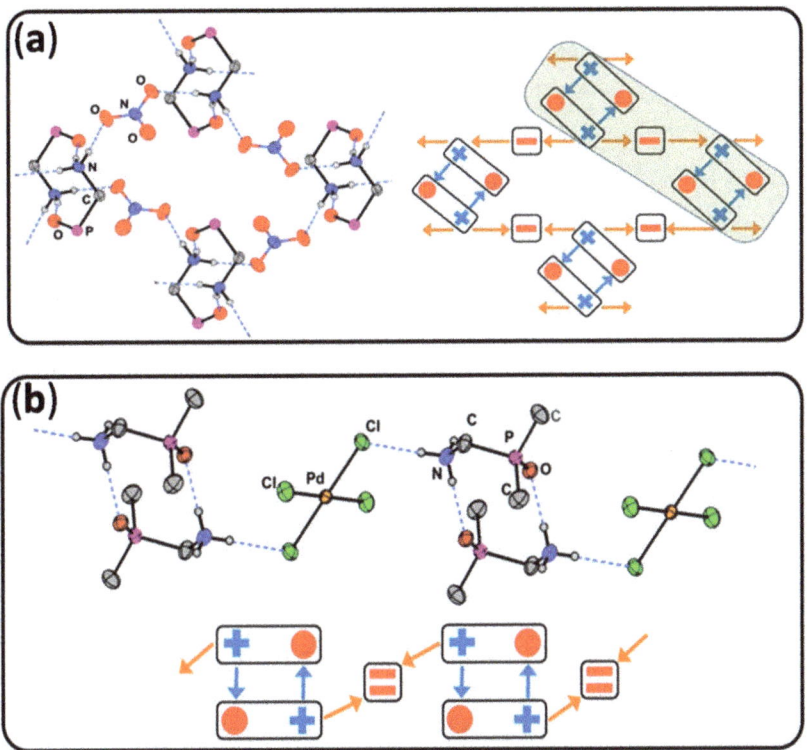

Figure 6. (a) the hydrogen bonded layered structure of *dpma*H[NO$_3$] and the corresponding constructor graph is shown; (b) the hydrogen bonded chain structure of (*dpma*H)$_2$[PdCl$_4$] and the corresponding modified constructor graph is shown.

3. Limitations

The use of constructor graphs, especially those with the informative icons, become cumbersome for more complex three-dimensional structures. There are similar limitations of use if there are several crystallographically independent building units and individual hydrogen bonding schemes for each of them.

4. Conclusions

In this contribution, it has been shown that the majority of *dpma*H$^+$ salts are closely structurally related. Their structural and hierarchical dependencies can easily be visualized by so-called constructor graphs. In contrast to the use of classical constructor graphs which consist of dots and arrows [10,33,34], informative icons and colored arrows are used here for the analysis and comparison of the hydrogen bonded architectures. As abstraction is ubiquitous used in chemistry, biochemistry and related fields, this modification of a mathematical tool towards a chemist-friendly design should broaden the usability of this method. Moreover, the potential of constructor graphs is illustrated by the integration of the structurally related α-aminomethylphosphinic acid into the structure systematology of simple *dpma*H+ salts.

Acknowledgments: Many of the *dpma*H+ salts were synthesized and characterized in the course of a student based research project support by the Lehrförderfond of the Heinrich-Heine-Universität Düsseldorf. I gratefully acknowledge support by the Ministry of Innovation, Science and Research of North-Rhine Westphalia and the German Research Foundation (DFG) for financial support (Xcalibur diffractometer; INST 208/533-1).

Conflicts of Interest: The author declares no conflict of interest.

References

1. Tsvetkov, E.N.; Kron, T.E.; Kabachnik, M.I. Synthesis of chloromethylphosphine oxides. *Izv. Akad. Nauk. Ser. Khim.* **1980**, *3*, 669–672.
2. Varbanov, S.G.; Agopian, G.; Borisov, G. Polyurethane foams based on dimethylaminomethylphosphine oxides adducts with ethylene and propylene oxides. *Eur. Polym. J.* **1987**, *23*, 639–642.
3. Kochel, A. Synthesis and magnetic properties of the copper(II) complex derived from dimethylaminomethylphosphine oxide ligand. X-ray crystal structure of DMAO and [Cu(NO$_3$)$_2$(POC$_3$H$_{10}$N)$_2$]. *Inorg. Chim. Acta* **2009**, *362*, 1379–1382.
4. Dodoff, N.; Macicek, J.; Angelova, O.; Varbanov, S.G.; Spassovska, N. Chromium(III), Cobalt(II), Nickel(II) and Copper(II) complexes of (dimethylphosphinyl)methanamine. Crystal structure of *fac*-tris{(dimethylphosphinyl)methanamine-*N,O*}nickel(II) chloride trihydrate. *J. Coord. Chem.* **1990**, *22*, 219–228.
5. Trendafilova, N.; Georgieva, I.; Bauer, G.; Varbanov, S.G.; Dodoff, N. IR and Raman study of Pt(II) and Pd(II) complexes of amino substituted phosphine oxides: Normal coordinate analysis. *Spectrochim. Acta A* **1997**, *53*, 819–828.
6. Vornholt, S.; Herrmann, R.; Reiss, G.J. Crystal structure of the trinuclear complex hexachlorido-1κ^3*Cl*,3κ^3*Cl*-bis(μ$_2$-dimethylphosphorylmethanamine-1:2κ^2*N:O*,3:2κ^2*N:O*)-bis(dimethylphosphorylmethanamine-2κ^2*N,O*)trizinc(II), C$_{12}$H$_{40}$Cl$_6$N$_4$O$_4$P$_4$Zn$_3$. *Z. Kristallogr. New Cryst. Struct.* **2014**, *229*, 440–442.
7. Borisov, G.; Varbanov, S.G.; Venanzi, L.M.; Albinati, A.; Demartin, F. Coordination of dimethyl(aminomethyl)phosphine oxide with Zinc(II), Nickel(II), and Palladium(II). *Inorg. Chem.* **1994**, *33*, 5430–5437.
8. Richert, M.E.; Helmbrecht, C.; Reiss, G.J. Synthesis, Spectroscopy and Crystal Structure of a New Copper Complex Built up by Cationic (Dimethylphosphoryl)methanaminium Ligands. *Mediterr. J. Chem.* **2014**, *3*, 3847–3853.
9. Etter, M.C. Encoding and decoding hydrogen-bond patterns of organic compounds. *Acc. Chem. Res.* **1990**, *23*, 120–126.
10. Grell, J.; Bernstein, J.; Tinhofer, G. Investigation of Hydrogen Bond Patterns: A Review of Mathematical Tools For the Graph Set Approach. *Crystallogr. Rev.* **2002**, *8*, 1–56.

11. Bernstein, J.; Davis, R.E.; Shimoni, L.; Chang, N.-L. Patterns in Hydrogen Bonding: Functionality and Graph Set Analysis in Crystals. *Angew. Chem. Int. Ed.* **1995**, *34*, 1555–1573.
12. Hursthouse, M.B.; Hughes, D.S.; Gelbrich, T.; Threlfall, T.L. Describing hydrogen-bonded structures; topology graphs, nodal symbols and connectivity tables, exemplified by five polymorphs of each of sulfathiazole and sulfapyridine. *Chem. Cent. J.* **2015**, *9*.
13. Knop, O.; Cameron, T.S.; James, M.A.; Falk, M. Alkylammonium hexachlorostannates(IV), $(R_nNH_{4-n})_2SnCl_6$: Crystal structure, infrared spectrum, and hydrogen bonding. *Can. J. Chem.* **1983**, *61*, 1620–1646.
14. Reiss, G.J. The pseudosymmetric structure of bis(diisopropylammonium) hexachloroiridate(IV) and its relationship to potassium hexachloroiridate(III). *Acta Crystallogr. E* **2002**, *58*, 47–50.
15. Bokach, N.A.; Pakhomova, T.B.; Kukushkin, V.Y.; Haukka, M.; Pombeiro, A.J.L. Hydrolytic metal-mediated coupling of dialkylcyanamides at a Pt(IV) center giving a new family of diimino ligands. *Inorg. Chem.* **2003**, *42*, 7560–7568.
16. Gillon, A.L.; Lewis, G.R.; Orpen, A.G.; Rotter, S.; Starbuck, J.; Wang, X.-M.; Rodriguez-Martin, Y.; Ruiz-Perez, C. Organic-inorganic hybrid solids: Control of perhalometallate solid state structures. *J. Chem. Soc. Dalton Trans.* **2000**, *21*, 3897–3905.
17. Coll, R.K.; Fergusson, J.E.; Penfold, B.R.; Rankin, D.A.; Robinson, W.T. The chloro and bromo complexes of iridium(III) and iridium(IV). III: The crystal structures of $K_3[IrCl_6]$ and $Cs_2[IrCl_5H_2O]$ and interpretation of the N.Q.R. data of chloroiridates(III). *Aust. J. Chem.* **1987**, *40*, 2115–2122.
18. Reiss, G.J.; Helmbrecht, C. Bis(diisopropylammonium) hexachloridostannate(IV). *Acta Crystallogr. E* **2012**, *68*, 1402–1403.
19. Bhattacharya, S.; Saraswatula, V.G.; Saha, B.K. Thermal Expansion in Alkane Diacids—Another Property Showing Alternation in an Odd–Even Series. *Cryst. Growth Des.* **2013**, *13*, 3651–3656.
20. Pienack, N.; Möller, K.; Näther, C.; Bensch, W. $(1,4-dabH)_2MnSnS_4$: The first thiostannate with integrated Mn^{2+} ions in an anionic chain structure. *Solid State Sci.* **2007**, *9*, 1110–1114.
21. Ratajczak-Sitarz, M.; Katrusiak, A.; Dega-Szafran, Z.; Stefański, G. Systematics in $NH^+\cdots N$-Bonded Monosalts of 4,4'-Bipyridine with Mineral Acids. *Cryst. Growth Des.* **2013**, *13*, 4378–4384.
22. Olejniczak, A.; Katrusiak, A.; Szafrański, M. Ten Polymorphs of $NH^+\cdots N$ Hydrogen-Bonded 1,4-Diazabicyclo[2.2.2]octane Complexes: Supramolecular Origin of Giant Anisotropic Dielectric Response in Polymorph V. *Cryst. Growth Des.* **2010**, *10*, 3537–3546.
23. Sheldrick, G.M. Crystal structure refinement with SHELXL. *Acta Crystallogr.* **2015**, *71*, 3–8.
24. Buhl, D.; Gün, H.; Jablonka, A.; Reiss, G.J. Synthesis, Structure and Spectroscopy of Two Structurally Related Hydrogen Bonded Compounds in the *dpma*/$HClO_4$ System; *dpma* = (dimethylphosphoryl)methanamine. *Crystals* **2013**, *3*, 350–362.

25. Reiss, G.J. (Dimethylphosphoryl)methanaminium iodide—(dimethylphosphoryl)methanamine (1:1). *Acta Crystallogr. E* **2013**, *69*, 1253–1254.
26. Czaikovsky, D.; Davidow, A.; Reiss, G.J. Crystal structure of (dimethylphosphoryl)methanaminium hydrogensulfate. *Z. Kristallogr. New Cryst. Struct.* **2014**, *229*, 29–30.
27. Glowiak, T.; Sawka-Dobrowolska, W. The crystal and molecular structure of α-aminomethylmethylphosphinic acid. *Acta Crystallogr. B* **1977**, *33*, 1522–1525.
28. Reiss, G.J.; Jörgens, S. (Dimethylphosphoryl)methanaminium chloride. *Acta Crystallogr. E* **2012**, *68*, 2899–2900.
29. Bianga, C.M.; Eggeling, J.; Reiss, G.J. (Dimethylphosphoryl)methanaminium nitrate. *Acta Crystallogr. E* **2013**, *69*, 1639–1640.
30. Kahrovic, E.; Orioli, P.; Bruni, B.; di Vaira, M.; Messori, L. Crystallographic evidence for decomposition of dimethylformamide in the presence of ruthenium(III) chloride. *Inorg. Chim. Acta* **2003**, *355*, 420–423.
31. Xu, W.; Lin, J.-L. Tetrakis(dimethylammonium) hexachlorotungstate(III) chloride. *Acta Crystallogr. E* **2007**, *63*, 767–769.
32. Reiss, G.J. Bis((dimethylphosphoryl)methanaminium) tetrachloridopalladate(II). *Acta Crystallogr. E* **2013**, *69*, 614–615.
33. Guzei, I.A.; Spencer, L.C.; Ainooson, M.K.; Darkwa, J. Constructor graph description of the hydrogen-bonding supramolecular assembly in two ionic compounds: 2-(pyrazol-1-yl) ethylammonium chloride and diaquadichloridobis (2-hydroxyethylammonium) cobalt(II) dichloride. *Acta Crystallogr. C* **2010**, *66*, 89–96.
34. Guzei, I.A.; Keter, F.K.; Spencer, L.C.; Darkwa, J. Constructor graph description of hydrogen bonding in a supramolecular assembly of (3,5-dimethyl-1H-pyrazol-4-ylmethyl)isopropylammonium chloride monohydrate. *Acta Crystallogr. C* **2007**, *63*, 481–483.

The Role of Hydrogen Bond in Designing Molecular Optical Materials

Leonardo H. R. Dos Santos and Piero Macchi

Abstract: In this perspective article, we revise some of the empirical and semi-empirical strategies for predicting how hydrogen bonding affects molecular and atomic polarizabilities in aggregates. We use p-nitroaniline and hydrated oxalic acid as working examples to illustrate the enhancement of donor and acceptor functional-group polarizabilities and their anisotropy. This is significant for the evaluation of electrical susceptibilities in crystals; and the properties derived from them like the refractive indices.

Reprinted from *Crystals*. Cite as: Dos Santos, L.H.R.; Macchi, P. The Role of Hydrogen Bond in Designing Molecular Optical Materials. *Crystals* **2016**, *6*, 43.

1. Introduction

Calculation and prediction of electrical properties of molecules and solids has for long been of interest [1]. These properties result from the forces that the electric charges mutually exert in a material, and being quantum-mechanical observables of the charge-density distribution, can be obtained, at least in principle, from both quantum-mechanical calculations and high-resolution X-ray diffraction. While the electrostatic moments or the electrostatic potential and its derivatives directly result from the ground-state distribution of electrons, other observables, for example the (hyper)polarizabilities, measure how the charge distribution changes under external stimuli (e.g., an electric field for the electric (hyper)polarizabilities). Therefore, these so-called "response properties" are consequence of the interplay between ground and excited states. When the material under study is molecular, thus ideally consisting of a large number of mutually perturbed closed-shell systems, the connection between the molecular charge density and the bulk electric susceptibilities is relevant [2]. Even more important is identifying those functional groups that induce a particular property of the bulk, either alone or through a cooperative effect.

Because functional materials are most likely to find applications in the solid state, and frequently in crystalline phases, it is also of relevance understanding and predicting the role played by the strongest intermolecular interactions on the molecular or functional-group charge distributions, and how this perturbation affects the properties of the molecules embedded in crystals [3]. In this respect, a good strategy is calculating accurate (hyper)polarizabilities in model systems, using highly correlated Hamiltonians and complete basis-sets, and transferring these quantities to larger molecules or extended structures. The latter are typically too challenging or even prohibitive for accurate first-principles simulations. Envisaging very efficient strategies, an intermediate step would consist in benchmarking density functionals and finite basis-sets against the most correlated Hamiltonians and the complete basis-set limit [4].

Having this in mind, in this article, we analyze the most relevant effects of intermolecular interactions, in particular hydrogen bonds, for the estimation of linear optical properties of molecular materials. We also discuss the state-of-the-art procedures to estimate polarizabilities of atoms and functional groups in molecules or crystals, and to derive bulk optical properties from these "microscopic" quantities.

2. Distributed Atomic and Functional-Group Polarizabilities

In order to identify the most important atoms and functional groups that determine the optical properties of a molecule or a crystalline material, we need a partitioning of the electrostatic moments and (hyper)polarizabilities into atomic contributions. In experimental electron-density determination, a formalism based on pseudo-atoms is widely adopted, which enables the refinement of electron-density models against measured structure-factor amplitudes [5]. In principle, the refined coefficients may be used to compute atomic electrostatic moments. However, there are two main limitations: (a) the multipole expansion is a fitting procedure that therefore returns only approximated quantities; and (b) the atomic multipole parameters may strongly correlate within a refinement procedure, therefore while the total electron density is quite accurately reconstructed, the atomic partitioning may be biased. A better estimation of atomic moments derives from space-partitioning of the total electron density reconstructed from the refined coefficients of the multipole expansion. In fact, the total electron density is less affected by the correlation among parameters of the multipolar expansion. For the partition, one can use the quantum theory of atoms in molecules (QTAIM) [6], which is more generally applicable to a theoretical electron density as well. An advantage of QTAIM is that the partitioning is *exact* (meaning that the sum of atomic contributions exactly reconstruct the total property) and the atomic basins are discrete, making it easier the definition of an important quantity like the charge transfer (see Supplementary Materials). Furthermore, QTAIM-basins are typically very well transferable among systems, thus allowing their accurate estimation in small, model molecules and exportation to larger ones. Similar schemes could be for example the Voronoi polyhedra or the Hirshfeld surface partitioning. Both are exclusive partitioning of the space, although the latter leaves some small volumes unassigned [7]. An alternative approach is the Hirshfeld atom: here the partitioning is complete, but the atomic basins overlap [8], which may create problems for assigning correctly the interatomic charge transfer.

Some Hilbert-space partitionings, such as Stone's [9], could also be used to obtain atomic or functional-group moments and (hyper)polarizabilities. However, with these methods, the partitioning suffers from being largely dependent on the atomic basis-sets and from a limited transferability. Notably, the computationally cheap Mulliken's partition [10] holds for atomic charges only. Although these can reconstruct a molecular dipole, the lack of terms describing the atomic intrinsic polarization (see Supplementary Materials) does not enable computing a fundamental part of the atomic polarizability.

Some partitioning schemes have been applied to estimate linear optical properties of atoms and functional groups in molecules or crystals. Several approaches have been proposed for the calculation of *distributed atomic polarizabilities*, i.e., the atomic polarizability tensors within a molecule or molecular aggregate (for a recent account, the reader is referred to [4] and references therein). Among those, Bader and coworkers developed QTAIM-partitioned polarizabilities and used them to evaluate intermolecular interaction energies and transferability of electric properties [11]. Keith's generalization of Bader's method removed the origin dependence from the QTAIM definition of atomic dipoles and polarizabilities, thus making them perfectly transferable to other systems [12]. Within this approach, the component $\alpha_{ij}(\Omega)$ of the atomic first-order polarizability tensor can be simply evaluated through the numerical derivative of the corresponding atomic dipole moments, with respect to an externally applied electric field:

$$\alpha_{ij}(\Omega) = \lim_{E_i \to 0} \frac{\mu_j^{E_i}(\Omega) - \mu_j^o(\Omega)}{E_i} \qquad (1)$$

where $\mu_j^{E_i}(\Omega)$ is the dipole moment component of the atomic basin Ω along the j direction computed with an electric field applied along the direction i. Keith's approach has been recently modified and implemented in a program called *PolaBer* [13], used to calculate the distributed polarizability tensors reported in this paper (for details of the implementation, as well as for the methodological procedures used in this work, see Supporting Information). The total, molecular or crystal polarizability can be straightforwardly obtained through summation of the atomic counterparts, as typical for any molecular property estimated using QTAIM. Because the polarizabilities have dimensions of volume, atomic and molecular tensors can be visualized as ellipsoids, see, for example, Figure 1, which shows *p*-nitroaniline, calculated at the CAM-B3LYP/cc-pVDZ level of theory. Most of the calculations in this work were performed using the modest-size cc-pVDZ basis-set. Even though augmentation with diffuse functions is highly desirable to accurately estimate electric moments and polarizabilities of molecules in an infinitely diluted gas, these functions are less relevant for estimating properties in aggregates or crystals, due to the existence of a "basis-set superposition" effect that takes place among basis-sets of vicinal groups or molecules in aggregation, see for example [14].

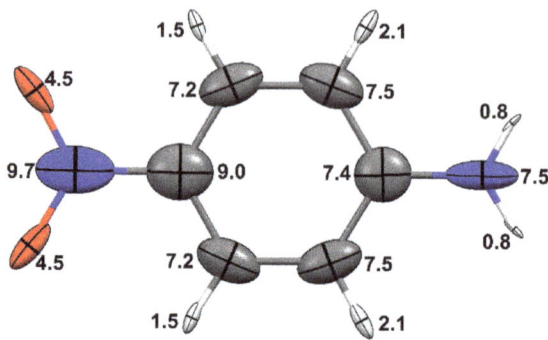

Figure 1. Distributed atomic polarizability ellipsoids after QTAIM partitioning of the molecular p-nitroaniline electron density, as obtained from a CAM-B3LYP/cc-pVDZ calculation. The isotropic polarizabilities, calculated as the arithmetical average of the main diagonal tensor components, are shown in au. The scaling factor for the ellipsoids is 0.3 Å^{-2}.

Otero *et al.* [15] criticized the use of Equation (1) because the ground-state molecular electron density $\rho^0(\mathbf{r})$ and the corresponding field-perturbed densities $\rho^{E_j}(\mathbf{r})$ are "independently" partitioned into their atomic contributions, without guaranteeing that the corresponding atomic basins are kept constant in the process. They developed, instead, a methodology to partition the polarizability distribution $\alpha(\mathbf{r})$ using Hirshfeld scheme. $\alpha(\mathbf{r})$ enables visualizing regions of negative polarizability in a molecule, *i.e.*, regions that would respond counterintuitively to an applied field (with negative charges moving in the direction of the negative field).

The criticism of Otero *et al.* [15] is, however, not fully correct. First of all, the numerical approach in Equation (1), in the limit of an infinitesimally small field, gives the exact definition of the dipolar electron density derivative that coincides with the analytical definition. Therefore, there is no bias in applying Equation (1) and the numerical solution is used for the sake of simplicity in the complex step of the atomic basin integration and the successive derivation with the field. If the values obtained through the numerical approach do not sum up perfectly to the total molecular polarizability (typically in the range of 0.1%), this can be easily overcome by using more precise integration inside the atomic basins and smaller electric fields. The integration is anyway the most crucial step, because of the difficult definition of QTAIM atomic basins. The fact that the atomic tensors are not symmetrical, instead, depends primarily on the fact that the atomic basins are themselves not symmetrical, which does not enable Equation (1) to be invariant with respect to exchange of i and j. Noteworthy, this is true with any discrete partitioning (e.g., Voronoi or Hirshfeld surface), unless the atom lies on a symmetric site. It holds also for the Hirshfeld atomic partitioning, because, although the volume in which atoms are integrated is infinite, the weight function that defines each atom is not symmetric (again unless the atom sits on a symmetric position). Therefore, the tensor symmetrization is a necessary step, used for visualizing the atomic polarizability ellipsoid. The non-symmetrized tensors would be the exact QTAIM-derived atomic tensors, but would also be more difficult to use. The tensor symmetrization has two important properties: (a) the isotropic atomic polarizability is invariant (because only the out of diagonal components are touched); and (b) the sum of the atomic tensors gives anyway the total (symmetric and exact) molecular polarizability, because the anti-symmetric part of the atomic tensors mutually cancel each other. Otero *et al.* [15] also claimed that in the calculation of integrated atomic polarizabilities, the regions of negative $\alpha(\mathbf{r})$ are "explicitly omitted". On the contrary, the integration inside an atomic basin implies that regions of negative polarizability be overwhelmed (not omitted) by those of positive polarizability, producing overall the expected result that on average the electron density responds intuitively to the applied field. Interestingly, however, the remark of Otero *et al.* [15] raises the question whether some atoms in a molecule may in principle have negative atomic tensors. Albeit possible (and anyway partitioning dependent), this is unlikely to occur for all atoms with a core, which necessarily implies a sufficiently large and almost symmetric basin. Therefore, the only atoms that may display a negative polarizability are H atoms, especially if involved in strong HB (where they exhibit small and irregular atomic basins).

Despite the above discussed criticism, it has been demonstrated that QTAIM distributed polarizabilities obtained using *PolaBer* are remarkably exportable among some series of molecules, including amino acids and organic optical materials [16]. Databanks of atomic and functional-group polarizabilities have been created and used to estimate the refractivity of a few materials [4]. Furthermore, this approach has been successfully applied to understand the origin of the refractive indices of molecular crystals in terms of their most fundamental building blocks, *i.e.*, atoms and functional groups [17], as well as to quantify the effect of hydrogen bonding on their optical behavior (see next section).

3. From Functional Groups and Molecules to Crystals

High-resolution X-ray diffraction is an established technique for determination of molecular electrostatic moments of atoms and molecules in crystals, with the advantage of providing not only magnitudes, but also all their tensor components. Some time ago, Fkyerat *et al.* [18,19] and Hamzaoui *et al.* [20,21] proposed a correlation between the molecular first-order polarizability α and first hyperpolarizability β with, respectively, the quadrupole and octupole electrostatic moments, derivable from the X-ray experiments. However, application of this method to organic non-linear optical materials revealed that many components of α and β tensors, calculated from X-ray diffraction, differ by more than an order of magnitude from theoretical results. Indeed, it was later shown that the one-electron density $\rho(\mathbf{r})$, obtained from the usual multipolar pseudo-atom formalisms, does not yield accurate response properties because electronic correlation is included only partially [22]. Instead, Jayatilaka and Cole [23] have pursued an X-ray constrained wavefunction approach to derive much more accurate $\rho(\mathbf{r})$ distributions for a few optical materials, including a metal-organic non-linear optical compound. The "experimental" wavefunctions yielded remarkably accurate electric properties, indicating the possibility to use X-ray constrained molecular orbitals for engineering this kind of materials. In the case a scheme is further assumed to localize the constrained canonical orbitals in particular atoms or functional groups [24], they can be exported even to much larger systems, thus allowing an inexpensive, though accurate, prediction of their electric properties as well, which is currently challenging for both theory and experiment.

The estimation of electrical properties from X-ray diffraction or fully periodic quantum-mechanical calculations [25] is indeed highly desirable, as these techniques inherently account for the entire crystal field. However, the need of high-quality single-crystals and the high computational costs, respectively, often hamper the application of these methods for large systems. Therefore, calculations on isolated molecules have been extensively applied to estimate electrostatic moments and (hyper)polarizabilities [3]. This requires including an empirical perturbation of the molecular electron density mimicking the crystal electric field. In fact, if the molecular electrostatic moments are calculated in an infinitely diluted gas, the agreement with experimental results in crystals (or with periodic calculations) are obviously poor.

The big challenge in computational chemistry is how to simulate the crystal electric field at the lowest computational costs. We should first of all consider that molecular electric moments are typically affected by short- and long-range effects [4]. Short-range effects are, for example, hydrogen bonds or other medium-strength intermolecular interactions. They may cause changes of the electron distribution and the molecular geometry as well, but they are possible only within the first coordination sphere around a molecule, therefore they are few. On the other hand, long-range effects, due mainly to electrostatic forces produced by interaction between electron-density clouds, are weaker (as their magnitude decrease as a power function of the intermolecular distance), but because they are potentially infinite in number their contribution may be of significance.

Hydrogen bonds tend to line up the molecules in order to minimize the energy [2,17]. This often, though not always, enhances their dipole moments due to induced polarization of the electron-density distributions [26]. This *dipole moment enhancement* observed when going from a single-molecule to an aggregated environment has been investigated in quite some detail for 2-methyl-4-nitroaniline, one of the molecular prototypes for linear and non-linear optical materials. A careful diffraction study by Spackman [27] reported a significant enhancement of 30%–40%, nevertheless smaller than that reported earlier by Howard *et al.* [28]. Spackman [26] has investigated more systematically a large number of crystals, calculating dipole enhancements up to *ca.* 30% for hydrogen-bonded species, whereas many diffraction experiments result in enhancements greater than 100%. However, it is recognized that molecular moments are highly dependent on the multipole model fitted against the X-ray diffracted intensities, and several studies reporting incredibly large enhancements were performed treating the thermal motion of hydrogen atoms as isotropic and without incorporating neutron diffraction estimates of X–H distances. Molecular dipole moments may be very difficult to estimate from experimental multipolar expansion, because they strongly depend on the contraction/expansion of atomic shells, often sensitive to even minor systematic effects in the datasets. Moreover, in non-centrosymmetric crystals, special attention is necessary to avoid model ambiguities, which often affect the estimation of molecular dipole moment. Last but not least, it should also be considered that some supramolecular packings may produce dipole depletion instead of enhancement.

The redistribution of electronic charge that molecules undergo upon aggregation has also important consequences for the estimation of optical properties. Studies on organic non-linear optical materials revealed that the hyperpolarizability of a molecule typically grows by a factor of three when a hydrogen-bonded cluster is considered in its surrounding [29]. Instead, the first-order polarizability α is much less susceptible, thus being determined to a great extent by the intramolecular connectivity (the chemical bonds) rather than the intermolecular forces. As an example, we consider the atomic polarizabilities corresponding to a head-to-tail aggregation of two *p*-nitroaniline molecules (see Figure 2), and compare these values with those of Figure 1. The moderate-to-weak hydrogen bond characterizing this system increases the polarizability of the donor ($-NH_2$) and acceptor ($-NO_2$) functional groups by 7% and 16%, respectively, whereas the polarizability of the phenyl ring enhances only slightly (approximately 1%). As shown in Table 1, the total molecular polarizability does not increase dramatically because of this interaction. In a real crystal, however, many more hydrogen bonds occur coupled with long-range interactions (electrostatics only). All these smaller contributions eventually sum up leading to a significant increase. Indeed, the molecular polarizability calculated under periodic boundary conditions (PBC) is much larger than that for the molecules composing the dimer (or for any other finite aggregate). As also shown in Table 1, the hydrogen bond has a significant effect on the polarizability anisotropy $\Delta\alpha$, a quantity directly related to the birefringence of a material.

Table 1. Components of the diagonal molecular polarizability tensor (au) for *p*-nitroaniline calculated in isolation (monomer), for the donor and acceptor molecules in the dimer, for the central molecule extracted from some hydrogen-bonded aggregates, and for a periodic B3LYP/cc-pVDZ calculation[a].

Aggregate	*p*-Nitroaniline					Donor Group (-NH$_2$)					Acceptor Group (-NO$_2$)				
	α_1	α_2	α_3	α_{ISO}	$\Delta\alpha$	α_1	α_2	α_3	α_{ISO}	$\Delta\alpha$	α_1	α_2	α_3	α_{ISO}	$\Delta\alpha$
Monomer	31.3	82.8	126.4	80.2	82.5	4.2	4.5	18.2	9.0	13.9	5.6	20.5	29.6	18.6	21.0
Dimer donor/acceptor	30.9/30.7	85.1/79.4	136.2/149.3	84.1/86.5	91.2/103.3	4.0/4.1	6.3/4.4	19.1/19.7	9.8/9.4	14.1/15.5	5.6/5.6	20.5/18.7	31.7/40.8	19.2/21.7	22.7/30.8
Trimer	30.5	81.8	161.5	91.2	114.3	3.9	6.2	20.8	10.3	15.9	5.5	18.6	43.9	22.7	33.8
Pentamer	30.0	85.9	183.1	99.7	134.2	3.7	6.8	25.1	11.9	20.0	5.7	23.6	48.2	25.8	37.0
Crystal (PBC)	70.0	128.6	200.3	133.0	113.0	-	-	-	-	-	-	-	-	-	-

Note: [a] Isotropic polarizability: $\alpha_{ISO} = \frac{1}{3}Tr\boldsymbol{\alpha} = \frac{1}{3}(\alpha_1 + \alpha_2 + \alpha_3)$. Anisotropy of the polarizability tensor: $\Delta\alpha = \{\frac{1}{2}[3Tr(\boldsymbol{\alpha}^2) - (Tr\boldsymbol{\alpha})^2]\}^{1/2}$.

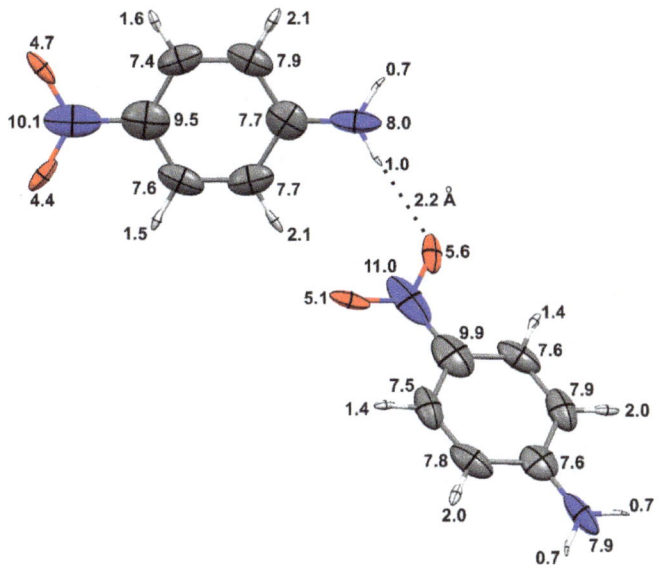

Figure 2. Distributed atomic polarizability ellipsoids of a *p*-nitroaniline dimer extracted from the crystal structure, calculated at the CAM-B3LYP/cc-pVDZ level of theory. The scaling factor for the ellipsoids is as in Figure 1.

As anticipated, calculations under periodic boundary conditions are in principle the correct approaches to model the crystalline effects, and therefore, to accurately estimate electrostatic and response properties of molecules in crystals because they might be dramatically influenced by intermolecular interactions. However, some problems affect the crystal-orbital or plane-wave-based methods: (a) the amount of electronic correlation that one can introduce is limited; (b) for crystal orbitals, convergence often fails when Bloch-type functions use diffuse atomic basis functions; and (c) plane-wave calculations, instead, exclude localized core-orbital functions and therefore their contribution to the properties, which can be quite substantial for hyperpolarizabilities, for example. In order to overcome these drawbacks, the so-called *supermolecule* or *cluster method* was proposed [3]. Within this approach, the properties of interest of several interacting molecules are evaluated as a whole, just like in standard molecular calculations. The goal is then extracting the properties featured by the "central" molecule around which the cluster is constructed. In fact, by comparing the electric properties of an isolated molecule and that of the molecular cluster, one gains insight into the role of short and medium-range intermolecular interactions, crucial for the design of optical materials [4]. However, this approach includes only partially the crystal field effects, because long-range interactions are not accounted. Furthermore, as the choice of the molecular aggregate is usually not unique, a computationally costly pre-screening of at least a few clusters is typically necessary [30].

Table 1 summarizes the polarizability of *p*-nitroaniline, as obtained from different aggregates constructed in order to progressively saturate the hydrogen-bond sites of the "central" molecule (see Figures 3 and 4). When going from the monomer to the trimer, the molecular isotropic polarizability increases by 12%, and a further enhancement of 9% is observed for the central molecule in the pentamer. Although the latter corresponds to a "hydrogen-bond saturation", this by no means indicates convergence of the polarizability because, as previously mentioned, long-range electrostatic interactions may still be of significance. Noteworthy, the polarizabilities of the hydrogen-bonded donor and acceptor groups in the central molecule of the pentamer are up to 30% larger than in the isolated molecule, whereas the enhancements in the phenyl rings are negligible. These results show that the distortions caused by the hydrogen bonds are largely localized in the functional groups directly involved in the bonding.

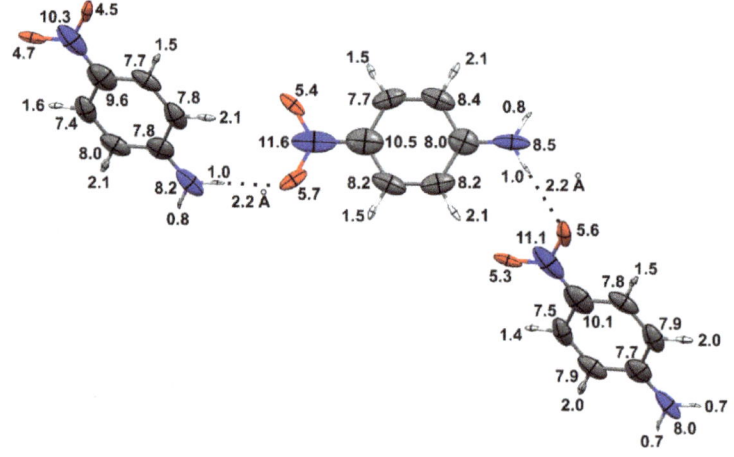

Figure 3. Distributed atomic polarizability ellipsoids of a *p*-nitroaniline trimer extracted from the crystal structure, calculated at the CAM-B3LYP/cc-pVDZ level of theory.

The cluster approach has been previously used to investigate the effect of hydrogen-bond formation on the (hyper)polarizabilities of a few other organic materials. For example, Balakina and coworkers [29,31], studied theoretically a series of non-linear optical chromophores and demonstrated that the formation of weak to moderate N–H···O bonds (with donor-acceptor distances in the range 2.8–3.5 Å) results in an increment of only 10%–15% of the molecular polarizability α, whereas the first hyperpolarizability β may increase up to three times. It is anyway noteworthy that even small changes in α will significantly affect the calculation of the electric susceptibility through a lattice summation.

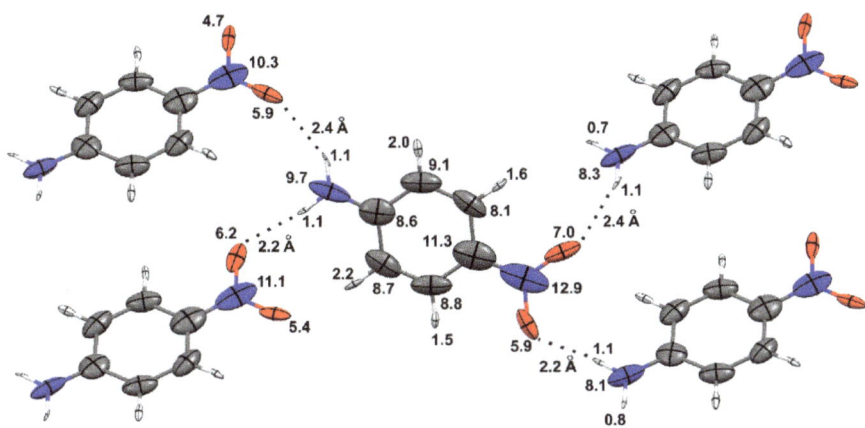

Figure 4. Distributed atomic polarizability ellipsoids of a *p*-nitroaniline pentamer extracted from the crystal structure, calculated at the CAM-B3LYP/cc-pVDZ level of theory.

The importance of explicitly taking into account hydrogen bonds in the estimations of optical properties becomes rather evident when the supermolecule approach is compared with calculations of single-molecules embedded in a polarizable continuum dielectric medium (PCM). This approach somewhat introduces averaged electrostatic interactions, improving calculations of isolated molecules, but it is not capable of predicting the correct polarization of the electron density caused by short-range interactions with partial covalent component, such as hydrogen bonds. Thus, the PCM approach often underestimates the (hyper)polarizabilities. Nevertheless, we are currently working [32] on a semi-continuum method, which combines PCM with calculations on a moderately large cluster, that can be quite effective to estimate optical properties of organic and metal-organic materials.

Although many studies investigated the effect of hydrogen bonding on QTAIM topological parameters, not much work has been devoted to study such influence on the corresponding atomic and functional-group polarizabilities. We have recently investigated many aggregates of amino acids in their zwitterionic configuration with the purpose to quantify the changes of α for the hydrogen-bond donor ($-NH_3^+$) and acceptor ($-COO^-$) functional groups [4,17]. While the isotropic polarizabilities of the groups do not change more than 10% upon hydrogen-bond formation, the tensors are quite significantly reoriented towards the hydrogen-bond direction. This means that these interactions may be useful for controlling the anisotropy of crystalline optical properties, rather than enhancing them. In fact, the anisotropy of the functional-group tensors was found to increase significantly (20%–30%) upon the formation of hydrogen bonds.

An alternative strategy to include crystal field effects in a single-molecule calculation or long-range effects in a molecular aggregate, is firstly to construct an *oriented-gas model*, one in which non-interacting identical molecules or aggregates lie side-by-side and the solid-state properties are simply appropriate combinations of the molecular ones. Afterwards, the effects of the surroundings are semi-empirically approximated using local-field factors [33]. Since molecular crystals feature in general only non-covalent intermolecular interactions, classical electrostatic models have been successfully adopted to estimate dipole moment enhancements and, ultimately, the electric response properties of crystals starting from a simple gas-phase calculation [30].

For the purpose of defining the local-field factor, a crystal may be represented as an array of equal dipoles distributed over a space lattice. A molecule embedded in the crystal experiences a local-electric field $\mathbf{E}_{local}$ that, in general, is a sum of any externally applied field and the internal field that results from the dipole moments of all other molecules in the unit cell. For relatively small values of $\mathbf{E}_{local}$, the dipole moment μ^{ind} induced at a particular molecule k depends only on its first-order polarizability, and can be computed as:

$$\mu^{ind}(k) = \alpha \cdot \mathbf{E}_{local} \qquad (2)$$

The dipole moment of the molecule embedded in a crystal or molecular cluster can be estimated as a simple sum of its permanent dipole moment μ, calculated in gas-phase, with the induced dipole μ^{ind}. Therefore, this approach allows the estimation of "in-crystal" dipole moments from the knowledge of the polarizability of the constituent molecules and the symmetry operations used to construct the aggregate. This formalism is usually called *rigorous local field theory* (RLFT) [34,35], and can be straightforwardly extended to compute atomic and functional-group induced dipole moments $\mu^{ind}(\Omega, k)$ in the k molecule of the crystal, provided that a partitioning scheme is used to calculate $\mu(\Omega)$ and $\alpha(\Omega)$. In this case, the *molecular point-dipole* realization of RLFT, *i.e.*, the one in which each molecule is considered as a point dipole (RLFT1), would be replaced by a *distributed atomic or functional-group point-dipole* treatment, in which n atomic dipoles are distributed over the molecule (RLFTn). The relative accuracy of these approximations depends on the size and shape of the molecules. For small compounds, such as urea or benzene [34], RLFT1 and RLFTn do not differ substantially, whereas RLFTn is typically more accurate to describe the anisotropies of larger systems, as shown for m-nitroaniline [35].

The first-order polarizabilities of the molecule embedded in its environment can be calculated through numerical differentiation of the induced dipole moments estimated through RLFT. Thus, the components $\alpha_{ij}(\Omega)$ of the polarizability calculated for the molecule in isolation are perturbed and become $\alpha'_{ij}(\Omega)$ in the crystal [4]. Analogously to Equation (1), $\alpha'_{ij}(\Omega)$ can be computed as:

$$\alpha'_{ij}(\Omega) = \lim_{E_i \to 0} \frac{[\alpha_{ij}(\Omega).E_{i,local}]^{E_i} - [\alpha_{ij}(\Omega).E_{i,local}]^0}{E_i} \qquad (3)$$

Although the local electric field of Equation (3) is just a zero-order approximation, only few works have attempted to iterate the process, using the dipole moment of the embedded molecule to compute an improved approximation to the electric field [4,26].

Our investigations on RLFT applied to amino acid clusters [4] suggested that in case extreme accuracy is desired, a full *ab-initio* treatment of the aggregate is necessary in order to satisfactory estimate functional-group or molecular polarizabilities because the semi-empirical approaches, although sometimes able to accurately predict dipole moment enhancements, are not capable to take into account the "re-accommodation" of the molecular electron density due to the interactions within the first coordination sphere around a molecule. The presence of the molecular vicinity manifests itself not only in terms of the polarization caused by the strongest intermolecular interactions, and of the charge-transfer along the hydrogen-bond directions, but also in terms of a volume contraction that the molecules undergo when going from isolation to the molecular aggregate. We are currently performing work in these directions, but RLFT still remains to be satisfactorily corrected for these factors.

Within RLFT, the first-order electric susceptibility tensor $Ø$ of a molecular crystal can simply be calculated from the $\alpha'(\Omega)$ tensors by applying well-known lattice-summation schemes [36,37]. The permittivity tensor is then obtained as $\varepsilon = \mathbf{I} + Ø$ and the three crystalline refractive indices n_i can be calculated from the eigenvalues ε_i of the permittivity, $n_i = \sqrt{\varepsilon_i}$. Table 2 summarizes the results for *p*-nitroaniline by comparing RLFT (within a molecular point-dipole implementation) against the simple oriented-gas model, a coupled-perturbed Kohn–Sham wave function using PBC, and a few estimations using intermediate levels of approximation (namely, X-ray constrained Hartree-Fock wavefunction, and an oriented-gas model using the molecule extracted from the pentamer), as well as with experimental values. Noteworthy, the experimental refractive indices are larger than all theoretically predictions because the finite frequency at which they are measured increases the refractivity with respect to the calculations performed with non-oscillating fields. An extrapolation to zero frequency would be possible if measurements were known with different wavelengths, which is not the case. Therefore, the periodic calculation can be used as benchmark for the efficiency of the other theoretical approaches.

In this respect, we observe that RLFT substantially improves over the oriented-gas, but still underperforms the periodic first-principles calculation. Interestingly, the electric susceptibility tensor and refractive indices of crystalline urea have been estimated from a variety of approaches [13]. Among them, RLFT resulted in values very close to the periodic first-principles methods, as well as to the experimental values extrapolated to zero frequency. The application of RLFT to a urea molecule extracted from a relatively large cluster slightly overcorrects the refractive indices.

Table 2. Main ($n_{i=1,2,3}$) and crystallographic ($n_{i=a,b,c}$) refractive indices for p-nitroaniline calculated using different models for simulating the crystal field [a].

Crystal Field Model	n_1	$n_{2\equiv b}$	n_3	n_a	n_c	n_{ISO}	Δn
Oriented-gas	1.22	1.41	1.55	1.43	1.36	1.40	0.33
RLFT	1.32	1.55	1.58	1.55	1.33	1.48	0.26
Pentamer	1.21	1.43	1.76	1.59	1.42	1.47	0.55
PBC [b]	1.36	1.60	1.85	1.69	1.55	1.60	0.49
XC-HF [c]	1.42	1.63	1.97	-	-	1.67	0.55
Exptl. [d]	1.525(2)	1.756(3)	1.788(4)	-	-	1.69(5)	0.26(7)

Notes: [a] Isotropic refractive index: $n_{ISO} = \frac{1}{3}\sum_i n_i$. The anisotropy Δn corresponds to the difference $n_3 - n_1$. [b] B3LYP/cc-pVDZ level of theory. [c] X-ray constrained Hartree–Fock calculation [38]. [d] Measured at 588 nm, see [38].

Finally, we consider the example of three aggregates composed of oxalic acid and water molecules (see Table 3 and Figure 5). Overall, the polarizability of the carboxylate groups are not dramatically affected by the hydrogen bonds formed with the solvent molecules, even though such interactions may be considered rather strong. Indeed, the –COOH isotropic polarizability increases only 4% when going from the oxalic acid in isolation to the tetra-hydrate cluster. The enhancement is more significant along the direction of the hydrogen-bond axes. Interestingly, the shorter hydrogen bonds (where oxalic acid acts as a donor) change less the oxalic acid polarizability than the longer ones (where oxalic acid is the acceptor). In this respect, it should also be taken into account that the volume associated with the –COOH group in the $C_2H_2O_4 \cdot 2\,H_2O$ (sHB) aggregate is inherently smaller than that in the $C_2H_2O_4 \cdot 2\,H_2O$ (mHB) aggregate (sHB is the aggregate produced by the shorter HB, mHB is the medium length HB adduct). Therefore, the fact that the α_3 tensor component of the –COOH polarizability in sHB is smaller than that in mHB is a manifestation not only of the extra electron-density polarization caused by the presence of the water molecules, but also of the change in volume undergone by the oxalic acid when hydrated, with respect to the anhydrous form. The α_{ISO}/V ratio for the –COOH group (with V calculated using a 0.001 au isodensity surface) is in fact constant for the whole series of aggregates.

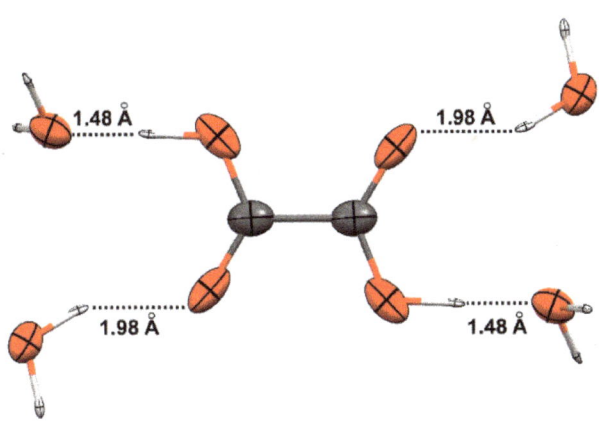

Figure 5. Distributed atomic polarizability ellipsoids after QTAIM partitioning of the molecular aggregate $C_2H_2O_4 \cdot 4H_2O$ electron density, as obtained from a CAM-B3LYP/aug-cc-pVDZ calculation. The scaling factor for the ellipsoids is 0.3 Å^{-2}.

Table 3. Diagonal polarizability tensor (au) for the carboxylic groups in oxalic acid, calculated in isolation and in some hydrogen-bonded aggregates at the CAM-B3LYP/aug-cc-pVDZ level of theory [a].

$C_2H_2O_4 \cdot x\, H_2O$	α_1	α_2	α_3	α_{ISO}	$\Delta\alpha$
0	13.0	23.5	23.9	20.1	10.7
2 (mHB) [b]	12.3	22.5	26.5	20.4	12.7
2 (sHB) [b]	13.0	23.6	26.1	20.9	12.0
4	12.2	23.2	27.2	20.9	13.5

Notes: [a] Isotropic polarizability: $\alpha_{ISO} = \frac{1}{3}Tr\alpha = \frac{1}{3}(\alpha_1 + \alpha_2 + \alpha_3)$. Anisotropy of the polarizability tensor: $\Delta\alpha = \{\frac{1}{2}[3Tr(\alpha^2) - (Tr\alpha)^2]\}^{1/2}$. [b] For $C_2H_2O_4 \cdot 2\, H_2O$, two aggregates were considered, mHB include only the water molecules forming the two longer hydrogen bonds in Figure 3 ($C_2H_2O_4$ is the hydrogen-bond acceptor), while sHB include only those forming the shorter bonds ($C_2H_2O_4$ is the donor).

4. Conclusions

In the framework of a long-term project dedicated to rationally designing optical materials, this article explored the role of hydrogen bonds in electrical response properties of some molecular crystals. In particular, we have discussed procedures to estimate the first-order electric susceptibility and the crystalline refractive indices from QTAIM-partitioned atomic and functional-group polarizabilities calculated from simple model systems, here exemplified by p-nitroaniline and oxalic acid. Figure 6 summarizes the main approaches discussed in this paper. The efficiency of the various methods strongly depends on the system under investigation, in particular on the chemical nature and strength of the intermolecular interactions.

For the examples taken into account in this work, the hydrogen bonds are usually responsible for an enhancement of the functional-group polarizabilities in the range of 10%–20%, a value that might be negligible when small aggregates are to be investigated, but that is of significance when bulk properties are estimated, as many hydrogen bonds are typically present in the crystalline phase. Hydrogen bonds also have the additional effect of reorienting the polarizability ellipsoids, thus affecting the anisotropy of the susceptibilities. This is a feature to be carefully considered when optical manifestations of this anisotropy (such as birefringence) are to be avoided or optimized. Furthermore, our analysis has shown that the most significant perturbations caused by the formation of hydrogen bonds are often concentrated in the atoms or groups directly involved in the intermolecular bonding. Therefore, while accurate estimation of these perturbations are typically necessary for predicting crystalline properties, the "inner" functional groups in a molecule may be safely approximated as the unperturbed ones, calculated from isolated model molecules.

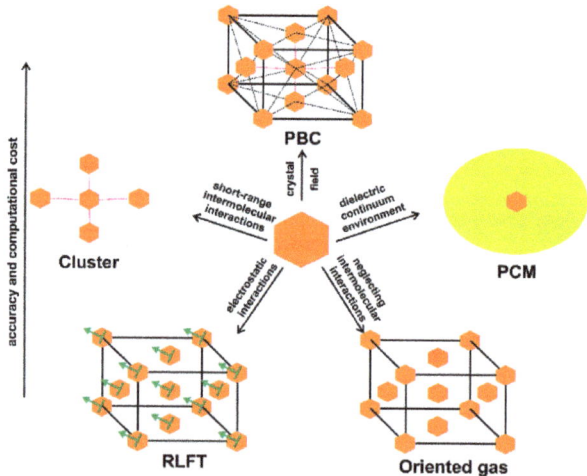

Figure 6. The approaches considered in this work to estimate linear optical properties of organic crystalline materials from their smallest building blocks, either molecules or functional groups.

These findings, along with previous works [4,17], pave the way for deriving databanks of transferable atomic or functional group polarizabilities that could be explored not only for estimating optical properties of complex materials, but also to estimate the dispersive component of the interaction energy between molecules in aggregates, or to map the most reactive sites in a material. Finally, the analyses carried out here should be extended for the relevant cases of non-linear optical materials, whose corresponding non-linear susceptibilities are much more challenging to estimate, either from first-principles or semi-empirically, because the effects of intermolecular interactions are typically much more pronounced.

Supplementary Materials: The Supplementary Materials are available online at www.mdpi.com/2073-4352/6/4/43/s1.

Acknowledgments: We thank the Swiss National Science Foundation (SNF-141271 and 160157) for financially supporting this work.

Author Contributions: Piero Macchi conceived the project and Leonardo H. R. Dos Santos carried out the calculations. Both authors analyzed the data and wrote the paper.

Conflicts of Interest: The authors declare no conflict of interest.

References

1. Medved, M.; Champagne, B.; Noga, J.; Perpète, E.A. *Computational Aspects of Electric Polarizability Calculations: Atoms, Molecules and Clusters*; Maroulis, G., Ed.; IOS Press: Amsterdam, The Netherlands, 2006; pp. 17–31.
2. Macchi, P. Crystallographic Approaches for the Investigation of Molecular Materials: Structure Property Relationships and Reverse Crystal Engineering. *CHIMIA Int. J. Chem.* **2014**, *68*, 31–37.
3. Champagne, B.; Bishop, D.M. Calculations of Nonlinear Optical Properties for the Solid State. *Adv. Chem. Phys.* **2003**, *126*, 41–92.
4. Dos Santos, L.H.R.; Krawczuk, A.; Macchi, P. Distributed Atomic Polarizabilities of Amino Acids and their Hydrogen-Bonded Aggregates. *J. Phys. Chem. A* **2015**, *119*, 3285–3298.
5. Coppens, P. *X-ray Charge Densities and Chemical Bonding*; Oxford University Press: New York, NY, USA, 1997.
6. Bader, R.F.W. *Atoms in Molecules: A Quantum Theory*; Oxford University Press: Oxford, UK, 1990.
7. Spackman, M.A.; Jayatilaka, D. Hirshfeld Surface Analysis. *CrystEngComm* **2009**, *11*, 19–32.
8. Jayatilaka, D.; Dittrich, B. X-ray structure refinement using aspherical atomic density functions obtained from quantum mechanical calculations. *Acta Crystallogr. Sect. A* **2001**, *64*, 383–393.
9. Stone, A.J. Distributed Polarizabilities. *Mol. Phys.* **1985**, *56*, 1065–1082.
10. Jensen, F. *Introduction to Computational Chemistry*; John Wiley and Sons: Chichester, UK, 2007.
11. Bader, R.F.W.; Keith, T.A.; Gough, K.M.; Laidig, K.E. Properties of atoms in molecules: Additivity and transferability of group polarizabilities. *Mol. Phys.* **1992**, *75*, 1167–1189.
12. Keith, T.A. *The Quantum Theory of Atoms in Molecules*; Matta, C.F., Boyd, R.J., Eds.; Wiley-VCH: Weinheim, Germany, 2007; pp. 61–94.

13. Krawczuk, A.; Pérez, D.; Macchi, P. PolaBer: A program to calculate and visualize distributed atomic polarizabilities based on electron density partitioning. *J. Appl. Crystallogr.* **2014**, *47*, 1452–1458.
14. Hammond, J.R.; Govind, N.; Kowalski, K.; Autschbach, J.; Xantheas, S.S. Accurate dipole polarizabilities for water clusters n = 2–12 at the coupled-cluster level of theory and benchmarking of various density functionals. *J. Chem. Phys.* **2009**, *131*, 214103.
15. Otero, N.; Van Alsenoy, C.; Pouchan, C.; Karamanis, P. Hirshfeld-based intrinsic polarizability density representations as a tool to analyze molecular polarizability. *J. Comput. Chem.* **2015**, *36*, 1831–1843.
16. Krawczuk-Pantula, A.; Pérez, D.; Stadnicka, K.; Macchi, P. Distributed atomic polarizabilities from electron density. I. Motivations and Theory. *Trans. Am. Cryst. Ass.* **2011**, *42*, 1–25.
17. Chimpri, A.S.; Gryl, M.; Dos Santos, L.H.R.; Krawczuk, A.; Macchi, P. Correlation between Accurate Electron Density and Linear Optical Properties in Amino Acid Derivatives: L-Histidinium Hydrogen Oxalate. *Cryst. Growth Des.* **2013**, *13*, 2995–3010.
18. Fkyerat, A.; Guelzim, A.; Baert, F.; Paulus, W.; Heger, G.; Zyss, J.; Périgaud, A. Electron density study by X-ray and neutron diffraction of an NLO compound: N-(4-nitrophenyl)-L-prolinol. Description of quadratic hyperpolarizability. *Acta Crystallogr. Sect. B-Struct. Sci.* **1995**, *51*, 197–209.
19. Fkyerat, A.; Guelzim, A.; Baert, F.; Heger, G.; Zyss, J.; Périgaud, A. Assessment of the polarizabilities (α, β) of a nonlinear optical compound [N-(-4nitrophenyl)-L-prolinol] from an experimental electronic density study. *Phys. Rev. B* **1996**, *53*, 16236.
20. Chouaih, A.; Hamzaoui, F.; Vergoten, G. Capability of X-ray diffraction to the determination of the macroscopic linear susceptibility in a crystalline environment: The case of 3-methyl 4-nitropyridine N-oxide (POM). *J. Mol. Struct.* **2005**, *738*, 33–38.
21. Hamzaoui, F.; Zanoun, A.; Vergoten, G. The molecular linear polarizability from X-ray diffraction study. The case of 3-methyl 4-nitropyridine N-oxide (POM). *J. Mol. Struct.* **2004**, *697*, 17–22.
22. Whitten, A.E.; Jayatilaka, D.; Spackman, M.A. Effective molecular polarizabilities and crystal refractive indices estimated from X-ray diffraction data. *J. Chem. Phys.* **2006**, *125*, 174505.
23. Hickstein, D.D.; Cole, J.M.; Turner, M.J.; Jayatilaka, D. Modeling electron density distributions from X-ray diffraction to derive optical properties: Constrained wavefunction *versus* multipole refinement. *J. Chem. Phys.* **2013**, *139*, 064108.
24. Dos Santos, L.H.R.; Genoni, A.; Macchi, P. Unconstrained and X-ray constrained extremely localized molecular orbitals: Analysis of the reconstructed electron density. *Acta Crystallogr. Sect. A-Found. Adv.* **2014**, *70*, 532–551.
25. Ferrero, M.; Rérat, M.; Orlando, R.; Dovesi, R.; Bush, I. Coupled Perturbed Kohn-Sham Calculation of Static Polarizabilities of Periodic Compounds. *J. Phys. Conf. Ser.* **2008**, *117*, 012016.

26. Spackman, M.; Munshi, P.; Jayatilaka, D. The use of dipole lattice sums to estimate electric fields and dipole moment enhancement in molecular crystals. *Chem. Phys. Lett.* **2007**, *443*, 87–91.
27. Whitten, A.E.; Turner, P.; Klooster, W.T.; Piltz, R.O.; Spackman, M.A. Reassessment of Large Dipole Moment Enhancements in Crystals: A Detailed Experimental and Theoretical Charge Density Analysis of 2-Methyl-4-nitroaniline. *J. Phys. Chem. A* **2006**, *110*, 8763–8776.
28. Howard, S.T.; Hursthouse, M.B.; Lehmann, C.W.; Mallinson, P.R.; Frampton, C.S. Experimental and theoretical study of the charge density in 2-methyl-4-nitroaniline. *J. Chem. Phys.* **1992**, *97*, 5616–5630.
29. Balakina, M.Y.; Fominykh, O.D. The quantum-chemical study of small clusters of organic chromophores: Topological analysis and non-linear optical properties. *Int. J. Quantum. Chem.* **2008**, *108*, 2678–2692.
30. Wu, K.; Snijders, J.G.; Lin, C. Reinvestigation of hydrogen bond effects on the polarizability and hyperpolarizability of urea molecular clusters. *J. Phys. Chem. B* **2002**, *106*, 8954–8958.
31. Balakina, M.Y.; Nefediev, S.E. Solvent effect on geometry and nonlinear optical response of conjugated organic molecules. *Int. J. Quantum Chem.* **2006**, *106*, 2245–2253.
32. Ernst, M.; Dos Santos, L.H.R.; Macchi, P. Optical properties of metal organic networks from the analysis of distributed atomic polarizabilities. **2016**. (In Preparation).
33. Dunmur, D.A. The Local Electric Field in Anisotropic Molecular Crystals. *Mol. Phys.* **1972**, *23*, 109–115.
34. Reis, H.; Raptis, S.; Papadopoulos, M.G.; Janssen, R.H.C.; Theodorou, D.N.; Munn, R.W. Calculation of macroscopic first- and third-order optical susceptibilies for the benzene crystal. *Theor. Chem. Acc.* **1998**, *99*, 384–390.
35. Reis, H.; Papadopoulos, M.G.; Calaminici, P.; Jug, K.; Köster, A.M. Calculation of macroscopic linear and nonlinear optical susceptibilities for the naphthalene, anthracene and meta-nitroaniline crystals. *Chem. Phys.* **2000**, *261*, 359–371.
36. Munn, R.W. Electric dipole interactions in molecular crystals. *Mol. Phys.* **1988**, *64*, 1–20.
37. Cummins, P.G.; Dunmur, D.A.; Munn, R.W.; Newham, R.J. Applications of the Ewald method. I. Calculation of multipole lattice sums. *Acta Crystallogr. Sect. A* **1976**, *32*, 847–853.
38. Jayatilaka, D.; Munshi, P.; Turner, M.J.; Howard, J.A.K.; Spackman, M.A. Refractive indices for molecular crystals from the response of X-ray constrained Hartree-Fock wavefunctions. *Phys. Chem. Chem. Phys.* **2009**, *11*, 7209–7218.

Diffusivity and Mobility of Adsorbed Water Layers at TiO$_2$ Rutile and Anatase Interfaces

Niall J. English

Abstract: Molecular-dynamics simulations have been carried out to study diffusion of water molecules adsorbed to anatase-(101) and rutile-(110) interfaces at room temperature (300 K). The mean squared displacement (MSD) of the adsorbed water layers were determined to estimate self-diffusivity therein, and the mobility of these various layers was gauged in terms of the "swopping" of water molecules between them. Diffusivity was substantially higher within the adsorbed monolayer at the anatase-(101) surface, whilst the anatase-(101) surface's more open access facilitates easier contact of adsorbed water molecules with those beyond the first layer, increasing the level of dynamical inter-layer exchange and mobility of the various layers. It is hypothesised that enhanced ease of access of water to the anatase-(101) surface helps to rationalise experimental observations of its comparatively greater photo-activity.

Reprinted from *Crystals*. Cite as: English, N.J. Diffusivity and Mobility of Adsorbed Water Layers at TiO$_2$ Rutile and Anatase Interfaces. *Crystals* **2016**, *6*, 1.

1. Introduction

In 1972, Fujishima and Honda first established that titania (TiO$_2$) could lead to splitting of water when exposed to visible light, thereby producing gaseous oxygen and hydrogen [1]. Since then, there has been considerable scrutiny of the properties of aqueous solutions in contact with titania surfaces. More generally, beyond titania, there are a great deal of renewable-energy applications potentially of interest involving photo-electrochemical splitting of water in dye-sensitised solar cells. Naturally, though, the non-toxic, inexpensive and abundant nature of titania render it especially attractive in this regard. Potential photo-active materials may involve support-metal-support-interaction (SMSI) changes to photo-catalytic properties [2]. In any event, given that titania is one of the most scrutinised oxides, there is somewhat of a paucity in our understanding of the characteristics of interfacial water molecules, together with their potential reactivity at surfaces. This is in spite of progress recently [3,4], including via molecular-simulation approaches and theoretical techniques [5], although the outlook in this regard is surely improved in recent years [4].

Titania-water interfaces allow for a detailed study of confined water molecules' dynamical properties. This is very important for the case where hydrogen-bonded molecules play an part in stabilising solutes by solvent interactions, and also forming thereby "cages" [3,4]. For instance, Inelastic Neutron Scattering (INS) measurements have led to vibrational spectra of adsorbed water on anatase powder and rutile rods [6,7]; this led to insight that confined adsorbed water molecules have dynamical and vibrational behaviour more redolent of less mobile ice vis-à-vis to a liquid [6,7]. This adsorbed-water vibrational behaviour has been studied in detail via molecular dynamics (MD) at interfaces of anatase-(101) and rutile-(110) with water [8,9]. Further, *ab initio* MD (AIMD) has led to important results in interesting studies recently [9,10] on librational and higher-frequency adsorbed-water modes. The rutile-(110)-water interface has been particularly scrutinised by MD [11–16]. Ion adsorption has been studied by Zhang *et al.* [11]. Importantly, Predota *et al.* have investigated structure of the electric double layer [12–14], establishing the underlying structural nature of water layers [12], adsorption of ions [13], and diffusivity and viscosity properties of water layers [14]. Crucially, Predota *et al.* concluded that diffusivity increases farther away from the interface, increasing to values synonymous with the bulk state [14]. The effects of protonation on surface properties and characteristics have been considered by Machesky *et al.* [15].

We have performed classical MD to characterise strain within adsorbed water molecules at a variety of titania-water interfaces, as well as considering orientations of their dipoles relative to the normal to the surface [16]. For a wide range of titania surfaces, we have also considered hydrogen-bonding kinetics between bridging oxygen atoms and water molecules adsorbed physically [17]. For confined water molecules, we have reproduced well vibrational-spectra data with respect to INS spectra [18]. For these rather confined layers, a particularly important feature in [18] was computing of self-diffusivity within these layers. We concluded recently that spatial distribution functions (SDF), in three dimensions, of adsorbed water show there is a more open-like topography for anatase-101 relative to rutile-(110) [19], facilitating easier contact of adsorbed water with that beyond the first layer. This boosts the degree of hydrogen bonding with water molecules outside the adsorbed layer. We concluded tentatively that this may rationalise different values for mobility in water layers [19], *i.e.*, lower mobility for rutile-110 interfaces, and rather higher for anatase-(101), providing extremes in the self-diffusivity of the adsorbed monolayer [18]. In any event, an empirical-potential model treatment of titania-water interfaces may well fail to capture the subtleties of their physico-chemical characteristics, together with hydroxylation properties [19]. To clarify this singularly important question, we performed AIMD of rutile-(110) surfaces with partial hydroxylation to determine vibrational properties of hydrogen bonds between bridging oxygen atoms and water, and of orientations of water molecules' dipoles with respect to the surface normal, for surfaces featuring oxygen-atom vacancies, as well as pristine ones [20].

Quite apart from the interest in rutile- and anatase- water interfaces' physico-chemical characteristics *per se*, there is the pertinent question of the link between these ground-state characteristics and photo-activity. Titania's most photo-active polymorph tends to be anatase, possessing greater stability vis-à-vis rutile for nanoparticles [21]. Pan *et al.* have measured and assessed a wide variety of facets of titania crystals, determining that clean anatase-(101) surfaces are more photo-active than their (001) analogues, contrary to many previous findings [22]; it was also determined that anatase-(101) surfaces afford greater photo-activity than rutile-(110). Naturally, in the context of the previous discussion of DFT-based and classical- MD modelling of titania-water interfaces, this raises the tantalising question of how greater photo-activity in anatase-(101) may be rationalised with respect to rutile-(110) from the perspective of structure, hydrogen-bonding arrangements and kinetics, and also, intriguingly, in terms of ease of access of water beyond the adsorbed layer to the titania surface. Given the tentative evidence of [19] from SDF considerations of anatase-(101)'s more accessible and open architecture facilitating hydrogen bonding with water beyond the adsorbed layer, this suggests that greater levels of mobility, or self-diffusivity, in the adsorbed layer of water at the anatase-(101) surface, observed from classical MD in [18], allow for greater scope of more "promiscuous" water contact with the surface. Naturally, this would increase photo-activity and water-splitting rates [22]. However, the contention of inter-layer water "swopping", or exchange, to allow for penetration of water molecules into the adsorbed layer in a dynamic equilibrium between layers, has not been explored in the literature to any extent, although Predota *et al.* have indeed studied self-diffusivity (via MD) increasing towards bulk-like values further away from the rutile-(110) surfaces [14]. In the present work, motivated by tackling these open questions of water self-diffusivity in layers for "extremes" of adsorbed-layer diffusional behaviour in anatase-(101) (higher) and rutile-(110) (low) [18], we study this in various layers, as well as exchange "events" into the adsorbed layer and outer ones, allowing conjecture as to the influence of these water-mobility characteristics, together with architecture of surfaces, on experimentally observed trends in titania-facet photo-activity.

2. Simulation Methodology

We carried out a 1 ns NVT MD simulation [23] for rutile-(110) and anatase-(101) TiO_2 surfaces, using in-house code, using classical dynamics under equilibrium conditions. We applied three-dimensional Ewald treatment for non-bonded interactions with a relative precision of 10^{-5} (in terms of variation of the number of reciprocal-space wavevectors and the real-space contributions [23]). A Nosé-Hoover NVT ensemble was employed at 300 K, in conjunction with a Velocity-Verlet integration with a 0.33 fs timestep. Bulk liquid water was relaxed for around 200 ps using the Anderson-Hoover NPT ensemble (300 K and 1 bar pressure), prior to running for 1 ns under NVT at 300 K with mild thermostat coupling (period of 0.5 ps). The necessary number of water molecules were added to realise an appropriate bulk-like density of circa 1 g/cm^3 between the titania surface and its periodically-imaged couterpart. The initial MD relaxation was performed for liquid water under NPT conditions, so as to achieve a box density corresponding to 1 bar pressure. The Matsui-Akaogi (MA) [24] model was applied for titania, whilst a flexible-SPC (FwSPC) [25] potential was used for water. Ti-Ow parameters were established using the Buckingham potential and O-Ow LJ potential (*cf.* Table 1) [16]. The application of these potentials has led to good accord between computed and experimental vibrational density of states, and also hydrogen-bonding characteristics on the presently-considered titania-water interfaces [18]. The MA model involves Buckingham-type interactions (see Table 1). All slabs used were free to move. The details of system size and box dimensions are specified in Table 2.

Table 1. Force-field parameters. Taken from [24] (titania) and [25] (water), with titania-water interactions as described in [16].

Buckingham Potential for TiO$_2$ and Water Oxygen: $A_{ij} \times \exp(-r_{ij}/\rho_{ij}) - C_{ij}/r_{ij}^6$			
i–j	A_{ij} (kcal·mol^{-1})	ρ_{ij} (Å)	C_{ij} (kcal·mol^{-1} Å^6)
Ti–O	391049.1	0.194	290.331
Ti–Ti	717647.4	0.154	121.067
O–O	271716.3	0.234	696.888
Ti–Ow	28593.0	0.265	148.000
Lennard-Jones potential for water: $\varepsilon_{ij}[(\sigma_{ij}/r_{ij})^1 - \sigma_{ij}/r_{ij})^6]$			
i–j	ε_{ij} (kcal·mol^{-1})		σ_{ij} (Å)
Ow–Ow	0.1554		3.165492
Harmonic potential for water: $k/2 \times (r_{ij} - r_0)^2$			
i–j	k_{ij} (kcal·mol^{-1} Å^{-2})		R^0_{ij} (Å)
Ow–Hw	1059.162		1.012
Harmonic angle bending potential for water: $k/2 \times (\theta - \theta_0)$			
i–j–k	θ_0 deg		k (kcal·mol^{-1} rad^{-2})
H–O–H	113.24		75.900

Atomic charges: q(Ti) = 2.196 e, q(O) = −1.098 e, q(Ow) = −0.82 e, q(Hw) = 0.41 e; Ow, Hw = water oxygen and hydrogen atoms

Table 2. Simulation-box dimensions and number of particles.

Phase (surface) X, Y, Z (Å)	System Size
Rutile (110) 26.26, 45.47, 69.490	(TiO$_2$)$_{630}$ (H$_2$O)$_{2000}$
Anatase (101) 71.46, 26.43, 72.680	(TiO$_2$)$_{1176}$ (H$_2$O)$_{3162}$

Surfaces (*cf.* Figure 1) were prepared from bulk rutile featuring lattice vectors $a_0 = b_0 = 4.593$ Å, $c_0 = 2.959$ Å (*P42/MNM*) and bulk anatase [$a_0 = b_0 = 3.776$ Å and $c_0 = 9.486$ Å (*I41/AMD*)]. Details of topography and construction of the surfaces are detailed elsewhere more completely [16]. The normal to the surface coincided with the z-axis. We estimated water self-diffusivity in the x-y plane (parallel to the surfaces) and z-direction (perpendicular thereto) via the mean squared displacement (MSD) over 1 ns, sampled in regions of increasing distance from the interface in 0.5 Å "bins", taking care with length of MSD and statistical sampling level to ensure establishment of the Fickian regime in each bin [23], whilst also monitoring swopping/exchange events between layers (*vide infra*).

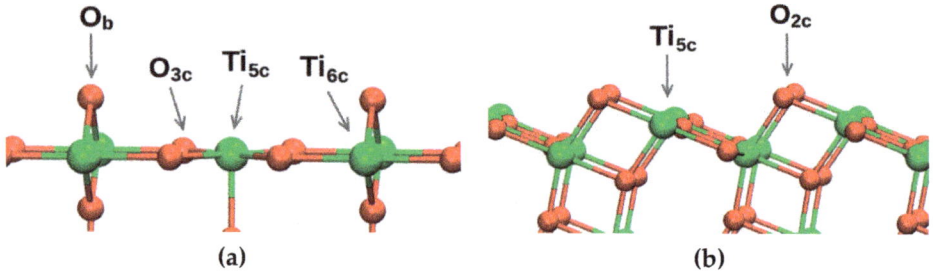

Figure 1. Structure of surfaces; laboratory z-direction is vertical. O_b stands for bridging oxygen, O_{3c} a three-coordinated surface oxygen, Ti_{5c} a penta-coordinated surface Ti atom, and Ti_{6c} denotes a hexa-coordinated Ti atom. (a) rutile-110, and (b) anatase-101.

3. Results and Discussion

The water-density "profiles", in terms of z-axis displacement from interfaces are very similar to those reported previously in [16] (*cf.* Figure 2). In this case, the first minimum is at ~2.9 Å distance from the plane of "uppermost" titanium atoms in the case of both surfaces (*cf.* Figure 2). This minimum is the *de facto* "border" between the adsorbed monolayer (referred to hereinafter as "ML") and layers further out from the surfaces, and, in both cases, uniform density is achieved within ~9 Å of the surfaces. There is a second, less distinct adsorbed layer (termed "2L") between ~2.9 and 4.5 Å distance which is not in any direct contact with the surface at any instant, and beyond this distance, the density layers become less clear-cut as more bulk-like behaviour is achieved, with a third density layer (dubbed "3L") partly evident between 4.5 and 6.3 Å.

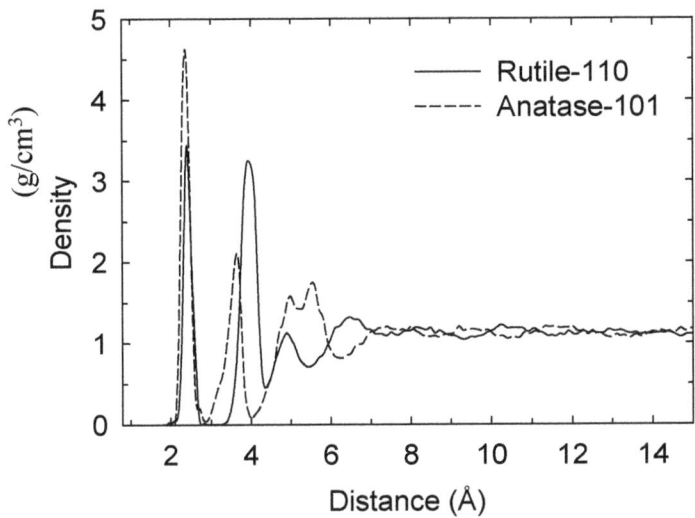

Figure 2. Density profile along z-axis of water molecules from topmost layers of titanium atoms.

The self-diffusivities of each of the three labelled density layers are specified in Table 3. The self-diffusivity of bulk water was found to be ~2.8 × 10^{-9} $m^2 \cdot s^{-1}$ (or ~0.93 × 10^{-9} $m^2 \cdot s^{-1}$ in each x-, y-, z-direction given its isotropic nature), whilst the experimental bulk-water value is 2.3 × 10^{-9} $m^2 \cdot s^{-1}$ [26]. The diffusivity values in the ML exhibit a certain anisotropy due to preferential motion along the local "ridged" architecture (*cf.* Figure 1), and are consistent with those in [18] (determined from a Green-Kubo integral of centre-of-mass velocity autocorrelation functions [23]). In any event, they tend towards bulk-like values with increasing distance from the surfaces (*cf.* Table 3). Those evaluated in 0.5 Å bins along the z-direction from the surfaces also show a gradual increase to recover bulk values within ~9–10 Å, consistent with recovering a bulk-like density profile (*cf.* Figure 3 of [16]). Predota *et al.* observed this in MD simulations for water in contact with rutile-110 [14], while Mamontov *et al.* have studied multi-layer water absorption via neutron scattering and MD, reporting also that the self-diffusivity of water increases away from the surface [27]. The markedly low value for ML water at rutile-(110) results from the atomistic architecture of the surface (*cf.* Figure 1a): adsorbed molecules are confined relatively rigidly in the region between O_b atoms, in contrast to the more "terraced" anatase-101 (*cf.* Figure 1b). The latter surface is more accessible to water molecules and allows for greater hydrogen-bonding interactions of somewhat localised "bound" water molecules with those beyond this layer. In [19], it was found that there are essentially two hydrogen bonds with molecules outside the adsorbed layer in anatase-(101), whilst there is around one such bond per adsorbed water molecule in rutile-(110). This disparity originates from the anatase-(101) surface's more accessible architecture. The present work confirms the tentative conclusion of [19] that this open surface structure and greater hydrogen-bond interaction and "communication" of anatase-(101) 1L water molecules with those in the second "2L" layer facilitates a greater self-diffusivity in the ML for anatase-(101), in stark contrast with the much less mobile, "ice-like" and "trapped" ML water molecules at rutile–(110). Indeed, [18] has determined these more ice-like vibrational features for ML water molecules at rutile-(110) via MD from velocity autocorrelation functions (to obtain the ML's vibrational density of states) with inelastic neutron scattering spectra.

Table 3. Self-diffusivities [$\times 10^{-9}$ m$^2 \cdot$s^{-1}] (x,y,z) in adsorbed layer (ML) and second and third layers (2L & 3L, respectively) from each surface (*cf.* density profiles in Figure 3 of [16]. Note that the sum of the different laboratory directions gives the total self-diffusivity. That of bulk water is ~2.8 $\times$ 10^{-9} m$^2 \cdot$s^{-1} (or ~0.93 $\times$ 10^{-9} m$^2 \cdot$s^{-1} in x,y,z), whilst the experimental bulk-water value is 2.3 $\times$ 10^{-9} m$^2 \cdot$s^{-1}.

Surface	ML	1L	2L
Rutile-(110)	0.011 $\pm$ 0.002, 0.063 $\pm$ 0.007, 0.021 $\pm$ 0.004	(0.31,0.37,0.34) $\pm$ 0.03	(0.58,0.60,0.66) $\pm$ 0.06
Anatase-(101)	0.70 $\pm$ 0.05, 0.75 $\pm$ 0.06, 0.67 $\pm$ 0.05	(0.80, 0.81, 0.76) $\pm$ 0.07	(0.83,0.85,0.80) $\pm$ 0.08

We now turn to the key point of inter-layer mobility, with special focus on exchange events between the ML and 2L density layers. It was found over 1 ns runs that the probability of an exchange event was only ~0.3% from the ML molecules to transition to the 2L layer at the rutile-(110) surface, but was markedly higher at ~4.2% at anatase-(101), with a similar number of reverse exchanges. Given the greater level of hydrogen-bond interactions (essentially double) for ML with 2L water molecules at anatase-(101), this "communication", coupled with a substantially more mobile ML layer (*cf.* Table 3), affords a much greater likelihood of such exchanges. It was found that there were a great deal more such events between 2L and 3L, approaching levels seen in bulk water, with less disparity between the two surfaces, as one would expect for a transition towards bulk-like diffusive behaviour (*cf.* Table 3).

4. Conclusions

MD has been carried out to study water self-diffusivity in quasi-distinct density layers, in contact with rutile-(110) and anatase-(101) surfaces at room temperature, via computation of MSD. Water diffusivity was substantially higher within the anatase-(101) ML, coupled with increasing the level of dynamical inter-layer (ML-1L) exchange and mobility of the various layers. An interesting hypothesis would be that this enhanced ease of access of water to the anatase-(101) surface helps to rationalise experimental observations of its comparatively greater photo-activity in relation to rutile-(110) [22,23], given the greater probability of access of water to the titania surface, allowing greater interaction possibilities with photo-excited holes at the surface. However, it must be noted that this conjecture is somewhat tentative at this stage.

At 300 K, there may well be an appreciable extent of chemically adsorbed water at both surfaces. Naturally, this would serve to change the flavor of anisotropy in water-water structuring relative to physical adsorption studied via classical MD, as in the present work. In any event, even accepting the distinct likelihood of some level of water chemical adsorption, this present work offers useful semi-quantitative insights into water diffusive behaviour. Employing "reactive" models [28] to study these further open questions in the future is to be considered and recommended, given their propensity to handle the complex tapestry of physico-chemical quirks of these surfaces' topographies more accurately.

Acknowledgments: The author thanks Science Foundation Ireland for funding under grant SFI 15/ERC-I3142.

Conflicts of Interest: The author declares no conflict of interest.

References

1. Fujishima, A.; Honda, K. Electrochemical Photolysis of Water at a Semiconductor Electrode. *Nature* **1972**, *238*, 37–38.
2. Haller, G.L.; Resasco, D.E. Metal–support interaction: Group VIII metals and reducible oxides. *Adv. Catal.* **1989**, *36*, 173–235.
3. Diebold, U. The surface science of titanium dioxide. *Surf. Sci. Rep.* **2003**, *48*, 53–229.
4. Henderson, M.A. The interaction of water with solid surfaces: Fundamental aspects revisitied. *Surf. Sci. Rep.* **2002**, *46*, 1–308.
5. Sun, C.; Lui, L.-M.; Selloni, A.; Lu, G.Q.; Smith, S.C. Titania-water interactions: A review of theoretical studies. *J. Mater. Chem.* **2010**, *20*, 10319–10334.
6. Levchenko, A.A.; Kolesnikov, A.I.; Ross, N.L.; Boerio-Goates, J.; Woodfield, B.F.; Li, G.; Navrotsky, A. Dynamics of Water Confined on a TiO_2 (Anatase) Surf. *J. Phys. Chem. A* **2007**, *111*, 12584–12588.
7. Spencer, E.C.; Levchenko, A.A.; Ross, N.L.; Kolesnikov, A.I.; Boerio-Goates, J.; Woodfield, B.F.; Navrotsky, A.; Li, G. Inelastic Neutron Scattering Study of Confined Surface Water on Rutile Nanoparticles. *J. Phys. Chem. A* **2009**, *113*, 2796–2800.
8. Mattioli, G.; Filippone, F.; Caminiti, R.; Bonapasta, A.A. Short Hydrogen Bonds at the Water/TiO_2 (Anatase) Interface. *J. Phys. Chem. C* **2008**, *112*, 13579–13586.
9. Kumar, N.; Neogi, S.; Kent, P.R.C.; Bandura, A.V.; Kubicki, J.D.; Wesolowski, D.J.; Cole, D.; Sofo, J.O. Hydrogen Bonds and Vibrations of Water on (110) Rutile. *J. Phys. Chem. C* **2009**, *113*, 13732–13740.

10. Russo, D.; Teixeira, J.; Kneller, L.; Copley, J.R. D.; Ollivier, J.; Perticaroli, S.; Pellegrini, E.; Gonzalez, M.A. Vibrational Density of States of Hydration Water at Biomolecular Sites: Hydrophobicity Promotes Low Density Amorphous Ice Behavior. *J. Am. Chem. Soc.* **2011**, *133*, 4882–4888.

11. Zhang, Z.; Fenter, P.; Cheng, L.; Sturchio, N.C.; Bedzyk, M.J.; Předota, M.; Bandura, A.; Kubicki, J.D.; Lvov, S.N.; Cummings, P.T.; *et al.* Ion Adsorption at the Rutile–Water Interface: Linking Molecular and Macroscopic Properties. *Langmuir* **2004**, *20*, 4954–4969.

12. Predota, M.; Bandura, A.V.; Cummings, P.T.; Kubicki, J.D.; Wesolowski, D.J.; Chialvo, A.A.; Machesky, M.L. Electric Double Layer at the Rutile (110) Surface. 1. Structure of Surfaces and Interfacial Water from Molecular Dynamics by Use of ab Initio Potentials. *J. Phys. Chem. B* **2004**, *108*, 12049–12060.

13. Predota, M.; Cummings, P.T.; Zhang, Z.; Fenter, P.; Wesolowski, D.J. Electric Double Layer at the Rutile (110) Surface. 2. Adsorption of Ions from Molecular Dynamics and X-ray Experiments. *J. Phys. Chem. B* **2004**, *108*, 12061–12072.

14. Predota, M.; Cummings, P.T.; Wesolowski, D.J. Electric Double Layer at the Rutile (110) Surface. 3. Inhomogeneous Viscosity and Diffusivity Measurement by Computer Simulations. *J. Phys. Chem. C* **2007**, *111*, 3071–3079.

15. Machesky, M.L.; Predota, M.; Wesolowski, D.J.; Vlcek, L.; Cummings, P.T.; Rosenqvist, J.; Ridley, M.K.; Kubicki, J.D.; Bandura, A.V.; Kumar, N.; *et al.* Surface Protonation at the Rutile (110) Interface: Explicit Incorporation of Solvation Structure within the Refined MUSIC Model Framework. *Langmuir* **2008**, *24*, 12331–12339.

16. Kavathekar, R.; Dev, P.; English, N.J.; MacElroy, J.M.D. Molecular dynamics study of water in contact with the TiO$_2$ rutile-110, 100, 101, 001 and anatase-101, 001 surface. *Mol. Phys.* **2011**, *109*, 1649–1656.

17. English, N.J.; Kavathekar, R.; MacElroy, J.M.D. Hydrogen-bond dynamical properties of adsorbed liquid water monolayers with various TiO$_2$ interfaces. *Mol. Phys.* **2012**, *110*, 2919–2925.

18. Kavathekar, R.S.; English, N.J.; MacElroy, J.M.D. Study of translational, librational and intra-molecular motion of adsorbed liquid water monolayers at various TiO$_2$ interfaces. *Mol Phys.* **2011**, *109*, 2645–2654.

19. Kavathekar, R.; English, N.J.; MacElroy, J.M.D. Spatial distribution of water monolayers at the TiO$_2$ rutile- and anatase-titania interfaces. *Chem. Phys. Lett.* **2012**, *554*, 102–106.

20. English, N.J. Dynamical properties of physically adsorbed water molecules at the TiO$_2$ rutile-(110) surface. *Chem. Phys. Lett.* **2013**, *583*, 125–130.

21. Barnard, A.S.; Curtiss, L.A. Prediction of TiO$_2$ Nanoparticle Phase and Shape Transitions Controlled by Surface Chemistry. *Nano Lett.* **2005**, *5*, 1261–1266.

22. Pan, J.; Liu, G.; Lu, G.Q.; Cheng, H.-M. On the True Photoreactivity Order of {001}, {010}, and {101} Facets of Anatase TiO$_2$ Crystals. *Angew. Chem. Int. Ed.* **2011**, *50*, 2133–2137.

23. Allen, M.P.; Tildesley, D.J. *Computer Simulation of Liquids*; Oxford University Press: Oxford, UK, 1987.

24. Matsui, M.; Akaogi, M. Molecular dynamics simulations of the structural and physical properties of the four polymorphs of TiO_2. *Mol. Simul.* **1991**, *6*, 239–244.
25. Wu, Y.; Tepper, H.L.; Voth, G.A. Flexible simple point-charge water model with improved liquid-state properties. *J. Chem. Phys.* **2006**, *124*, 024503.
26. Lamanna, R.; Delmelle, M.; Cannistraro, S. Role of hydrogen-bond cooperativity and free-volume fluctuations in the non-Arrhenius behavior of water self-diffusion: A continuity-of-states model. *Phys. Rev. E* **1994**, *49*, 2841–2850.
27. Mamontov, E.; Vlcek, L.; Wesolowski, D.J.; Cummings, P.T.; Wang, W.; Anovitz, L.M.; Rosenqvist, J.; Brown, C.M.; Sakai, V.G. Dynamics and Structure of Hydration Water on Rutile and Cassiterite Nanopowders Studied by Quasielastic Neutron Scattering and Molecular Dynamics Simulations. *J. Phys. Chem. C* **2007**, *111*, 4328–4341.
28. Raju, M.; Kim, S.-Y.; van Duin, A.C.T.; Fichthorn, K.A. ReaxFF Reactive Force Field Study of the Dissociation of Water on Titania Surfaces. *J. Phys. Chem. C* **2013**, *117*, 10558–10572.

Dissection of the Factors Affecting Formation of a CH···O H-Bond. A Case Study

Steve Scheiner

Abstract: Quantum calculations are used to examine how various constituent components of a large molecule contribute to the formation of an internal CH···O H-bond. Such a bond is present in the interaction between two amide units, connected together by a series of functional groups. Each group is removed one at a time, so as to monitor the effect of each upon the H-bond, and thereby learn the bare essentials that are necessary for its formation, as well as how its presence affects the overall molecular structure. Also studied is the perturbation caused by change in the length of the aliphatic chain connecting the two amide groups. The energy of the CH···O H-bond is calculated directly, as is the rigidity of the entire molecular framework.

Reprinted from *Crystals*. Cite as: Scheiner, S. Dissection of the Factors Affecting Formation of a CH···O H-Bond. A Case Study. *Crystals* **2015**, *5*, 327–343.

---

1. Introduction

There are numerous factors that are known to influence the three-dimensional structure adopted by molecules within crystals [1–8]. In terms of short-range forces the strengths of the covalent bonds are reflected in their equilibrium internuclear distances, and the bond and dihedral angles are a product of orbital interactions and interelectronic repulsions. Coulombic forces are typically the most important component of longer range interactions, acting between buildups and depletions of electron density that are sometimes considered in terms of partial atomic charges. London forces are of much shorter range, serving as an attractive element although they are sometimes masked by stronger electrostatic interactions.

In addition to the aforementioned general sorts of forces, there are a range of more specific noncovalent forces. Examples from the recent literature include halogen [9–17], chalcogen [18–22], and pnicogen [23–30] bonds where an atom from one of these families is attracted to an electronegative atom, such as O, S, or even P. This attraction is facilitated by a highly anisotropic charge distribution. That is, a halogen atom X, for example, is characterized by an overall partial negative charge. But a deeper analysis reveals a distribution of electron density in which a belt of negative potential surrounds a crown of positive charge, usually opposite the C–X bond, which can be attracted to a negative atom on the other subunit. This

Coulombic attraction is supplemented by charge transfer from the O atom to a σ^* C–X antibonding orbital, as well as dispersive forces.

Perhaps the most well-known and best understood noncovalent force is the hydrogen bond (HB), wherein a H atom acts as a bridge [31,32] helping to hold together a pair of electronegative atoms, commonly conceived as F, O, or N. The proton acceptor atom donates a certain amount of electron density from one of its lone electron pairs to form what might be described as a 3c–4e bond. Analyses of crystal structures commonly account for these attractive forces, not only between neighboring molecules, but also HBs internal to each molecule. And indeed, the presence of such HBs are well known to produce crystal structures that differ from those that would occur in their absence.

However, a good deal of work has demonstrated that the HB is a more general phenomenon [26,33–41]. For example, the proton acceptor need not use a lone electron pair as an electron source, but can make use instead of individual π or σ bonds, or even a larger aromatic π system. Another acceptor can be a hydride atom [42–45], in what has come to be called a dihydrogen bond. Nor is it necessary that the bridging H atom be covalently bound to a highly electronegative atom, as SH and ClH can participate as well [46–51]. In fact, even the C atom with an electronegativity comparable to H, has been shown to be a potent proton donor in certain circumstances [52–59]. With specific regard to CH··O HBs, a good deal has been learned about them in recent years. Electron-withdrawing substituents placed near to the pertinent CH group enhance its potency as a proton donor. C atoms with sp hybridization are considerably stronger donors than are sp^2 or sp^3-hybridized systems. And CH···O HBs sometimes manifest an intriguing distinction from more conventional HBs: Rather than the usual red shift of the OH covalent bond stretching frequency that is commonly observed in the IR spectra of OH··O HBs, some (but not all) CH···O HBs display a blue shift of the analogous C–H stretch. Taking these issues into consideration, certain CH···O HBs have been shown to have strength comparable to, and sometimes even exceeding, those of conventional HBs.

Given the potential strength of CH···O HBs, it would be imprudent to ignore the effect that they can exert upon crystal structures. Indeed, the literature is replete with cases where the presence of one or more such bonds have influenced the structure [60–66]. Recent work [67] has shown for example, that a CH··O HB can override the normal trans-planar conformational preferences of α-fluoroamides. In the biological realm too, CH···O HBs play quite an important role [53,68–75] in systems varying from interhelical interactions in proteins to structures of oligosaccharides and carbohydrates and nucleic acids [76–79]. There is some evidence that CH··O HBs may play a previously overlooked role in the structure of such protein stalwarts as the β-sheet [80,81]. Their importance is clear as well in the catalytic mechanism of a family of lysine methyltransferases [82,83].

While much of the earlier work was strongly indicative of the influence of CH···O HBs, there has been little direct information concerning just how different the structure might be in their absence. Nor is there substantial data dealing with how the structure is altered by small changes in the nature of substituents, especially those that are not directly associated with the HB of interest. The present work attempts to open a window into some of the subtleties of the CH···O HB. A molecule is taken as a case study, for which there is a crystal structure available. In this way, the quantum calculations have a secure starting point to be sure that the computed structure matches experiment. A CH···O HB connects a pair of amide groups, which are separated from one another by a series of units, notably a substituted phenyl ring and an ether functionality. Then small alterations are made to various parts of the molecule, after which the geometry and energetics of the internal CH···O HB is carefully monitored. Some of these alterations include the removal of first an electron-withdrawing CF_3 substituent on the phenyl ring, and then the aromatic system itself. The length of an aliphatic system that links the proton donor and acceptor is tested, to see how the HB is affected by their degree of separation. And the application of quantum calculations allows a determination of the energy of the CH···O HB, in isolation from the accompanying NH···O bond, for each of the modified structures.

2. Computational Methods

All calculations were carried out using the Gaussian-03 [84] suite of programs. The polarized 6-31+G** basis set, augmented by diffuse functions, was applied to the system at both the B3LYP and MP2 levels of theory. Geometries were fully optimized except as noted below. Minima were verified via a lack of imaginary frequencies. Atoms in Molecules (AIM) calculations [85,86] supplied quantitative data about bond strengths.

3. Results

Molecule I, displayed in Figure 1, was taken as the starting point for this study. I contains a pair of amide units, which are capable of forming both a NH··O and CH··O HB with one another. The relative orientation of these two amide units is conditioned on the separating groups, including a phenyl ring with CF_3 substituent, and an ether group. The geometry of I was fully optimized at the B3LYP/6-31+G** level. The structure obtained is illustrated in Figure 1, which also includes the labeling that is used below, and the interatomic distances of a number of possible HBs. There are a number of intramolecular interactions that may influence the preferred geometry of this molecule. The aryl hydrogen atom, H_{C1}, may engage in a CH···O H-bond with the O atom of the upper amide group. The R(H···O) distance of 2.18 Å is certainly short enough to support such a supposition. This attraction may help guide the preferred orientation of this amide. With regard to the lower amide, there is clearly a conventional NH···O H-bond with the NH of the upper amide, which probably affects the orientation of both. There is also a possible CH_{C2}···O interaction, with R(H···O) = 2.517 Å, which would affect not only the amide orientation, but the rotational preference of the upper methyl group as well. The presence of such a bond is verified by the presence of a bond path between the two atoms, with a density of 0.008 au at the bond critical point, illustrated by the red number in Figure 1 (in units of 10^3 au). This bond is weaker than the nearby NH···O HB with a density of 0.012 au. It is notable in this respect that the methyl group does not seem to form an intra-amide CH···O H-bond, preferring instead the interamide variety. In other words, the dihedral angle $\varphi(OCCH_{C2})$ of 171° is such that neither of the dihedral angles involving the other two methyl hydrogens are less than 50°. Lastly, H_N lies within 2.14 Å of the ether O atom of the backbone.

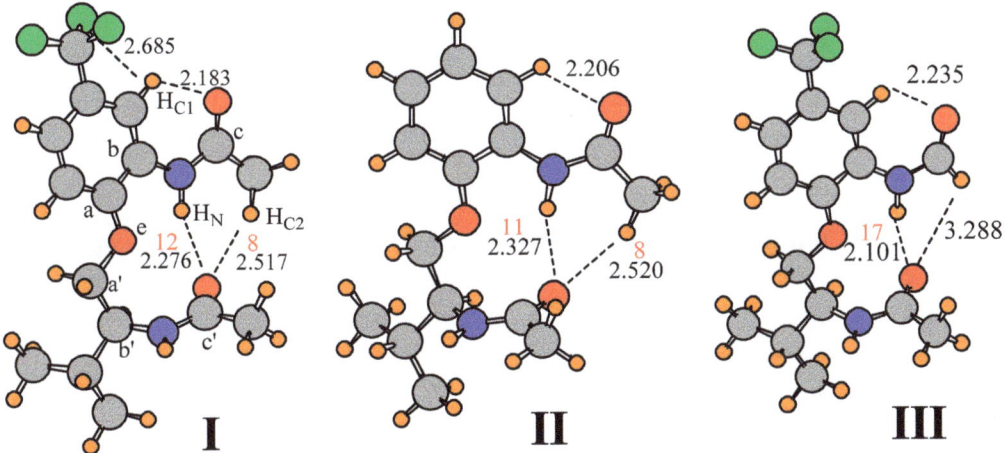

Figure 1. Optimized geometries of indicated molecules, showing atomic labels. Densities at Atoms in Molecules (AIM) bond critical points are reported in red, in units of 10^{-3} au.

Optimization of this structure at the MP2 level, with the same 6-31+G** basis set, leads to a very similar geometry. The H-bond lengths are displayed in the second column of Table 1 which shows only minor perturbations from the B3LYP details. There is a 0.01 Å lengthening of $R(H_{C1}···O)$ and a 0.10 shortening of the other putative CH···O H-bond. The computed distances are similar to those obtained in the crystal structure [87], reported in the last column of Table 1. These similarities lend support to the idea that the calculations properly capture the essential aspects of the structural components of I.

Table 1. H-bond distances (Å) and dihedral angles (degrees) in computed and X-ray structure of I. Calculated results with 6-31+G** basis set.

H-bond	B3LYP	MP2	X-ray
$H_N \cdots O$	2.276	2.116	2.364
$H_{C2} \cdots O$	2.517	2.417	2.538
$H_N \cdots O_e$	2.136	2.155	2.163
$H_{C1} \cdots O$	2.183	2.196	2.261
$\varphi(C_a C_b N C_c)$	180	−169	−171
$\varphi(C_{a'} C_{b'} N C_{c'})$	−92	−90	−93

With specific respect to the orientations of the upper and lower amide groups, the φ(CCNC) dihedral angles describe the rotations around the C-N bond of interest in each case. As reported in the last rows of Table 1, the upper amide group is approximately coplanar with the phenyl ring, with a $\varphi(C_a C_b N C_c)$ dihedral angle of approximately 180°. The lower amide, on the other hand, is approximately perpendicular to the relevant $C_{a'}$–$C_{b'}$ bond of the backbone, with a corresponding dihedral angle of around −90°. Given the nearly fully extended structure within the backbone between the two amide groups, these dihedral angles lead to a nearly perpendicular disposition of the two amides relative to one another which optimally positions the O of the lower amide to interact with the NH and CH donors of the upper amide. It may be noted that both the B3LYP and MP2 structures provide near coincidence with the X-ray values of the dihedral angles.

As noted above, the conformation of the upper terminal methyl group is such that there is no interaction of any of its H atoms with the O atom of the adjacent amide. It was wondered whether the optimized structure of I was perhaps not the global minimum, and that a rotation of the methyl group might lower its energy. The methyl group was therefore rotated by about 60° so that a dihedral angle φ(OCCH) was set equal to 0°, and then the structure reoptimized. However, the methyl group simply rotated back to that in Figure 1, indicating that a geometry of this altered type does not represent a minimum on the surface. This observation supports the idea that a $CH_{C2} \cdots O$ HB with the lower amide group is stronger than any intra-amide interaction that might develop. One might wonder how strong this preference is on a quantitative energetic level. It was found that the aforementioned 60° rotation of the terminal methyl group, which effectively breaks any interamide H-bond and forces an interaction between a methyl H and the adjacent O on the same amide unit, raised the energy of the system by 0.6 kcal/mol. This quantity may be taken as a crude estimate of the differing strengths of the inter- and intra-amide CH$\cdots$O interactions.

Another issue concerns the rigidity with which the upper amide group is held in the orientation of the global minimum of Figure 1. A quantitative measure of this rigidity was obtained by a forced rotation of the entire upper amide group,

NHCOCH$_3$, around the C$_b$–N bond. This rotation was followed by a geometry optimization, but holding this particular dihedral angle fixed. It should be stressed that such a rotation strains not only the CH$_{C2}$···O H-bond, but also another CH···O H-bond involving the aromatic H$_{C1}$, as well as the conventional NH···O H-bond. It was found that a rotation in one direction by 30° raised the energy of the complex by 1.9 kcal/mol, and the energy rose by 2.5 kcal for a rotation in the other direction, providing an indication of the rigidity with which this amide is held in place.

3.1. Related Model Molecules

One might wonder about the influence of the CF$_3$ group of the aromatic ring upon the structure of this molecule. As a strongly electron-withdrawing group, CF$_3$ might be expected to strengthen the H$_{C1}$···O interaction by making H$_{C1}$ more electropositive. Replacement of CF$_3$ by a simple H atom, leading to structure II in Figure 1, lengthens this HB but only by 0.02 Å, suggesting only a marginal weakening. Also, as may be seen in the first column of Table 2, a greater perturbation is observed in the NH···O HB which stretches by 0.05 Å even though it is further removed from the site of substitution. It is not only the HB length that influences its strength, but also the angles. Most particular in this regard is the N/CH···O angle, to which HB energies are most sensitive. The first row of Table 2 shows that the two H-bonds between the upper and lower amides are within 15–28° of linearity, but that there is a 60° deviation in the θ(CH$_{C1}$···O) angle. It is important to note that these distortion angles are essentially unchanged when the CF$_3$ of I is replaced by H in II.

Table 2. Calculated H-bond distances (Å) and angles (degrees) in I and its various derivatives.

Structure	r(H$_N$···O)	r(H$_{C2}$···O)	r(H$_N$···O$_e$)	r(H$_{C1}$···O)	θ(NH$_N$···O)	θ(CH$_{C2}$···O)	θ(NH$_N$···O$_e$)	θ(CH$_{C1}$···O)
I	2.276	2.517	2.136	2.183	165	152	106	120
II	2.327	2.520	2.135	2.206	165	154	106	120
III	2.101	–	2.161	2.235	166	–	104	120
IV	2.377	2.579	2.118	2.213	164	154	107	120
V	2.198	2.850	2.198	–	158	143	99	–
VI	2.234	2.784	2.449	–	163	145	97	–
VII	2.135	–	2.477	–	163	–	95	–

In concert with the geometries and AIM quantities, an alternate indicator of the strength of a given HB is the amount of charge transfer from the proton acceptor to the donor. Natural bond orbital (NBO) analysis [88,89] provides an energetic measure of this transfer in the form of the second-order perturbation energy E(2) into the σ* antibonding orbital of the proton donor. In contrast to the AIM bond critical point densities which are little changed, comparison of these quantities for structures I and II in Table 3 indicates a small but significant weakening of H-bonds of both NH···O and CH···O type, with the exception of the NH···O interaction involving the ether O atom O$_e$.

Table 3. Natural bond orbital (NBO) values of E(2) (kcal/mol) for potential H-bonds in structures I-VII.

Structure	$NH_N \cdots O$ [a]	$CH_{C2} \cdots O$	$NH_N \cdots O_e$	$CH_{C1} \cdots O$
I	3.80	1.49	2.65	3.62
II	3.17	0.97	2.68	3.25
III	7.30	–	2.38	3.19
IV	2.63	0.79	2.92	3.18
V	4.81	–	0.90	–
VI	4.28	–	0.59	–
VII	5.89	–	–	–

[a] includes electron donation from both O lone pair and CO π bond.

In addition to the H-bond geometries, there are a number of measures of the relative orientations of the two amide units. First of these are the two $\varphi(C_aC_bNC_c)$ dihedral angles that describe the rotation of each amide around the relevant C–N bond axis. As indicated in Table 4, these angles are respectively 180° and −92° for the upper and lower amide units in I. This difference would suggest the two amide units are roughly perpendicular to one another. And indeed the angle between the two amide planes, as defined by their O–C–N linkages, θ(a–a'), is rather close to perpendicular, at 74°. Note that this angle is unchanged in II after removal of the CF_3 group. Another means of considering the interamide orientation has to do with sighting down the C_b–C_a–O–$C_{a'}$–$C_{b'}$ backbone, and evaluating the separation of the two units around this axis. An angle of 180° would correspond to the two amides being on direct opposite sides of the backbone, and the angle would be 0° if precisely lined up behind one another. This dihedral angle φ(a–a') is defined as $\varphi(NC_bC_{b'}N)$ and is equal to 57° for both I and II. The geometric data, therefore, argue for very little influence of the CF_3 group upon the preferred structure.

Another perspective on the effect of the CF_3 group arises if one compares the energetics of rotation of the upper amide group both with and without this group. If the CF_3 indeed strengthens the $H_{C1} \cdots O$ attraction, then its removal ought to permit a freer rotation of the amide. However, as reported in Table 5, after this group was replaced by H, the energy of ±30° rotation of the amide was unaffected, further supporting the idea that this group has a minimal effect on any such CH⋯O H-bond. Just as rotation of the upper $NHCOCH_3$ group will strain the H-bonds between the upper and lower amides, so too will a rotation of the lower amide. The results in Table 5 indicate the degrees of flexibility of the two amide groups are comparable to one another. Further, the data indicate that the removal of the CF_3 group has only very minor effects on this flexibility.

Table 4. Angles (degrees) that define the relative orientation of the two amide groups. See Figure 1 for labeling.

Structure	$\varphi(C_aC_bNC_c)$	$\varphi(C_{a'}C_{b'}NC_{c'})$	$\theta(a-a')$ [a]	$\varphi(a-a')$ [b]
I	180	−92	73.8	57
II	180	−92	74.2	57
III	−178	−91	65.7	53
IV	178	−92	73.1	57
V	179	−92	90.5	57
VI	−136	−92	77.3	34
VII	−131	−92	78.7	33

[a] angle between amide planes (O-C-N links) [b] dihedral angle $\varphi(NC_bC_{b'}N)$.

Table 5. Distortion energies (kcal/mol) computed for ±30° rotations of upper and lower NHCOCH$_3$ groups for I and its derivatives.

Structure	Upper Amide		Lower Amide	
	+30	−30	+30	−30
I	1.94	2.51	2.57	1.98
II	1.90	2.46	2.73	1.69
III	2.21	2.49	2.51	2.07
IV	2.02	2.40	2.47	2.00
V	2.73	2.85	2.21	2.05
VI	1.98	2.82	2.12	2.14
VII	2.20	2.47	2.09	2.20

The interamide CH$_{C2}\cdots$O H-bond can be eliminated almost entirely by replacing the upper methyl group by a H atom. The optimized structure of III illustrated in Figure 1 shows that the distance from the carbonyl O of the lower amide to the substituted H is more than 3.2 Å, beyond the range where any such H-bond would contribute much to the energy. And indeed, both AIM and NBO analysis supports the absence of such a H-bond, as in the third row of Table 3. This replacement does have the effect of shortening the NH$\cdots$O H-bond by some 0.26 Å relative to I, so a strengthened NH$\cdots$O may compensate for the lost CH$\cdots$O. The NBO measure of NH$\cdots$O H-bond strength is in fact approximately doubled in III relative to I and II, and the AIM density grows to 0.017 au. Note also that the removal of the methyl group has little effect upon the geometry of the CH$_{C1}\cdots$O H-bond, or its value of E(2) in Table 3. The first two columns of Table 4 show that this replacement does not perturb the orientations of the two amide groups very much, less than 2°. On the other hand, there is a noticeable alteration in the angle between the planes of the two amides, as $\theta(a-a')$ drops from 74° to 66°. This small increase in the deviation from perpendicularity may reflect the increased importance of the NH⋯O H-bond. The rigidity parameters in Table 5 are virtually unaffected by removal of the $CH_{C2}\cdots O$ H-bond. One might conclude, then, that this particular $CH_{C2}\cdots O$ H-bond in and of itself has little influence upon the conformation; any loss is compensated by a strengthening of the other interamide H-bond, of the conventional NH⋯O type.

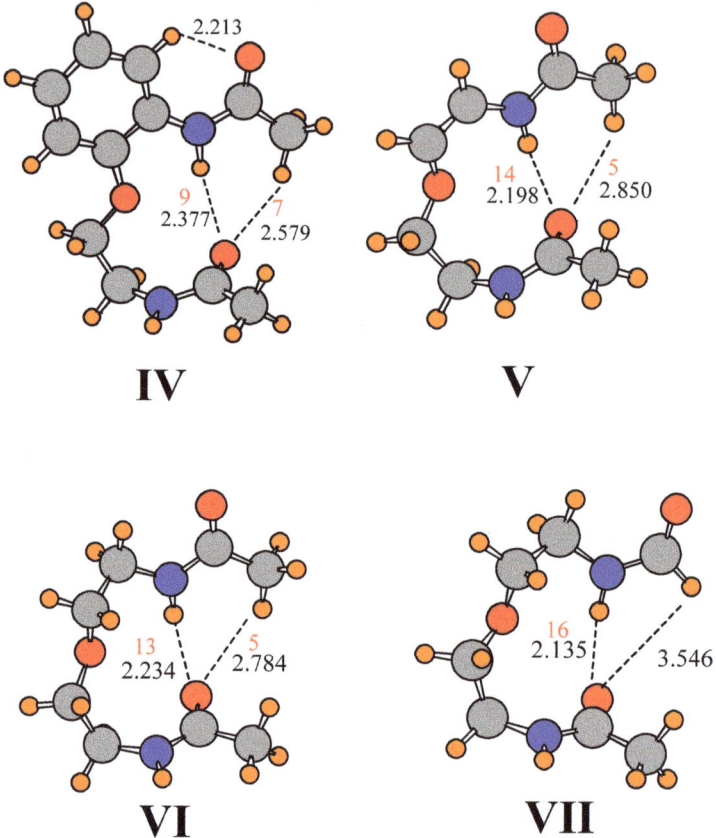

Figure 2. Optimized geometries. Distances in Å. Densities at AIM bond critical points are reported in red, in units of 10^{-3} au.

It is reasonable to presume that the isopropyl group of I has little effect upon the preferred orientations of the two amide groups, or the energetics of twisting them. When the isopropyl group was replaced by H, in addition to the removal of the CF_3 group, the geometrical parameters reported for IV in Tables 2 and 4 (see Figure 2) indicate only minor changes. There are short elongations of three H-bonds, accompanied by small reductions in E(2) and ρ_{BCP}, but the orientations of the two amides remain unchanged. With regard to energetics of distortion, there are only small differences caused by removal of the isopropyl group, on the order of 0.3 kcal/mol, which confirms the small changes of the structure, and the lack of influence of the isopropyl group.

One might anticipate that the largest contribution of the aromatic ring to the orientations of the two amide groups is associated with the presence of a possible $CH_{C1} \cdots O$ H-bond. Removing the phenyl group, and its potential H-bond entirely, replacing the aromatic linkage by a simple HC=CH linkage, leads to structure V illustrated in Figure 2. The dihedral angles that describe the orientations of the two amide groups relative to the backbone remain unchanged. The removal of a possible $CH_{C1} \cdots O$ H-bond elongates the other $CH \cdots O$ H-bond between the two amides, while shortening the $NH \cdots O$ H-bond to the same O atom. The E(2) and ρ_{BCP} values reflect these changes in H-bond length; indeed, the magnitude of E(2) for the $CH_{C2} \cdots O$ HB drops below the 0.5 kcal/mol threshold. At the same time, both of these H-bonds become a little less linear, differing from I by 7–11°. These small changes are accompanied by a very nearly perfectly perpendicular arrangement of the two amide planes with a $\theta(a–a')$ angle of 90.5° As may be seen in Table 5, this simplification has little effect on the energetics of rotation of the lower amide, but would appear to make it a bit more difficult to rotate the upper amide. This observation implies that, rather than holding the upper amide in place, any $CH_{C1} \cdots O$ H-bond involving the phenyl CH adds a small amount of flexibility, perhaps by countering the interactions with the lower amide.

The entire system can be made more symmetric by replacing the alkenyl spacer by alkane only, with a $CH_2CH_2OCH_2CH_2$ spacer, as in structure VI. This change leads to a small stretch in the $NH \cdots O$ H-bond which is accompanied by a contraction in the $CH \cdots O$ bond. But these changes have no substantive effect on the H-bond strengths relative to V as measured by ρ_{BCP}. There is also a 43° change in the dihedral angle associated with the upper amide, some of which arises from the different hybridization around the C_a and C_b atoms to which it is attached. This rotation of the upper amide group takes the two amide planes back to an angle which more closely approximates the value of $\theta(a–a')$ in the parent molecule I. However, the alkane spacer causes a 23° reduction in the relative positions of the two amides around the backbone axis, with $\varphi(a–a')= 34°$. The change from alkene to alkane spacer has minimal effect upon the energetics of rotating the lower amide, but does flatten out the potential for the upper amide, albeit in only one direction, toward positive direction of rotation.

Finally, in order to examine the possible effects of the $CH_{C2} \cdots O$ H-bond on this scaled down version of the molecule in VI, the methyl group was removed from the upper amide, replaced by a H atom. After reoptimization of this structure VII, the geometry obtained is illustrated in Figure 2, where a small contraction of the remaining $NH \cdots O$ H-bond may be seen, along with a growth in ρ_{BCP}. As in the earlier case of methyl replacement, the remaining H atom is too far from the O to pose a significant attraction, leaving NH···O as the only interamide attractive force. Nonetheless, there is only a small change in one of the dihedral angles, in that the upper amide rotates 5° relative to VI. The energetics of rotation of the two amide groups, reported in the last row of Table 5 are only slightly changed from VI. The data would thus support the supposition that it is primarily the NH···O H-bond, rather than either possible CH···O that is responsible for the mutual arrangements of the two amide groups in I, as well as for the energetics holding the two amide groups in place.

3.2. Direct Evaluation of H-Bond Energies

Given the foregoing findings, it would be desirable to have a more direct estimate of the energetics of each type of H-bond. However, there is no clear means of computing an interaction energy for an intramolecular interaction. In order to provide some estimate of the strength of the individual NH···O and CH···O interactions between the two amide units, a philosophy was followed that was used earlier for a similar problem involving amide units in the β-sheet of proteins [80,90] and other systems [91]. The fully optimized geometry of molecule I was taken as a starting point, Figure 3a, and the two amide units were extracted, leaving their relative orientations unchanged from that in Figure 3a. The amide dimer Figure 3b contains both the NH···O and CH···O interactions. In order to compute each separately, the replacement of the upper CH_3 group in Figure 3b by a H atom, leads to the pair in Figure 3c which leaves only the NH···O interaction. Likewise, the replacement of the upper NH_2 group in Figure 3b by H leads to dimer Figure 3d which contains only CH···O. It should be reiterated that the geometries of the subunits in each case were held in the structures in the full molecule Figure 3a, so the interaction energies of Figure 3c,d ought to represent at least a reasonable approximation of the NH···O and CH···O H-bond energies, respectively. These quantities were corrected for basis set superposition error [92] by the counterpoise procedure [93]. They were computed to be −4.74 kcal/mol for NH···O and −2.06 kcal/mol for CH···O, a little less than half the NH···O value. These values correspond nicely with other calculations designed to compute these HB energies in other systems [60,68,81,94–96].

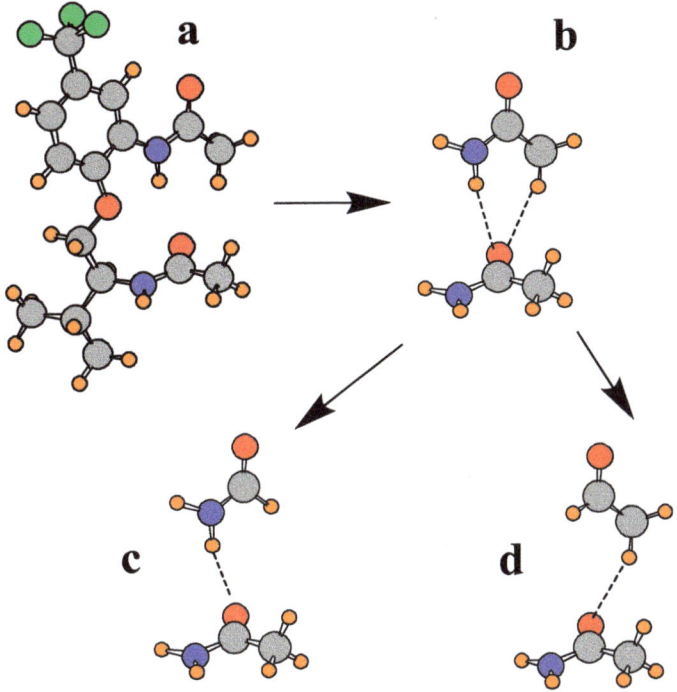

Figure 3. Scheme for partitioning total interaction energy between pair of amides into separate contributions from NH···O and CH···O.

The substantial interaction energy for NH···O comes even at the expense of a H-bond that suffers certain distortions from the intermolecular orientation it would adopt in the absence of intramolecular constraints associated with the full molecule I. The θ(NH···O) angle is fairly close to linearity, at 165°, which should not induce too much distortion energy. On the other hand, the bridging NH proton lies well out of the plane of the other amide which contains the two carbonyl O lone pairs. Specifically, its 78° deviation from this plane is largely responsible for reducing the H-bond energy from what it might be otherwise. In contrast, the bridging CH proton is only 40° from this same plane, although the θ(CH···O) angle is 28° from linearity. But the bottom line is that the CH··O H-bond energy is estimated to be roughly 40% of the NH···O interaction, a proportion that conforms to a number of other calculations of these two quantities [57,60,97–99].

3.3. Consideration of Other Possible Minima

In any study of this type, there is always the question as to whether the X-ray structure, which pertains to one molecule surrounded by others, which can interact with it in numerous ways, also represents the global minimum on the surface of a single molecule. In other words, are there other minima on the surface of the single molecule which might be lower in energy than that which closely resembles the X-ray geometry? In order to test this notion, a number of different structures were taken as starting points, from which a geometry optimization was carried out.

The first candidate was one in which both the upper and lower amide groups were rotated by roughly 180°, around the C_b–N and $C_{b'}$–N bonds, respectively. The structure contains a NH··O H-bond between the two amides, but this H-bond involves a C=O from the upper amide, and N–H from the lower, opposite to that in I above. Geometry optimization yielded a minimum which retains this H-bond, but this structure was 4.8 kcal/mol higher in energy than conformation I. A second possible starting point orients the two amide groups parallel to one another in a stacked arrangement, with no interamide H-bond, a structure which prior calculations have indicated might represent a minimum [100]. The minimum obtained for this structure was 6.0 kcal/mol higher in energy than I. Also explored was the possibility of a reorientation within the backbone. A ~120° rotation around the $C_{a'}$–$C_{b'}$ bond removes the two amide groups from the vicinity of one another, eliminating any interamide H-bonds. The optimized structure was found to be 2.1 kcal/mol higher than I. Likewise for a rotation around the O_e–$C_{a'}$ bond: a 120° rotation, followed by optimization, leads to a higher energy, in this case by 5.3 kcal/mol.

The probing for a possibly lower energy structure was not limited to rotations of only one bond at a time. For example, the rotations were also considered around the C_b–N and $C_{a'}$–$C_{b'}$ bonds, and in each case a range of different starting angles were considered. Also considered were different starting points for the C_b–N and $C_{a'}$–O_e dihedral angles. In all cases, the optimized structures were of higher energy than I. These various attempts lead to the conclusion that the X-ray structure represents the global minimum of this molecule.

3.4. Nature and Length of Spacer Group

Another question relates to the nature of the backbone that lies between the two amide groups. As described above, changing the alkene group of I to an alkane (V→VI) has only a modest effect on the system, the primary influence being a 40° change in the $\varphi(C_aC_bNC_c)$ dihedral angle between the chain and the upper amide. However, this change is due in part to the change in hybridization of the related C atoms, and does not reflect as much of a change in the relative orientation of the two amide units which readjusts by some 13–17°. Importantly, there are only small changes in the two H-bond geometries.

There are five atoms that separate the two amide groups in I; the central atom is an ether-type O atom. Changing that O atom to a CH_2 group takes the spacer to a simple alkane of length 5. The conformation of this molecule, denoted **5**, is illustrated in Figure 4, where it may be seen that the replacement of O by CH$_2$ causes the two amide groups to separate from one another, breaking any interamide H-bonds. The reason for this separation would appear to be that the replacement of O by CH$_2$ places one of the methylene H atoms in close coincidence, ~1.7 Å, with the NH proton of the upper amide. The steric repulsion between these two atoms outweighs any attraction of the NH to the lower carbonyl O, causing the geometry to move off to structure **5**. However, this factor is eliminated when the central methylene group is removed. Substitution of the 5-carbon spacer group in **5** by the shorter 4-C chain leads to geometry **4** in Figure 4, from which it may be seen that the two amide groups once again reestablish their H-bonds, both NH··O and CH··O, although the latter bond is long enough that one may question its contribution to the stability. The problem arising from overly short H···H contacts in **5** when the two amide groups lie within H-bonding distance is eliminated in **4**: The NH proton lies no closer than 2.3 Å to any of the methylene H atoms of the spacer chain.

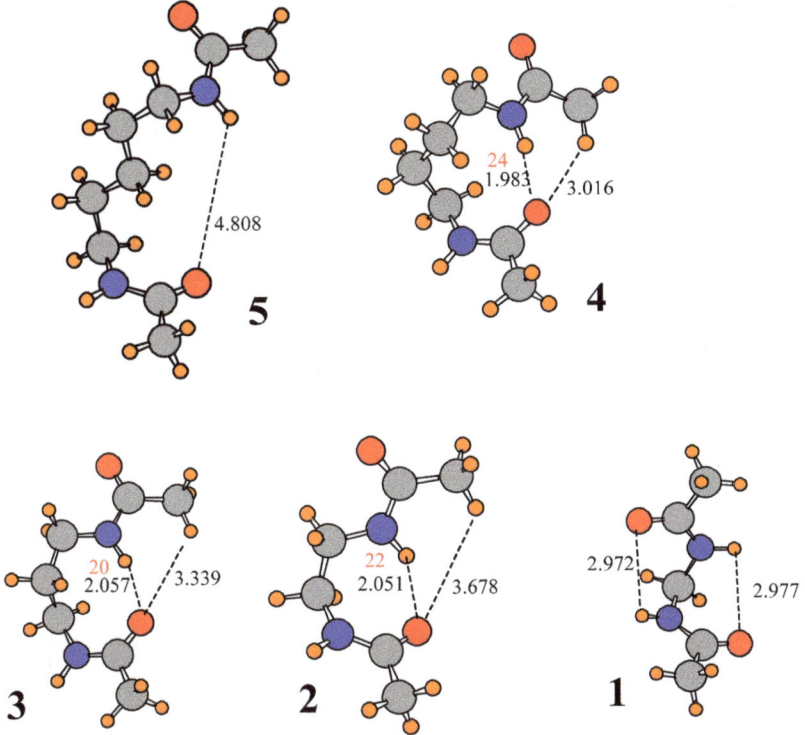

Figure 4. Optimized geometries of derivatives of I, where spacer groups consist of simple alkane chains of various lengths. Number of molecule is equal to number of methylene groups in spacer chain. Distances in Å. Densities at AIM bond critical points are reported in red, in units of 10^{-3} au.

The orientation of **4** remains, albeit with some small changes when the number of links goes down to 3 and 2, although the CH···O distance of >3.3 Å in **2** and **3** suggests the absence of a pertinent H-bond. Leaving only a single C link changes the structure a great deal. The very short spacer no longer allows the two amides to approach one another. The only options remaining in the way of H-bonds are the very strained geometries depicted for **1** in Figure 4, which may be categorized as 6-membered rings. The θ(NH···O) angles of 95° push the limits of allowable deviation from linearity for a true H-bond. Moreover, the H and O atoms are separated by nearly 3 Å, quite long for a H-bond.

Given the dramatic transition from **VII** to **5**, wherein the two amide groups separate from one another when the ether O_e is replaced by a methylene group, it would be natural to wonder if perhaps this O atom holds the two amide groups together by virtue of interaction with NH of the upper amide. However, there are two factors that argue against this idea. In the first place, any such NH···O H-bond is highly distorted. The θ(NH···O) angles in structures **I-VII** vary from 95° to 107°, all quite far from the optimal 180° of a linear H-bond. In the second place, if the loss of this putative NH···O H-bond were indeed the cause of the amide separation in **5**, then it is difficult to explain why the two amides are able to come together again in **4**, **3** and **2**, none of which contain an ether O atom. It is concluded that while one cannot argue with a certain amount of attraction between an ether O and the NH proton of the upper amide which is lost when the O_e is replaced by CH_2, the driving force for the amide separation in **5** is more likely to be steric repulsion between this NH proton and the methylene hydrogens.

4. Summary and Conclusions

The geometry of molecule I contains a pair of amide units that interact directly with one another. The upper amide is approximately coplanar with the neighboring phenyl ring, and the lower amide is roughly perpendicular to the first. The conformation is stabilized by a pair of HBs, an interamide NH···O, supplemented by a CH···O to the same O proton acceptor by the terminal methyl group. The latter HB prevents the normal HB between this methyl group and the adjacent O atom.

The optimized conformation, and its associated interamide HBs, are not significantly affected by the removal of the CF_3 group from the phenyl ring, nor is there an appreciable change in the rigidity with which the amide groups are held. Another group with little influence upon the molecular geometry is the pendant isopropyl group. Removal of the entire aromatic ring, on the other hand, induces a shortening of NH··O and a concomitant lengthening of CH···O. A small perturbation, opposite to that associated with removal of the phenyl ring, is introduced by saturation of the alkenyl spacer with a simple alkyl chain of the same length.

A direct evaluation of the interaction energies associated with the pertinent interamide NH··O and CH··O HBs yielded values of 4.7 and 2.1 kcal/mol, respectively. If the CH···O HB is destroyed by eliminating the proton donor methyl group, the remaining NH···O HB compensates by strengthening its own interaction with the O proton acceptor of the lower amide and shortening R(NH···O). This trend is true whether it is the entire molecule I under consideration, or the scaled down model lacking CF_3, the phenyl ring, or the alkenyl spacer.

If the ether O atom that separates the two amide units is replaced by a CH_2 methylene group, interatomic repulsions force the two amide groups to separate from one another, breaking their HBs. But these interamide HBs return when this methylene spacer group is removed, shortening the spacer group to four atoms. This same interamide conformation, with its NH···O and CH···O HBs persists when the spacer chain is shortened further, to three and even to two atoms, although the CH···O HB is somewhat longer in the latter two cases.

In summary, it appears that the CH···O HB is a real contributor to the geometry adopted by this molecule. This HB, along with the stronger NH···O HB, is not an artifact of the particular substituents, persisting even when the molecule is stripped down to its bare essentials.

Acknowledgments: The author is indebted to Martin D. Smith for suggesting this project.

Conflicts of Interest: The authors declare no conflict of interest.

References

1. Bartashevich, E.V.; Tsirelson, V.G. Interplay between non-covalent interactions in complexes and crystals with halogen bonds. *Russ. Chem. Rev.* **2014**, *83*, 1181–1203.
2. Jeffrey, G.A.; Maluszynska, H. A survey of hydrogen bond geometries in the crystal structures of amino acids. *Int. J. Biol. Macromol.* **1982**, *4*, 173–185.
3. Kroon, J.; Kanters, J.A. Non-linearity of hydrogen bonds in molecular crystals. *Nature* **1974**, *248*, 667–668.
4. Mukherjee, A.; Tothadi, S.; Desiraju, G.R. Halogen bonds in crystal engineering: Like hydrogen bonds yet different. *Acc. Chem. Res.* **2014**, *47*, 2514–2524.
5. Aakeröy, C.B.; Panikkattu, S.; Chopade, P.D.; Desper, J. Competing hydrogen-bond and halogen-bond donors in crystal engineering. *Cryst. Eng. Comm.* **2013**, *15*, 3125–3136.
6. Metrangolo, P.; Resnati, G.; Pilati, T.; Biella, S. Halogen bonding in crystal engineering. In *Halogen Bonding. Fundamentals and Applications*; Metrangolo, P., Resnati, G., Eds.; Springer: Berlin, Germany, 2008; Volume 126, pp. 105–136.
7. Steiner, T. C–H···O hydrogen bonding in crystals. *Cryst. Rev.* **2003**, *9*, 177–228.
8. Gould, R.O.; Gray, A.M.; Taylor, P.; Walkinshaw, M.D. Crystal environments and geometries of leucine, isoleucine, valine, and phenylalanine provide estimates of minimum nonbonded contact and preferred van der waals interaction distances. *J. Am. Chem. Soc.* **1985**, *107*, 5921–5927.

9. Zheng, Q.-N.; Liu, X.-H.; Chen, T.; Yan, H.-J.; Cook, T.; Wang, D.; Stang, P.J.; Wan, L.-J. Formation of halogen bond-based 2D supramolecular assemblies by electric manipulation. *J. Am. Chem. Soc.* **2015**, *137*, 6128–6131.
10. Politzer, P.; Murray, J.S. A unified view of halogen bonding, hydrogen bonding and other σ-hole interactions. In *Noncovalent Forces*; Scheiner, S., Ed.; Springer: Dordrecht, The Netherland, 2015; Volume 19, pp. 357–389.
11. Zierkiewicz, W.; Bieńko, D.C.; Michalska, D.; Zeegers-Huyskens, T. Theoretical investigation of the halogen bonded complexes between carbonyl bases and molecular chlorine. *J. Comput. Chem.* **2015**, *36*, 821–832.
12. Joseph, J.A.; McDowell, S.A.C. Comparative computational study of model halogen-bonded complexes of FKrCl. *J. Phys. Chem. A* **2015**, *119*, 2568–2577.
13. Robinson, S.W.; Mustoe, C.L.; White, N.G.; Brown, A.; Thompson, A.L.; Kennepohl, P.; Beer, P.D. Evidence for halogen bond covalency in acyclic and interlocked halogen-bonding receptor anion recognition. *J. Am. Chem. Soc.* **2015**, *137*, 499–507.
14. Grabowski, S.J. Halogen bond with the multivalent halogen acting as the lewis acid center. *Chem. Phys. Lett.* **2014**, *605–606*, 131–136.
15. Hauchecorne, D.; Herrebout, W.A. Experimental characterization of C–X···Y–C (X = Br, I; Y = F, Cl) halogen-halogen bonds. *J. Phys. Chem. A* **2013**, *117*, 11548–11557.
16. Adhikari, U.; Scheiner, S. Sensitivity of pnicogen, chalcogen, halogen and H-Bonds to angular distortions. *Chem. Phys. Lett.* **2012**, *532*, 31–35.
17. Bauzá, A.; Quiñonero, D.; Deyà, P.M.; Frontera, A. Halogen bonding *versus* chalcogen and pnicogen Bonding: A combined cambridge structural database and theoretical study. *Cryst. Eng. Comm.* **2013**, *15*, 3137–3144.
18. Rosenfield, R.E.; Parthasarathy, R.; Dunitz, J.D. Directional preferences of nonbonded atomic contacts with divalent sulfur. 1. Electrophiles and nucleophiles. *J. Am. Chem. Soc.* **1977**, *99*, 4860–4862.
19. Adhikari, U.; Scheiner, S. Effects of charge and substituent on the S···N chalcogen bond. *J. Phys. Chem. A* **2014**, *118*, 3183–3192.
20. Pang, X.; Jin, W.J. Exploring the halogen bond specific solvent effects in halogenated solvent systems by esr probe. *New J. Chem.* **2015**, *39*, 5477–5483.
21. Azofra, L.M.; Alkorta, I.; Scheiner, S. Chalcogen bonds in complexes of soxy (X, Y = F, Cl) with Nitrogen Bases. *J. Phys. Chem. A* **2015**, *119*, 535–541.
22. Nziko, P.N.Z.; Scheiner, S. Intramolecular S···O chalcogen bond as stabilizing factor in geometry of substituted phenyl-SF_3 molecules. *J. Org. Chem.* **2015**, *80*, 2356–2363.
23. Scheiner, S. The pnicogen bond: Its relation to hydrogen, halogen, and other noncovalent bonds. *Acc. Chem. Res.* **2013**, *46*, 280–288.
24. Scheiner, S. Effects of multiple substitution upon the P···N noncovalent interaction. *Chem. Phys.* **2011**, *387*, 79–84.
25. Del Bene, J.E.; Alkorta, I.; Elguero, J. Properties of cationic pnicogen-bonded complexes $F_{4-n}H_nP^+$:N-Base with F–P···N linear and n = 0–3. *J. Phys. Chem. A* **2015**, *119*, 5853–5864.

26. Scheiner, S. Detailed comparison of the pnicogen bond with chalcogen, halogen and hydrogen bonds. *Int. J. Quantum Chem.* **2013**, *113*, 1609–1620.
27. Del Bene, J.E.; Alkorta, I.; Elguero, J. The pnicogen bond in review: Structures, energies, bonding properties, and spin-spin coupling constants of complexes stabilized by pnicogen bonds. In *Noncovalent Forces*; Scheiner, S., Ed.; Springer: Dordrecht, The Netherland, 2015; Volume 19, pp. 191–263.
28. Sarkar, S.; Pavan, M.S.; Guru Row, T.N. Experimental validation of "pnicogen bonding" in nitrogen by charge density analysis. *Phys. Chem. Chem. Phys.* **2015**, *17*, 2330–2334.
29. Adhikari, U.; Scheiner, S. Substituent effects on Cl···N, S···N, and P···N noncovalent bonds. *J. Phys. Chem. A* **2012**, *116*, 3487–3497.
30. Scheiner, S. On the properties of X···N noncovalent interactions for first-, second- and third-row X atoms. *J. Chem. Phys.* **2011**, *134*.
31. Joesten, M.D.; Schaad, L.J. *Hydrogen Bonding*; Marcel Dekker: New York, NY, USA, 1974; p. 622.
32. Schuster, P.; Zundel, G.; Sandorfy, C. *The Hydrogen Bond. Recent Developments in Theory and Experiments*; North-Holland Publishing Co.: Amsterdam, The Netherlands, 1976.
33. Arunan, E.; Desiraju, G.R.; Klein, R.A.; Sadlej, J.; Scheiner, S.; Alkorta, I.; Clary, D.C.; Crabtree, R.H.; Dannenberg, J.J.; Hobza, P.; et al. Definition of the Hydrogen Bond. *Pure Appl. Chem.* **2011**, *83*, 1637–1641.
34. Falvello, L.R. The hydrogen bond, front and center. *Angew. Chem. Int. Ed. Engl.* **2010**, *49*, 10045–10047.
35. Latajka, Z.; Scheiner, S. Structure, energetics and vibrational spectrum of H$_2$O-HCl. *J. Chem. Phys.* **1987**, *87*, 5928–5936.
36. Sandoval-Lira, J.; Fuentes, L.; Quintero, L.; Höpfl, H.; Hernández-Pérez, J.M.; Terán, J.L.; Sartillo-Piscil, F. The stabilizing role of the intramolecular C–H···O hydrogen bond in cyclic amides derived from α-methylbenzylamine. *J. Org. Chem.* **2015**, *80*, 4481–4490.
37. Adhikari, U.; Scheiner, S. Competition between lone pair–Π, halogen bond, and hydrogen bond in adducts of water with perhalogenated alkenes C$_2$Cl$_n$F$_{4-n}$ (n = 0–4). *Chem. Phys.* **2014**, *440*, 53–63.
38. Latajka, Z.; Scheiner, S. Basis sets for molecular interactions. 2. Application to H$_3$N-HF, H$_3$N-HOH, H$_2$O-HF, (NH$_3$)$_2$, and H$_3$CH–OH$_2$. *J. Comput. Chem.* **1987**, *5*, 674–682.
39. Grabowski, S.J. Dihydrogen bond and X–H···σ interaction as sub-classes of hydrogen bond. *J. Phys. Org. Chem.* **2013**, *26*, 452–459.
40. Latajka, Z.; Scheiner, S. Structure, energetics and vibrational spectra of dimers, trimers, and tetramers of HX (X = Cl, Br, I). *Chem. Phys.* **1997**, *216*, 37–52.
41. Mundlapati, V.R.; Ghosh, S.; Bhattacherjee, A.; Tiwari, P.; Biswal, H.S. Critical assessment of the strength of hydrogen bonds between the sulfur atom of methionine/cysteine and backbone amides in proteins. *J. Phys. Chem. Lett.* **2015**, *6*, 1385–1389.
42. Matta, C.F.; Hernández-Trujillo, J.; Tang, T.-H.; Bader, R.F.W. Hydrogen-hydrogen bonding: A stabilizing interaction in molecules and crystals. *Chem. Eur. J.* **2003**, *9*, 1940–1951.

43. Hernández-Trujillo, J.; Matta, C. Hydrogen-hydrogen bonding in biphenyl revisited. *Struct. Chem.* **2007**, *18*, 849–857.
44. Orlova, G.; Scheiner, S. Intermolecular MH···HF bonding in monohydride Mo and W complexes. *J. Phys. Chem. A* **1998**, *102*, 260–269.
45. Kar, T.; Scheiner, S. Comparison between hydrogen and dihydrogen bonds among H_3BNH_3, H_2BNH_2, and NH_3. *J. Chem. Phys.* **2003**, *119*, 1473–1482.
46. Biswal, H.S.; Wategaonkar, S. Sulfur, not too far behind O, N, and C: SH···π hydrogen bond. *J. Phys. Chem. A* **2009**, *113*, 12774–12782.
47. Cabaleiro-Lago, E.M.; Rodríguez-Otero, J.; Peña-Gallego, Á. Characteristics of the interaction of azulene with water and hydrogen sulfide: A computational study. *J. Chem. Phys.* **2008**, *129*.
48. Bhattacherjee, A.; Matsuda, Y.; Fujii, A.; Wategaonkar, S. Acid-base formalism in dispersion-stabilized S–H···Y (Y=O, S) hydrogen-bonding interactions. *J. Phys. Chem. A* **2015**, *119*, 1117–1126.
49. Solimannejad, M.; Gharabaghi, M.; Scheiner, S. SH···N and SH···P blue-shifting H-bonds and N···P interactions in complexes pairing HSN with amines and phosphines. *J. Chem. Phys.* **2011**, *134*.
50. Minkov, V.S.; Boldyreva, E.V. Contribution of Weak S–H···O Hydrogen Bonds to the Side Chain Motions in D,L-Homocysteine on Cooling. *J. Phys. Chem. B* **2014**, *118*, 8513–8523.
51. Biswal, H.S. Hydrogen bonds involving sulfur: New insights from *ab initio* calculations and gas phase laser spectroscopy. In *Noncovalent Forces*; Scheiner, S., Ed.; Springer: Dordrecht, The Netherland, 2015; Volume 19, pp. 15–45.
52. Cybulski, S.; Scheiner, S. Hydrogen bonding and proton transfers involving triply bonded atoms. HC≡N and HC≡CH. *J. Am. Chem. Soc.* **1987**, *109*, 4199–4206.
53. Zierke, M.; Smieško, M.; Rabbani, S.; Aeschbacher, T.; Cutting, B.; Allain, F.H.-T.; Schubert, M.; Ernst, B. Stabilization of branched oligosaccharides: Lewis benefits from a nonconventional C–H···O hydrogen bond. *J. Am. Chem. Soc.* **2013**, *135*, 13464–13472.
54. Gu, Y.; Kar, T.; Scheiner, S. Comparison of the CH···N and CH···O interactions involving substituted alkanes. *J. Mol. Struct.* **2000**, *552*, 17–31.
55. Michielsen, B.; Verlackt, C.; van der Veken, B.J.; Herrebout, W.A. C–H···X (X = S, P) hydrogen bonding: The complexes of halothane with dimethyl sulfide and trimethylphosphine. *J. Mol. Struct.* **2012**, *1023*, 90–95.
56. Scheiner, S.; Kar, T. Effect of solvent upon CH···O hydrogen bonds with implications for protein folding. *J. Phys. Chem. B* **2005**, *109*, 3681–3689.
57. Sánchez-Sanz, G.; Trujillo, C.; Alkorta, I.; Elguero, J. Weak interactions between hypohalous acids and dimethylchalcogens. *Phys. Chem. Chem. Phys.* **2012**, *14*, 9880–9889.
58. Kryachko, E.; Scheiner, S. CH···F Hydrogen bonds. Dimers of fluoromethanes. *J. Phys. Chem. A* **2004**, *108*, 2527–2535.
59. Scheiner, S. Relative strengths of NH···O and CH···O hydrogen bonds between polypeptide chain segments. *J. Phys. Chem. B* **2005**, *109*, 16132–16141.

60. Rest, C.; Martin, A.; Stepanenko, V.; Allampally, N.K.; Schmidt, D.; Fernandez, G. Multiple CH···O interactions involving glycol chains as driving force for the self-assembly of amphiphilic Pd(II) complexes. *Chem. Commun.* **2014**, *50*, 13366–13369.
61. Sigalov, M.V.; Doronina, E.P.; Sidorkin, V.F. C_{ar}–H···O hydrogen bonds in substituted isobenzofuranone derivatives: Geometric, topological, and NMR characterization. *J. Phys. Chem. A* **2012**, *116*, 7718–7725.
62. Madura, I.D.; Zachara, J.; Hajmowicz, H.; Synoradzki, L. Interplay of carbonyl-carbonyl, C–H···O and C–H···π interactions in hierarchical supramolecular assembly of tartaric anhydrides—Tartaric acid and its O-acyl derivatives: Part II. *J. Mol. Struct.* **2012**, *1017*, 98–105.
63. You, L.-Y.; Chen, S.-G.; Zhao, X.; Liu, Y.; Lan, W.-X.; Zhang, Y.; Lu, H.-J.; Cao, C.-Y.; Li, Z.-T. C–H···O hydrogen bonding induced triazole foldamers: Efficient halogen bonding receptors for organohalogens. *Angew. Chem. Int. Ed.* **2012**, *51*, 1657–1661.
64. Lee, K.-M.; Chen, J.C.C.; Chen, H.-Y.; Lin, I.J.B. A triple helical structure supported solely by C–H···O hydrogen bonding. *Chem. Commun.* **2012**, *48*, 1242–1244.
65. Vibhute, A.M.; Gonnade, R.G.; Swathi, R.S.; Sureshan, K.M. Strength from weakness: Opportunistic CH···O hydrogen bonds differentially dictate the conformational fate in solid and solution states. *Chem. Commun.* **2012**, *48*, 717–719.
66. Sonoda, Y.; Goto, M.; Ikeda, T.; Shimoi, Y.; Hayashi, S.; Yamawaki, H.; Kanesato, M. Ntermolecular CH···O hydrogen bonds in formyl-substituted diphenylhexatriene, a [2 + 2] photoreactive organic solid: Crystal structure and IR, NMR spectroscopic evidence. *J. Mol. Struct.* **2011**, *1006*, 366–374.
67. Jones, C.R.; Baruah, P.K.; Thompson, A.L.; Scheiner, S.; Smith, M.D. Can a C–H···O interaction be a determinant of conformation. *J. Am. Chem. Soc.* **2012**, *134*, 12064–12071.
68. Derewenda, Z.S.; Lee, L.; Derewenda, U. The occurrence of C–H···O hydrogen bonds in proteins. *J. Mol. Biol.* **1995**, *252*, 248–262.
69. Mueller, B.K.; Subramanian, S.; Senes, A. A Frequent, GxxxG-mediated, transmembrane association motif is optimized for the formation of interhelical C_α–H hydrogen bonds. *Proc. Nat. Acad. Sci. USA* **2014**, *111*, E888–E895.
70. Moore, K.B.; Migues, A.N.; Schaefer, H.F.; Vergenz, R.A. Streptococcal hyaluronate lyase reveals the presence of a structurally significant C–H···O hydrogen bond. *Chem. Eur. J.* **2014**, *20*, 990–998.
71. Venugopalan, P.; Kishore, R. Unusual folding propensity of an unsubstituted b,g-hybrid model peptide: Importance of the C–H...O intramolecular hydrogen bond. *Chem. Eur. J.* **2013**, *19*, 9908–9915.
72. Yang, H.; Wong, M.W. Oxyanion hole stabilization by C–H···O interaction in a transition state-a three-point interaction model for cinchona alkaloid-catalyzed asymmetric methanolysis of meso-cyclic anhydrides. *J. Am. Chem. Soc.* **2013**, *135*, 5808–5818.
73. Sheppard, D.; Li, D.-W.; Godoy-Ruiz, R.; Brschweiler, R.; Tugarinov, V. Variation in quadrupole couplings of α deuterons in ubiquitin suggests the presence of C^α–H^α···O=C hydrogen bonds. *J. Am. Chem. Soc.* **2010**, *132*, 7709–7719.

74. Jones, C.R.; Qureshi, M.K.N.; Truscott, F.R.; Hsu, S.-T.D.; Morrison, A.J.; Smith, M.D. A nonpeptidic reverse turn that promotes parallel sheet structure stabilized by C–H···O hydrogen bonds in a cyclopropane γ-peptide. *Angew. Chem. Int. Ed.* **2008**, *47*, 7099–7102.
75. Yoneda, Y.; Mereiter, K.; Jaeger, C.; Brecker, L.; Kosma, P.; Rosenau, T.; French, A. van der waals *versus* hydrogen-bonding forces in a crystalline analog of cellotetraose: Cyclohexyl 4'-O-cyclohexyl β-D-cellobioside cylohexane solvate. *J. Am. Chem. Soc.* **2008**, *130*, 16678–16690.
76. Grunenberg, J. Direct assessment of interresidue forces in watson-crick base pairs using theoretical compliance constants. *J. Am. Chem. Soc.* **2004**, *126*, 16310–16311.
77. Brovarets', O.O.; Yurenko, Y.P.; Hovorun, D.M. Intermolecular CH···O/N H-Bonds in the biologically important pairs of natural nucleobases: A thorough quantum-chemical study. *J. Biomol. Struct. Dyn.* **2013**, *32*, 993–1022.
78. Brovarets', O.O.; Yurenko, Y.P.; Hovorun, D.M. The significant role of the intermolecular CH··O/N hydrogen bonds in governing the biologically important pairs of the DNA and RNA modified bases: A comprehensive theoretical investigation. *J. Biomol. Struct. Dyn.* **2014**, *33*, 1624–1652.
79. Yurenko, Y.P.; Zhurakivsky, R.O.; Samijlenko, S.P.; Hovorun, D.M. Intramolecular CH . . . O hydrogen bonds in the Ai and Bi DNA-like conformers of canonical nucleosides and their watson-crick pairs. Quantum chemical and aim analysis. *J. Biomol. Struct. Dyn.* **2011**, *29*, 51–65.
80. Scheiner, S. Contributions of NH···O and CH···O H-Bonds to the stability of β-sheets in proteins. *J. Phys. Chem. B* **2006**, *110*, 18670–18679.
81. Adhikari, U.; Scheiner, S. First steps in growth of a polypeptide toward β-sheet structure. *J. Phys. Chem. B* **2013**, *117*, 11575–11583.
82. Horowitz, S.; Dirk, L.M.A.; Yesselman, J.D.; Nimtz, J.S.; Adhikari, U.; Mehl, R.A.; Scheiner, S.; Houtz, R.L.; Al-Hashimi, H.M.; Trievel, R.C. Conservation and functional Importance of carbon-oxygen hydrogen bonding in adomet-dependent methyltransferases. *J. Am. Chem. Soc.* **2013**, *135*, 15536–15548.
83. Horowitz, S.; Adhikari, U.; Dirk, L.M.A.; Del Rizzo, P.A.; Mehl, R.A.; Houtz, R.L.; Al-Hashimi, H.M.; Scheiner, S.; Trievel, R.C. Manipulating unconventional CH-based hydrogen bonding in a methyltransferase via noncanonical amino acid mutagenesis. *ACS Chem. Biol.* **2014**, *9*, 1692–1697.
84. Frisch, M.J.; Trucks, G.W.; Schlegel, H.B.; Scuseria, G.E.; Robb, M.A.; Cheeseman, J.R.; Zakrzewski, V.G.; Montgomery, J.J.A.; Stratmann, R.E.; Burant, J.C.; *et al. Gaussian03*; D.01; Gaussian, Inc.: Pittsburgh, PA, USA, 2003.
85. Bader, R.F.W. *Atoms in Molecules, A Quantum Theory*; Clarendon Press: Oxford, UK, 1990; Volume 22, p. 438.
86. Carroll, M.T.; Bader, R.F.W. An analysis of the hydrogen bond in base-HF complexes using the theory of atoms in molecules. *Mol. Phys.* **1988**, *65*, 695–722.
87. Smith, M.D.; Oxford University, Oxford, UK. Personal Communication, 2013.

88. Reed, A.E.; Weinhold, F. Natural bond orbital analysis of near hartree-fock water dimer. *J. Chem. Phys.* **1983**, *78*, 4066–4073.
89. Reed, A.E.; Weinhold, F.; Curtiss, L.A.; Pochatko, D.J. Natural bond orbital analysis of molecular interactions: Theoretical studies of binary complexes of HF, H_2O, NH_3, N_2, O_2, F_2, CO and CO_2 with HF, H_2O, and NH_3. *J. Chem. Phys.* **1986**, *84*, 5687–5705.
90. Guo, H.; Gorin, A.; Guo, H. A peptide-linkage deletion procedure for estimate of energetic contributions of individual peptide groups in a complex environment: Application to parallel β-sheets. *Interdiscip. Sci. Comput. Life Sci.* **2009**, *1*, 12–20.
91. Guo, H.; Beahm, R.F.; Guo, H. Stabilization and destabilization of the C^δ–H$\cdots$O=C hydrogen bonds involving proline residues in helices. *J. Phys. Chem. B* **2004**, *108*, 18065–18072.
92. Latajka, Z.; Scheiner, S. Primary and secondary basis set superposition error at the SCF and MP2 Levels: H_3N–Li^+ and H_2O–Li^+. *J. Chem. Phys.* **1987**, *87*, 1194–1204.
93. Boys, S.F.; Bernardi, F. The calculation of small molecular interactions by the differences of separate total energies. Some procedures with reduced errors. *Mol. Phys.* **1970**, *19*, 553–566.
94. Nepal, B.; Scheiner, S. Angular dependence of hydrogen bond energy in neutral and charged systems containing CH and NH proton donors. *Chem. Phys. Lett.* **2015**, *630*, 6–11.
95. Scheiner, S. Weak H-Bonds. Comparisons of CH$\cdots$O to NH$\cdots$O in proteins and PH$\cdots$N to direct P$\cdots$N interactions. *Phys. Chem. Chem. Phys.* **2011**, *13*, 13860–13872.
96. Scheiner, S. The strength with which a peptide group can form a hydrogen bond varies with the internal conformation of the polypeptide chain. *J. Phys. Chem. B* **2007**, *111*, 11312–11317.
97. Gu, Y.; Kar, T.; Scheiner, S. Fundamental properties of the CH$\cdots$O interaction: Is it a true hydrogen bond? *J. Am. Chem. Soc.* **1999**, *121*, 9411–9422.
98. Scheiner, S.; Kar, T.; Gu, Y. Strength of the $C^\alpha H\cdots O$ hydrogen bond of amino acid residues. *J. Biol. Chem.* **2001**, *276*, 9832–9837.
99. Kar, T.; Scheiner, S. Comparison of cooperativity in CH$\cdots$O and OH$\cdots$O hydrogen bonds. *J. Phys. Chem. A* **2004**, *108*, 9161–9168.
100. Adhikari, U.; Scheiner, S. Preferred configurations of peptide-peptide interactions. *J. Phys. Chem. A* **2013**, *117*, 489–496.

Very Strong Parallel Interactions between Two Saturated Acyclic Groups Closed with Intramolecular Hydrogen Bonds Forming Hydrogen-Bridged Rings

Jelena P. Blagojević, Goran V. Janjić and Snežana D. Zarić

Abstract: Saturated acyclic four-atom groups closed with a classic intramolecular hydrogen bond, generating planar five-membered rings (hydrogen-bridged quasi-rings), in which at least one of the ring atoms is bonded to other non-ring atoms that are not in the ring plane and, thus, capable to form intermolecular interactions, were studied in this work, in order to find the preferred mutual positions of these species in crystals and evaluate strength of intermolecular interactions. We studied parallel interactions of these rings by analysing crystal structures in the Cambridge Structural Database (CSD) and by quantum chemical calculations. The rings can have one hydrogen atom out of the ring plane that can form hydrogen bonds between two parallel rings. Hence, in these systems with parallel rings, two types of hydrogen bonds can be present, one in the ring, and the other one between two parallel rings. The CSD search showed that 27% of the rings in the crystal structures form parallel interactions. The calculations at very accurate CCSD(T)/CBS level revealed strong interactions, in model systems of thiosemicarbazide, semicarbazide and glycolamide dimers the energies are -9.68, -7.12 and -4.25 kcal/mol. The hydrogen bonds between rings, as well as dispersion interactions contribute to the strong interaction energies.

Reprinted from *Crystals*. Cite as: Blagojević, J.P.; Janjić, G.V.; Zarić, S.D. Very Strong Parallel Interactions between Two Saturated Acyclic Groups Closed with Intramolecular Hydrogen Bonds Forming Hydrogen-Bridged Rings. *Crystals* **2016**, *6*, 34.

1. Introduction

The stacking interactions between aromatic rings are one of the well examined non-covalent interactions [1–21]. However, not only aromatic, but other planar molecules and fragments can also form stacking (parallel) interactions [22–46]. Interestingly, several studies showed that stacking interactions of other planar molecules can be even stronger than stacking between aromatic molecules [22–46], indicating the importance of these interactions. Analysis of the crystal structures from Cambridge Structural Database (CSD) have shown that planar chelate rings with delocalized π-bonds can form stackinginteractions with C_6-aromatic rings, and with other chelate rings [24–27]. The interaction energies of nickel and copper six-membered chelate rings with benzene, calculated at very acurate CCSD(T)/CBS level, are −4.77 and −6.39 kcal/mol respectively [43,44], remarkably stronger than stacking interaction between two benzene molecules, −2.73 kcal/mol [10].

Supramolecular chemistry of hydrogen-bridged quasi-ring species becomes very interesting from the fundamental point of view, however, it could be also useful in molecular recognition, crystal engineering or biochemistry. Interactions of hydrogen-bridged rings in supramolecular arrangements are similar to interactions of classical rings formed by covalent bonding. For example, quasi-chelate rings can participate in C–H···π interactions similar to aromatic organic molecules [47]. Rings formed by resonance-assisted hydrogen bonding [45,48–50] can form π-stacking interactions. Moreover, hydrogen-bridged rings with saturated bonds can also form stacking interactions [46]. In our previous study, it was shown that, in crystal structures from Cambridge Structural Database (CSD), 27% of five-membered hydrogen-bridged rings form intermolecular stacking interactions [46]. Interaction energy calculated at very accurate CCSD(T)/CBS level are −4.89 kcal/mol and −2.95 kcal/mol, dependent on ring structure. These interactions are stronger than stacking between two benzene molecules (−2.73 kcal/mol) [10].

The aim of this work was to deepen the knowledge about intermolecular interactions of hydrogen-bridged rings by analyzing interaction of the saturated acyclic four-atom groups closed by a classic intramolecular hydrogen bond, generating planar five-membered quasi-rings, with at least one non-ring atom attached to the ring and situated out of the ring plane. This type of rings is observed quite frequently in crystal structures from the CSD. One example of such ring is presented in Figure 1a. The rings can have one hydrogen atom out of the ring plane that can form hydrogen bonds between two parallel rings. Hence, in these systems, two types of hydrogen bonds can be present, one in the ring, and the other one between two parallel rings. The parallel interactions of the rings were studied by searching Cambridge Structural Database (CSD) and by quantum chemical calculations.

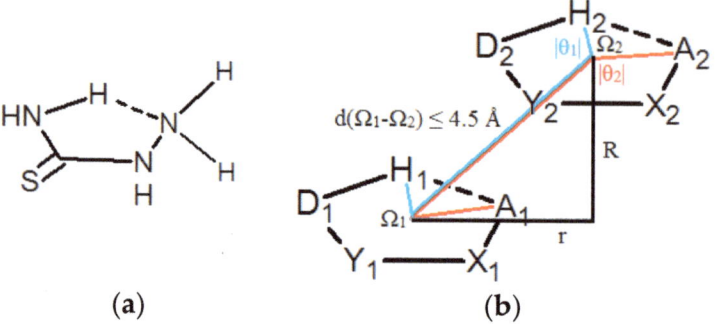

Figure 1. (a) An example of a molecule with non–planar groups in the ring, that have hydrogen capable for hydrogen bond between two parallel rings. (b) Geometric parameters and atom labeling scheme used for the description of intermolecular interactions of saturated hydrogen-bridged rings, studied in this work; Ω marks the centroid of the ring, X and Y letters stand for any atoms adjacent to acceptor (A) and donor (D) atoms, respectively, R and r mark normal distance and offset value, respectively, θ_1 and θ_2-torsion angles $H_1\Omega_1\Omega_2H_2$ and $A_1\Omega_1\Omega_2A_2$, respectively; non–planar groups are omitted for simplicity.

2. Results

2.1. Search and Analysis of Crystal Structures from the CSD

To get an insight into the interactions of the hydrogen-bridged rings without multiple bonds that also have non-planar groups in the ring (Figure 1a), search and analysis of the data in crystal structures from the CSD were performed. Structures studied in this work are not limited to small molecules similar to the one presented in Figure 1a, but larger molecules which have structural fragments, as shown in Figure 1b, are included.

There are 1068 rings that satisfy Criteria 1–6, listed below in the Materials and Methods Section. There are 352 (33%) contacts of the two hydrogen-bridged quasi-rings that have distances between two centroids 4.5 Å or less, while 289 contacts (27%) have parallel orientation of the rings (interplanar angles smaller than 10°). Interestingly, in our previous search of hydrogen-bridged rings, with all planar groups in the ring, we found 978 rings in the CSD, while 27% (264) of the rings also formed parallel interactions [46].

The interplanar angle (π) distribution for contacts that have distances between two centroids less than 4.5 Å, indicating preferred parallel orientation of the middle ring planes, is given in Figure 2.

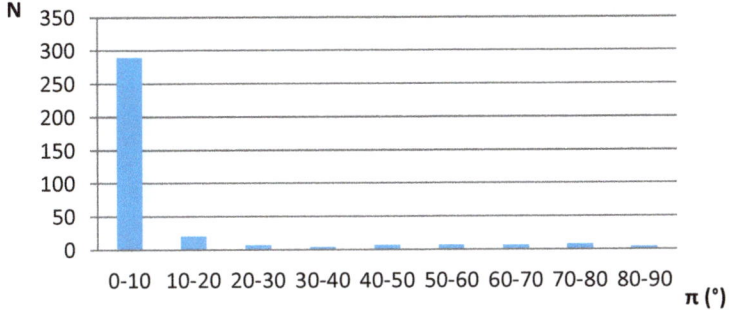

Figure 2. Interplanar angle distributions of five-membered saturated hydrogen-bridged rings with non-planar groups in the rings in crystal structures. N—number of contacts.

The data in Figure 3 show that most of the parallel contacts have the normal distances between 3.0 and 3.5 Å. These are distances typical for stacking of organic aromatic rings (3–4 Å) [17–22] and other planar rings that form stacking interactions [24–27,43–46].

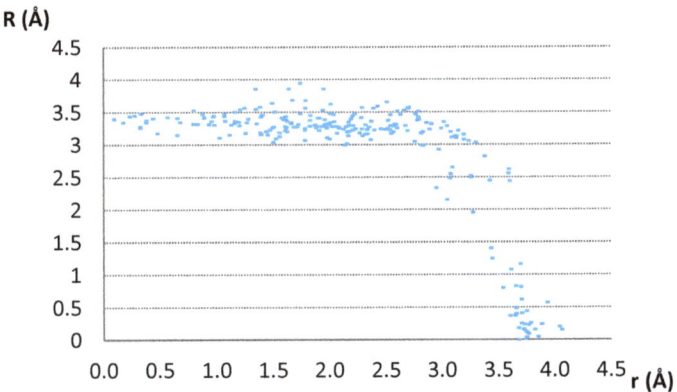

Figure 3. Interplanar distances (R) of contacts having parallel ring planes, plotted as a function of offset values.

Analysis of distributions of torsion angles $H_1\Omega_1\Omega_2H_2$ (θ_1) and $A_1\Omega_1\Omega_2A_2$ (θ_2) (Figure 1) shows that the large majority of contacts has absolute torsion angles around 180° (Figure 4), indicating that the "head to tail" orientations are preferred.

Statistical analysis of elemental composition of the hydrogen-bridged rings studied in this work is given in Table 1.

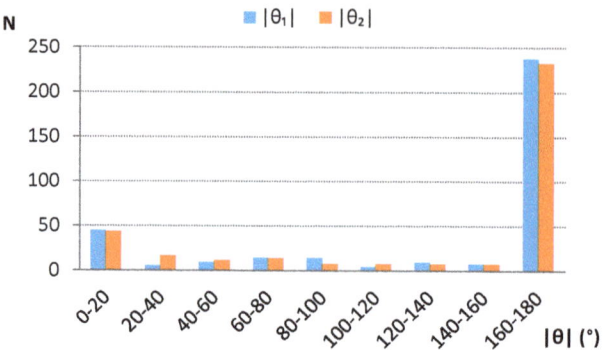

Figure 4. Absolute torsion angle θ_1 ($H_1\Omega_1\Omega_2H_2$) and θ_2 ($A_1\Omega_1\Omega_2A_2$) distributions. Centroid and atom labeling is consistent with the scheme given in Figure 1. N—number of contacts.

Table 1. Statistical analysis of elemental composition; the mark Z stands for halogen.

D,Y,X,A	Percentage of Total Number of Contacts
N,C,N,N	49.1
N,C,C,O	18.8
N,C,C,Z	8.0
N,C,N,O	4.3
other	19.8

Many contacts (49.1%, or 173 contacts) involve rings containing N (as D, donor atom), C (as Y, atom adjacent to the donor), N (as X, atom adjacent to the acceptor), and N (as A acceptor atom), using atom labeling given in Figure 1. The majority of the contacts are between aminoguanidinium cation derivatives (55.5%, or 96 contacts), but crystal structures of aminoguanidinium salts are largely affected by ionic forces, so model systems containing aminoguanidinium cation were not used in the calculations.

Among the structures with NCNN sequence, relatively abundant are thiosemicarbazide derivatives (31.8%, or 55 contacts), with a sulfur atom doubly bonded to the carbon atom and semicarbazide derivatives (8.7%, or 15 contacts), with an oxygen atom doubly bonded to the carbon atom. In all contacts of these two groups, coordination number of D and X atoms is 3, *i.e.*, these atoms belong to the planar groups (Figure 5). If a coordination number of D, Y or X atoms is 3, it means that these atoms belong to planar groups, since the ring itself is defined to be planar by constraints applied in CSD search (Materials and Methods Section). Coordination numbers of D, Y or X atoms larger than 3 indicate that these atoms belong to non-planar groups, mostly tetrahedral (or other geometries when Y or X are metals). Coordination number of A larger than 2 indicates that A belongs to the non-planar group, as well as in structures where coordination number of A is 2, A is oxygen atom and the substituent is hydrogen atom. Molecules 4 and 5, shown in Figure 5, are studied in our previous work, where all atoms in the ring belong to planar groups. In this work, however, at least one, or more, atoms in the ring should belong to non-planar groups. In molecules **1** and **2**, A is part of non-planar group, while in **3**, X is part of non-planar group.

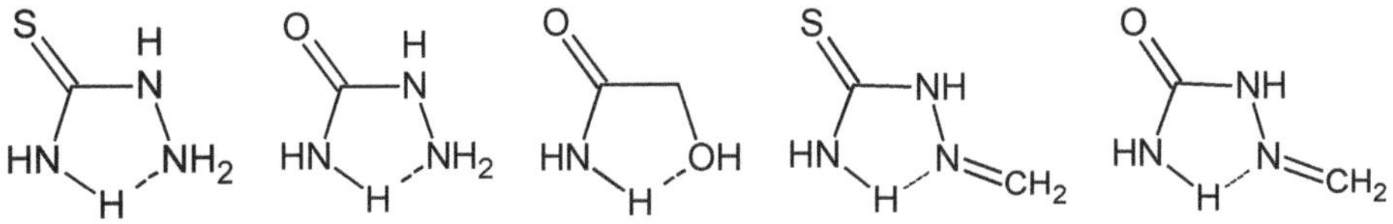

Figure 5. Molecules chosen for the model systems for quantum chemical calculations of parallel interaction energies; 1–3 systems studied in this work; 4–5 similar systems from our previous work.

When A belongs to a non-planar group (Figure 1), it has a hydrogen atom as a substituent in almost all cases (98% of contacts of thiosemicarbazide (**1**) derivatives and 93% of contacts of semicarbazide (**2**) derivatives).

Another relatively numerous group of rings that form parallel interactions (18.8% of total number of contacts, or 66 contacts) has N, C, C and O atoms as D, X, Y and A atoms, respectively (Table 1). Most of them (87.9% or 58 contacts) are glycolamide (**3**) derivatives. In these structures, D atom always belongs to a planar group (coordination number is always 3), while X atom always belongs to a non-planar group with coordination number 4. Both substituents at X atom are hydrogen atoms in 79% of contacts. A large majority of contacts of this group (98%) has an A atom that belongs to a planar group or hydrogen atom as a substituent on A atom. This hydrogen atom is directed away from the area of the other molecule in the dimer, so it is not capable to form hydrogen bond with it (Figure 5).

It should be noted that D and A are donor and acceptor for intramolecular hydrogen bond, while they can have different roles in intermolecular hydrogen bonds between two rings (Figures 5–7).

2.2. Quantum Chemical Calculations

Hydrogen-bridged rings, thiosemicarbazide (**1**), semicarbazide (**2**) and glycolamide (**3**), presented in Figure 5, were used in model systems to calculate stacking interactions. These rings were chosen since their derivatives occur quite frequently in crystal structures (Table 1). In order to show that these rings are appropriate representatives for whole set of rings studied in this work, we analyzed geometric parameters for derivatives of thiosemicarbazide, semicarbazide and glycolamide rings separately.

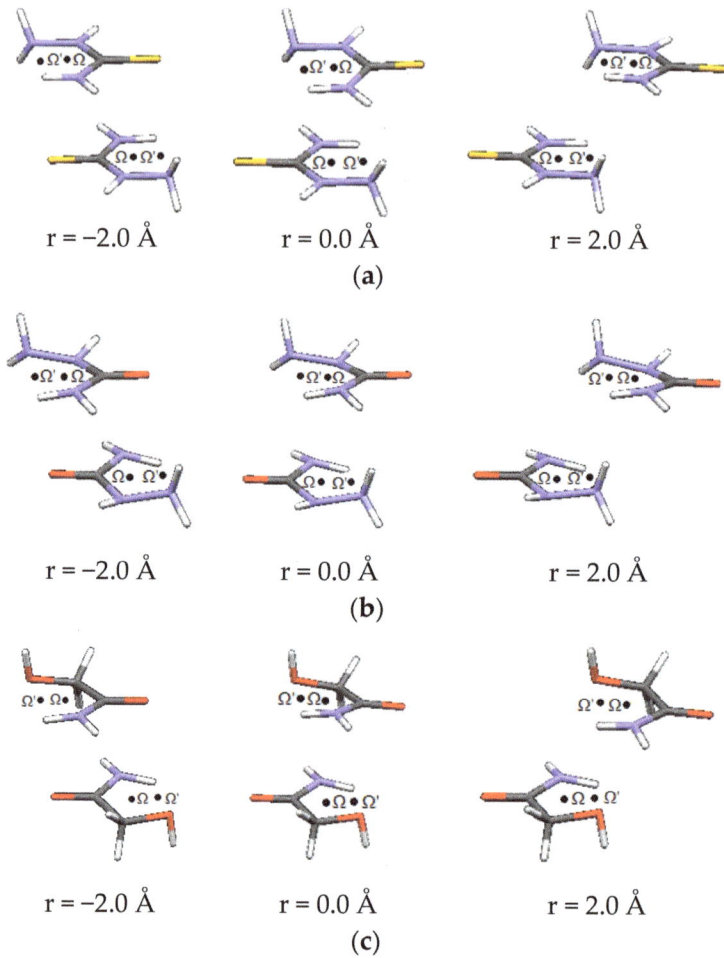

Figure 6. Parallel interactions of (a) thiosemicarbazide, (b) semicarbazide and (c) glycolamide dimers, at three offset values along Ω-Ω'direction.

Trends concerning interactions of their derivatives in crystal structures are the same (Figures S1–S4) as overall trends (Figures 2–4), justifying the use of these molecules in model system for quantum chemical calculations. Geometries of optimized dimers of thiosemicarbazide, semicarbazide and glycolamide in the gas phase, as well as some typical structural patterns in the crystals of thiosemicarbazide and derivatives of semicarbazide and glycolamide are also given in ESI (Figures S5 and S6).

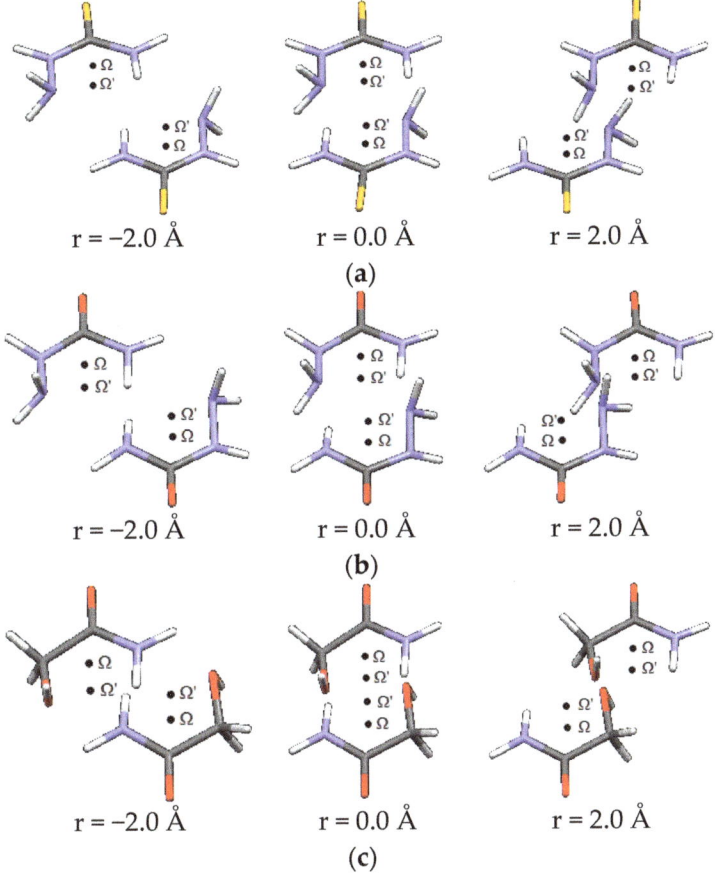

Figure 7. Parallel interactions of (a) thiosemicarbazide, (b) semicarbazide and (c) glycolamide dimers, at three offset values along direction orthogonal to Ω-Ω'.

In Figure 5, structures of two hydrogen-bridged rings, **4** and **5**, that we used as model systems to calculate interaction energies in our previous study [46] are also presented. Model systems used for calculations are composed of two quasi-cyclic molecules in the antiparallel position (parallel alignment of the ring planes and torsion angles $H_1\Omega_1\Omega_2H_2$ and $A_1\Omega_1\Omega_2A_2$ of 180°) since data from the crystal structures showed antiparallel orientation of interacting rings (Figure 4).

Potential curves of interaction energies are obtained by moving molecules along two directions. The first is Ω-Ω' direction, where Ω represents the centroid of the molecule, while Ω' represents the centroid of the hydrogen-acceptor bond (Figure 6).

The second direction is orthogonal to the Ω-Ω' direction. Representative geometries are shown in Figure 7.

The potential curves were calculated by varying offset values (r) in steps of 0.5 Å, while normal distances (R) were varied for every particular offset value in order to obtain the strongest energy.

For calculating potential curves, we used DFT methods, which are in good agreement with very accurate CCSD(T)/CBS method. Benchmark calculations are presented in Tables S1–S5. Model systems used for the benchmark analysis are shown in Figure S7. Potential curves are shown in Figure 8, while corresponding normal distances are shown in Figure S8, ESI.

Potential curves in the Ω-Ω' direction show minima for thiosemicarbazide, semicarbazide and glycolamide model systems at negative offset values of −1.5, −2.0, and −3.5 Å, respectively, with interaction energies, calculated at CCSD(T)/CBS level, of −7.66, −4.90 and −4.25 kcal/mol, respectively. The minima on potential curves orthogonal to the Ω-Ω' direction are at positive offset values of 2.5 Å for thiosemicarbazide and semicarbazide model systems with energies of −9.68 and −7.12 kcal/mol, respectively (Figure 8, Table 2). The minimum for glycolamide model system is at offset of −2.0 Å with the energy of −2.21 kcal/mol.

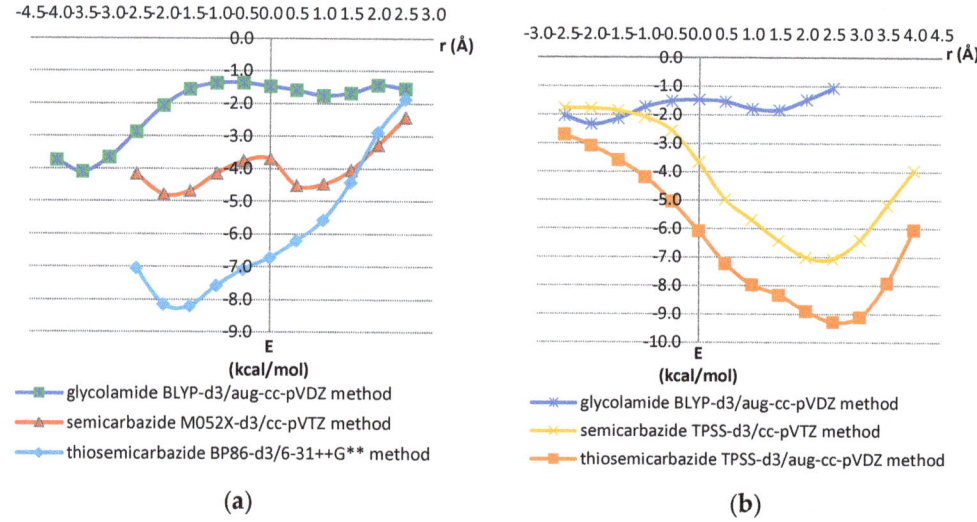

Figure 8. (a) Potential curves for offset values varied in the Ω-Ω' direction; (b) Potential curves for offset values varied orthogonal to Ω-Ω' direction. The interactions energies for each offset value r were calculated by varying the normal distance (R) between two molecules in a series of single point calculations. The strongest calculated energy for each offset value is presented. The selected geometries are shown in Figures 6 and 7.

Table 2. Interaction energies and offsets of potential curves minima for molecules in Figure 5. Energies are in kcal/mol, distances are in Å.

Model System	Ω-Ω' Direction			Orthogonal to Ω-Ω' Direction		
	r	E_m	CCSD(T)/CBS	r	E_m	CCSD(T)/CBS
1	-1.5	-8.19	-7.66	2.5	-9.27	-9.68
2	-2.0	-4.79	-4.90	2.5	-7.07	-7.12
3	-3.5	-4.09	-4.25	-2.0	-2.33	-2.21
4	0.0	-4.49	-4.84	-1.0	-4.91	-4.89
5	0.0	-2.88	-2.95	-1.0	-3.19	-2.95

Note: E_m—interaction energy at minima on potential curve in Ω-Ω' direction calcualated at: 1: BP86-d3/6-31++G**; 2: M052X-d3/cc-pVTZ; 3: BLYP-d3/aug-cc-pVDZ; 4: blyp-d3/cc-pVTZ; 5: blyp-d3/cc-pVTZ levels. E_m—interaction energy at minima on potential curve orthogonal to Ω-Ω' direction calculated at: 1: tpss-d3/aug-cc-pVDZ; 2: tpss-d3/cc-pVTZ; 3: BLYP-d3/aug-cc-pVDZ; 4: M06HF-d3/cc-pVDZ; 5: M052X-d3/6-31++G** levels.

Potential curves for thiosemicarbazide and semicarbazide model systems have similar shape, because of the similar molecular structure, however, the interaction energies are stronger for thiosemicarbazide, for almost all offsets. For all three model systems, the most stable interactions are calculated for the geometries with very favorable hydrogen bonds between rings, as shown in Figure 9. In dimers of thiosemicarbazide and semicarbazide (Figure 9a,b) intermolecular hydrogen bonds are bifurcated. Nitrogen atoms in position A (Figure 1b) are acceptor atoms of intramolecular hydrogen bonds and donors for the intermolecular hydrogen bonds. Acceptor atoms for the intermolecular hydrogen bonds are sulfur (Figure 9a) or oxygen atoms (Figure 9b), as well as nitrogen atoms in the position X (Figures 1b and 9a,b). In the case of glycolamide, tetrahedral carbon atoms (position X, Figure 1b) are donors, while oxygen atoms of carbonyl groups are acceptors (Figure 9c) for intermolecular hydrogen bonds.

The data of potential curves and interaction energies calculated at very accurate CCSD(T)/CBS level show that interactions are quite strong (Figure 8, Table 2). The strongest are interactions in thiosemicarbazide dimer −9.68 kcal/mol, while interaction in semicarbazide dimer is weaker, however, still quite strong, −7.12 kcal/mol. Among the model systems used in this study, the weakest interaction was calculated in glycolamide dimer, −4.25 kcal/mol.

In the geometries of the minima on potential curves, there are two simultaneous hydrogen bonds between two quasi-rings that are in dimers of thiosemicarbazide and semicarbazide bifurcated (Figure 9). In order to evaluate the contribution of single hydrogen bond to the interaction energies, we performed calculations on model systems: thiosemicarbazide and semicarbazide molecules with ammonia and of glycolamide molecule with methanol (Figure 10). The position of NH and CH bonds that form intermolecular hydrogen bonds are the same as the position of the corresponding NH and CH bonds in geometries of minima shown in Figure 9, with identical bond lengths and valence angles. Energies are calculated using the same methods that are used for calculating potential curves (Figure 8), *i.e.*, tpss-d3/aug-cc-pVDZ for thiosemicarbazide/ammonia, tpss-d3/cc-pVTZ for semicarbazide/ammonia, and blyp-d3/aug-cc-pVDZ for glycolamide/methanol. Evaluated energies of hydrogen bonds are −3.89, −1.80 and −1.67 kcal/mol, respectively. Although contributions of intermolecular hydrogen bonds to total interaction energies are considerable, the total interaction energies in Table 2 are more than energies of two hydrogen bonds, indicating the existence of additional contributions to total energies.

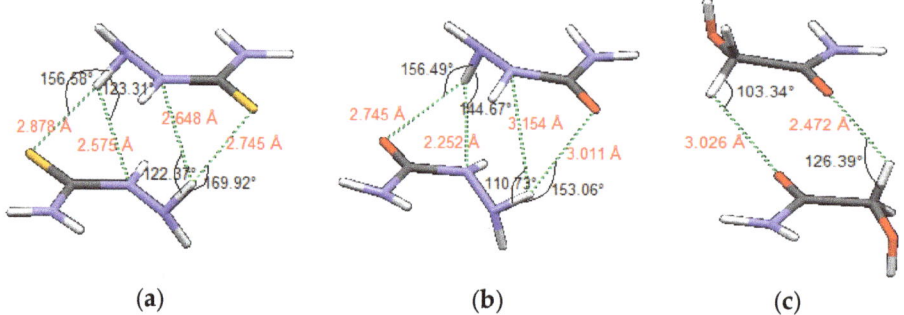

Figure 9. Geometries of the strongest calculated interactions of (**a**) thiosemicarbazide, (**b**) semicarbazide and (**c**) glycolamide dimers, with presented intermolecular hydrogen bonds and CH–O interactions.

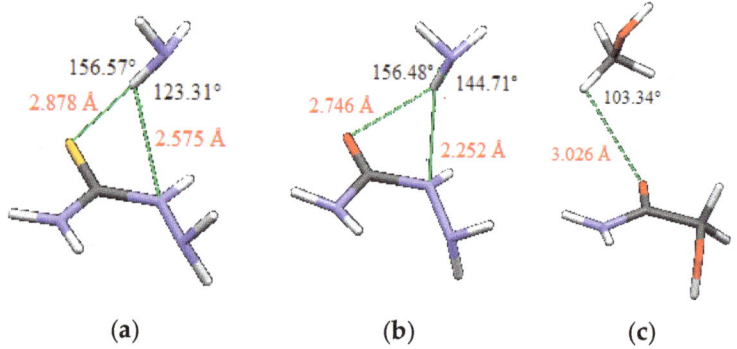

Figure 10. Model systems for evaluating energies of hydrogen bonds (**a**) thiosemicarbazide and ammonia, (**b**) semicarbazide and ammonia, and (**c**) glycolamide and methanol. Positions and geometries of intermolecular hydrogen bonds are identical as in Figure 9. The calculated energies are (**a**) −3.89 kcal/mol, (**b**) −1.80 kcal/mol and (**c**) −1.67 kcal/mol.

Maps of electrostatic potentials for thiosemicarbazide, semicarbazide and glycolamide molecules (Figure 11) show that electron density is more localized on oxygen atom of semicarbazide than on the sulfur atom of thiosemicarbazide. In addition, the hydrogen atom in tiosemicarbazide is more positive than in semicarbazide. Electrostatic potential maps of glycolamide molecule show that hydrogen and oxygen involved in intermolecular interactions are less positive and less negative, respectively, than hydrogens and sulfur/oxygen atoms in thiosemicarbazide and semicarbazide.

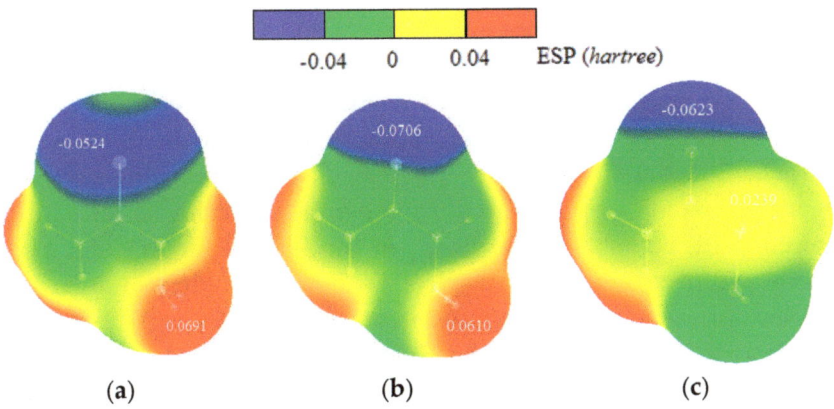

Figure 11. Maps of electrostatic potentials for (**a**) thiosemicarbazide, (**b**) semicarbazide and (**c**) glycolamide molecules. The values of ESP maxima and minima in hartrees are indicated onto the surfaces of the maps.

3. Discussion

Orientations of the rings in crystal structures correspond to antiparallel arrangement (Figures 4, 6 and 7). In our previous work, we studied interactions of similar hydrogen-bridged rings, however, with all planar groups in the ring [46], as was mentioned above. The analysis of the data from crystal structures showed very similar trends as observed in this study.

Stacking interactions of similar systems without intermolecular hydrogen bonds (model systems of molecules 4 and 5, Figure 5) that were calculated in our previous work, −4.89 and −2.95 kcal/mol [46], are significantly weaker than interactions in model systems **1** and **2**, studied in this work, −9.68 and −7.12 kcal/mol, at CCSD(T)/CBS level (Table 2). The geometries in the dimers are also quite different because of the hydrogen bonds. This indicates that the presence of hydrogen atoms between the ring planes that form hydrogen bonds, significantly influence the geometries and strengthen interactions between two parallel rings. In systems **1** and **2**, there are double bifurcated intermolecular hydrogen bonds, NH–N and NH–S in the case of thiosemicarbazide (**1**) and NH–N and NH–O in case of semicarbazide (**2**) (Figure 9), which contribute to the strength of the interactions.

Evaluated energies of single hydrogen bonds (Figure 10) indicate significant contribution of hydrogen bonds to total interaction energy between two quasi-rings, however, indicate also other contributions to total energy. Namely total interaction energies for thiosemicarbazide, semicarbazide, and glycolamide (-9.27, -7.07 and -4.09 kcal/mol respectively) are larger than energies of two hydrogen bonds (-7.78, -3.60 and -3.34 kcal/mol, respectively). One can also notice that interaction energy in thiosemicarbazide and semicarbazide dimers is less than sum of hydrogen bonds (Figure 10) and interaction energies in dimers of **4** and **5** (Figure 5, Table 2), without intermolecular hydrogen bonds between quasi-rings.

The dimers of thiosemicarbazide and semicarbazide have similar geometries, so energy differences can be attributed mostly to the nature of sulfur and oxygen atoms. Stronger interaction in case of thiosemicarbazide is probably a consequence of stronger dispersion component that can be anticipated for larger sulfur atom. Indeed, electrostatic potential maps, discussed above, show larger sulfur atom with more delocalized negative charge (Figure 11). Besides, hydrogen atom involved in intermolecular interaction is slightly more acidic (more partially positive) in thiosemicarbazide than in semicarbazide (Figure 11), so that also contributes to the energy difference. Electrostatic potential maps can explain weaker interactions in glycolamide dimer. One reason for weaker interaction in the case of glycolamide is the type of hydrogen bonds between rings, which are in this case relatively weak CH–O interactions (Figure 10). Another reason is that CH–O bonds in glycolamide dimer are not bifurcated as in dimers of thiosemicarbazide and semicarbazide.

Curves of changing normal distances for various offset values presented in Figure S8, show that normal distances are shorter for thiosemicarbazide rings, in accordance with the strongest interaction energies.

The calculations on correlation energies show that correlation component is higher for thiosemirarbazide than for semicarbazide (Table S6), indicating that dispersion is also important in these interactions.

4. Materials and Methods

A CSD search (CSD version 5.35, November 2013. and updates, May 2014) is performed by using ConQuest 1.16 [51]. Constraints applied in search were: (1) distances between donor (D) and acceptor (A) atoms within the ring less than 4.0 Å; (2) angles between donor (D), hydrogen, and acceptor (A) atoms within the ring from 90° to 180°; (3) absolute torsions AXYD and XYDH (Figure 1) from 0 to 10°; (4) donor and acceptor atoms include N, O, Cl, S and F atoms, due to their considerable electronegativities; (5) all covalent bonds within the ring are set to be single acyclic; (6) structures where all atoms in the ring belong to planar groups (coordination number of D, Y and X atoms less than four and coordination number of A atom less than three are excluded from further considerations, since noncovalent interactions of these species were analyzed in our previous work [46]); and (7) intermolecular contacts having distances between two centroids 4.5 Å or less are considered in this work. The crystallographic R factor is set to be less than 10%, the error-free coordinates according to the criteria used in the CSD, the H-atom positions were normalized using the CSD default X–H bond lengths (O–H = 0.983 Å; C–H = 1.083 Å and N–H = 1.009 Å), no polymer structures and no powder structures were included.

Single-point calculations were used to evaluate interaction energy between two rings. Optimizations of monomers, thiosemicarbazide, semicarbazide, and glycolamide, are done at MP2/cc-pVTZ [52,53] level. The methods for calculation of interaction energy potential curves were chosen based on good agreement with CCSD(T)/CBS method [54] (Tables S1–S5, ESI). Interaction energy was determined as a difference of the dimer energy and the sum of energies of monomers, having included correction of basis set superposition error (BSSE) [55]. All calculations are done by using Gaussian09 series of programs [56].

Maps of electrostatic potentials for thiosemicarbazide, semicarbazide and glycolamide molecules, calculated and visualized from wavefunction files using the Wavefunction Analysis Program (WFA-SAS) [57,58]. The wavefunctions were calculated on MP2/cc-pVTZ level of theory.

5. Conclusions

Search and analysis of the crystal structures in the Cambridge Structural Database (CSD) show that saturated acyclic four-atom groups closed with a classic intramolecular hydrogen bond, generating planar five-membered rings (hydrogen-bridged quasi-rings), in which at least one of the ring atoms is bonded to other non-ring atoms that are not in the ring plane and, thus, capable to form intermolecular interactions, were observed quite frequently in crystal structures from the CSD. In the crystal structures, we found 1068 rings, while 289 (27%) of them form parallel interactions.

The rings can have one hydrogen atom out of the ring plane that gives possibility for hydrogen bonds between two parallel rings. Hence, in these systems, two types of hydrogen bonds can be present, one in the ring, and the other one between two parallel rings.

The strongest interaction energies, calculated at very accurate CCSD(T)/CBS level, are -9.68 and -7.12 kcal/mol for thiosemicarbazide and semicarbazide interactions, respectively. Similar rings that do not have possibility for hydrogen bonds between rings have significantly weaker interaction energies of -4.89 and -2.95 kcal/mol [46], indicating the importance of hydrogen bonds for the strength of the interactions. Evaluated energies of hydrogen bonds in thiosemicarbazide and semicarbazide are −7.78 and −3.60 kcal/mol, respectively.

The results presented in this paper recognize that interactions, and their strong interaction energies, can be important in all supramolecular systems.

Supplementary Materials: The supplementary materials are available online at http://www.mdpi.com/ 2073-4352/6/4/34/s1.

Acknowledgments: This work was supported by the Serbian Ministry of Education, Science and Technological Development (Grant 172065). The HPC resources and services used in this work were partially provided by the IT Research Computing group in Texas A&M University at Qatar. IT Research Computing is funded by the Qatar Foundation for Education, Science and Community Development (http://www.qf.org.qa).

Author Contributions: Snežana D. Zarić and Goran V. Janjić conceived and designed this study; Jelena P. Blagojević and Goran V. Janjić performed the study; Jelena P. Blagojević, Goran V. Janjić and Snežana D. Zarić analyzed the data; Jelena P. Blagojević and Snežana D. Zarić wrote the paper.

Conflicts of Interest: The authors declare no conflict of interest.

Abbreviations

The following abbreviations are used in this manuscript:

CSD	Cambridge Structural Database
CCSD(T)	Coupled-Cluster with Single and Duble and Perturbative Triple excitations
CBS	Complete Basis Set
DFT	Density Functional Theory
ESI	Electronic Supporting Information
ESP	Electrostatic Potential
MP2	Møller-Plesset Perturbation Theory of second order
cc-pVTZ	Correlation-Consistent Polarized Valence-only Triple-zeta basis set
BSSE	Basis Set Superposition Error
WFA-SAS	Wavefunction Analysis-Surface Analysis Suite

References

1. Salonen, L.M.; Ellermann, M.; Diederich, F. Aromatic rings in chemical and biological recognition: Energetics and structures. *Angew. Chem. Int. Ed.* **2011**, *50*, 4808–4842.
2. O'Sullivan, M.C.; Durham, T.B.; Valdes, H.E.; Dauer, K.L.; Karney, N.J.; Forrestel, A.C.; Bacchi, C.J.; Baker, J.F. Dibenzosuberyl substituted polyamines and analogs of clomipramine as effective inhibitors of trypanothione reductase; molecular docking, and assessment of trypanocidal activities. *Bioorgan. Med. Chem.* **2015**, *23*, 996–1010.
3. Woziwodzka, A.; Gołuński, G.; Wyrzykowski, D.; Kaźmierkiewicz, R.; Piosik, J. Caffeine and other methylxanthines as interceptors of food-borne aromatic mutagens: Inhibition of Trp-P-1 and Trp-P-2 mutagenic activity. *Chem. Res. Toxicol.* **2013**, *26*, 1660–1673.

4. Thio, Y.; Toh, S.W.; Xue, F.; Vittal, J.J. Self-assembly of a 15-nickel metallamacrocyclic complex derived from the L-glutamic acid Schiff base ligand. *Dalton Trans.* **2014**, *43*, 5998–6001.
5. Ma, M.; Kuang, Y.; Gao, Y.; Zhang, Y.; Gao, P.; Xu, B. Aromatic-aromatic interactions induce the self-assembly of pentapeptidic derivatives in water to form nanofibers and supramolecular hydrogels. *J. Am. Chem. Soc.* **2010**, *132*, 2719–2728.
6. Schneider, H.-J. Binding mechanisms in supramolecular complexes. *Angew. Chem. Int. Ed.* **2009**, *48*, 3924–3977.
7. Schweizer, W.B.; Dunitz, J.D. Quantum Mechanical Calculations for Benzene Dimer Energies: Present Problems and Future Challenges. *J. Chem. Theory Comput.* **2006**, *2*, 288–291.
8. Tsuzuki, S.; Honda, K.; Uchimaru, T.; Mikami, M.; Tanabe, K. Origin of Attraction and Directionality of the π/π Interaction: Model Chemistry Calculations of Benzene Dimer Interaction. *J. Am. Chem. Soc.* **2002**, *124*, 104–112.
9. Sinnokrot, M.O.; Valeev, E.F.; Sherrill, C.D. Estimates of the Ab Initio Limit for $\pi-\pi$ Interactions: The Benzene Dimer. *J. Am. Chem. Soc.* **2002**, *124*, 10887–10893.
10. Řezáč, J.; Riley, K.E.; Hobza, P. S66: A Well-balanced Database of Benchmark Interaction Energies Relevant to Biomolecular Structures. *J. Chem. Theory Comput.* **2011**, *7*, 2427–2438.
11. Janowski, T.; Pulay, P. High accuracy benchmark calculations on the benzene dimer potential energy surface. *Chem. Phys. Lett.* **2007**, *447*, 27–32.
12. Bludský, O.; Rubes, M.; Soldán, P.; Nachtigall, P. Investigation of the benzene-dimer potential energy surface: DFT/CCSD(T) correction scheme. *J. Chem. Phys.* **2008**, *128*, 114102.
13. Hohenstein, E.G.; Sherrill, C.D. Effects of heteroatoms on aromatic π-π interactions: Benzene-pyridine and pyridine dimer. *J. Phys. Chem. A* **2009**, *113*, 878–886.
14. Geronimo, I.; Lee, E.C.; Singh, N.J.; Kim, K.S. How Different are Electron-Rich and Electron-Deficient π Interactions? *J. Chem. Theory Comput.* **2010**, *6*, 1931–1934.
15. Ninković, D.B.; Janjić, G.V.; Veljković, D.Ž.; Sredojević, D.N.; Zarić, S.D. What are the preferred horizontal displacements in parallel aromatic-aromatic interactions? Significant interactions at large displacements. *Chemphyschem* **2011**, *12*, 3511–3514.
16. Ninković, D.B.; Andrić, J.M.; Malkov, S.N.; Zarić, S.D. What are the preferred horizontal displacements of aromatic-aromatic interactions in proteins? Comparison with the calculated benzene-benzene potential energy surface. *Phys. Chem. Chem. Phys.* **2014**, *16*, 11173–11177.
17. Sinnokrot, M.O.; Sherrill, C.D. High-accuracy quantum mechanical studies of π-π interactions in benzene dimers. *J. Phys. Chem. A* **2006**, *110*, 10656–10668.
18. Podeszwa, R.; Bukowski, R.; Szalewicz, K. Potential energy surface for the benzene dimer and perturbational analysis of π-π interactions. *J. Phys. Chem. A* **2006**, *110*, 10345–10354.
19. Pitoňák, M.; Neogrády, P.; Řezáč, J.; Jurečka, P.; Urban, M.; Hobza, P. Benzene Dimer: High-Level Wave Function and Density Functional Theory Calculations. *J. Chem. Theory Comput.* **2008**, *4*, 1829–1834.

20. Ninković, D.B.; Janjić, G.V.; Zarić, S.D. Crystallographic and ab Initio Study of Pyridine Stacking Interactions. Local Nature of Hydrogen Bond Effect in Stacking Interactions. *Cryst. Growth Des.* **2012**, *12*, 1060–1063.
21. Ninković, D.B.; Andrić, J.M.; Zarić, S.D. Parallel interactions at large horizontal displacement in pyridine-pyridine and benzene-pyridine dimers. *Chemphyschem* **2013**, *14*, 237–243.
22. Craven, E.; Zhang, C.; Janiak, C.; Rheinwald, G.; Lang, H. Synthesis, Structure and Solution Chemistry of (5, 5′-Dimethyl-2, 2′-bipyridine)(IDA)copper(II) and Structural Comparison With Aqua(IDA)(1, 10-phenanthroline)copper(II) (IDA = iminodiacetato). *Z. Anorg. Allg. Chem.* **2003**, *629*, 2282–2290.
23. Janjić, G.V.; Veljković, D.Z.; Zarić, S.D. Water/Aromatic Parallel Alignment Interactions. Significant Interactions at Large Horizontal Displacements. *Cryst. Growth Des.* **2011**, *11*, 2680–2683.
24. Sredojević, D.; Bogdanović, G.A.; Tomić, Z.D.; Zarić, S.D. Stacking *vs.* CH–π interactions between chelate and aryl rings in crystal structures of square-planar transition metal complexes. *CrystEngComm* **2007**, *9*, 793–798.
25. Sredojević, D.N.; Tomić, Z.D.; Zarić, S.D. Evidence of Chelate−Chelate Stacking Interactions in Crystal Structures of Transition-Metal Complexes. *Cryst. Growth Des.* **2010**, *10*, 3901–3908.
26. Tomić, Z.D.; Leovac, V.M.; Pokorni, S.V.; Zobel, D.; Zarić, S.D. Crystal Structure of Bis[acetone-1-naphthoylhydrazinato(−1)]copper(II) and Investigations of Intermolecular Interactions. *Eur. J. Inorg. Chem.* **2003**, *6*, 1222–1226.
27. Tomić, Z.D.; Sredojević, D.; Zarić, S.D. Stacking Interactions between Chelate and Phenyl Rings in Square-Planar Transition Metal Complexes. *Cryst. Growth Des.* **2006**, *6*, 29–31.
28. Wang, X.-J.; Jian, H.-X.; Liu, Z.-P.; Ni, Q.-L.; Gui, L.-C.; Tang, L.-H. Assembly molecular architectures based on structural variation of metalloligand [Cu(PySal)2] (PySal = 3-pyridylmethylsalicylideneimino). *Polyhedron* **2008**, *27*, 2634–2642.
29. Granifo, J.; Vargas, M.; Garland, M.T.; Ibáñez, A.; Gaviño, R.; Baggio, R. The novel ligand 4′-phenyl-3,2′:6′,3″-terpyridine (L) and the supramolecular structure of the dinuclear complex [Zn2(μ-L)(acac)4]·H2O (acac = acetylacetonato). *Inorg. Chem. Commun.* **2008**, *11*, 1388–1391.
30. Philip, V.; Suni, V.; Prathapachandra Kurup, M.R.; Nethaji, M. Structural and spectral studies of nickel(II) complexes of di-2-pyridyl ketone N4,N4-(butane-1,4-diyl)thiosemicarbazone. *Polyhedron* **2004**, *23*, 1225–1233.
31. Konidaris, K.F.; Tsipis, A.C.; Kostakis, G.E. Shedding Light on Intermolecular Metal-Organic Ring Interactions by Theoretical Studies. *ChemPlusChem* **2012**, *77*, 354–360.
32. Konidaris, K.F.; Morrison, C.N.; Servetas, J.G.; Haukka, M.; Lan, Y.; Powell, A.K.; Plakatouras, J.C.; Kostakis, G.E. Supramolecular assemblies involving metal–organic ring interactions: Heterometallic Cu(II)–Ln(III) two-dimensional coordination polymers. *CrystEngComm* **2012**, *14*, 1842–1849.

33. Konidaris, K.F.; Powell, A.K.; Kostakis, G.E. Peculiar structural findings in coordination chemistry of malonamide–N,N'-diacetic acid. *CrystEngComm* **2011**, *13*, 5872–5876.
34. Tiekink, E.R.T. Molecular crystals by design? *Chem. Commun.* **2014**, *50*, 11079–11082.
35. Baul, T.S.B.; Kundu, S.; Mitra, S.; Höpfl, H.; Tiekink, E.R.T.; Linden, A. The influence of counter ion and ligand methyl substitution on the solid-state structures and photophysical properties of mercury(II) complexes with (*E*)-*N*-(pyridin-2-ylmethylidene)arylamines. *Dalton Trans.* **2013**, *42*, 1905–1920.
36. Khavasi, H.R.; Sadegh, B.M.M. Influence of N-heteroaromatic π-π stacking on supramolecular assembly and coordination geometry; effect of a single-atom change in the ligand. *Dalton Trans.* **2015**, *44*, 5488–5502.
37. Hosseini-Monfared, H.; Pousaneh, E.; Sadighian, S.; Ng, S.W.; Tiekink, E.R.T. Syntheses, Structures, and Catalytic Activity of Copper(II)-Aroylhydrazone Complexes. *Z. Anorg. Allg. Chem.* **2013**, *639*, 435–442.
38. Ni, Q.-L.; Jiang, X.-F.; Gui, L.-C.; Wang, X.-J.; Yang, K.-G.; Bi, X.-S. Synthesis, structures and characterization of a series of Cu(I)-diimine complexes with labile N,N'-bis((diphenylphosphino) methyl)naphthalene-1,5-diamine: Diverse structures directed by π–π stacking interactions. *New J. Chem.* **2011**, *35*, 2471–2476.
39. Molčanov, K.; Jurić, M.; Kojić-Prodić, B. Stacking of metal chelating rings with π-systems in mononuclear complexes of copper(II) with 3,6-dichloro-2,5-dihydroxy-1,4-benzoquinone (chloranilic acid) and 2,2'-bipyridine ligands. *Dalton Trans.* **2013**, *42*, 15756–15765.
40. Akine, S.; Varadi, Z.; Nabeshima, T. Synthesis of Planar Metal Complexes and the Stacking Abilities of Naphthalenediol-Based Acyclic and Macrocyclic Salen-Type Ligands. *Eur. J. Inorg. Chem.* **2013**, *35*, 5987–5998.
41. Melnic, E.; Coropceanu, E.B.; Kulikova, O.V.; Siminel, A.V.; Anderson, D.; Rivera-Jacquez, H.J.; Masunov, A.E.; Fonari, M.S.; Kravtsov, V.C. Robust Packing Patterns and Luminescence Quenching in Mononuclear [Cu(II)(phen)$_2$] Sulfates. *J. Phys. Chem. C* **2014**, *118*, 30087–30100.
42. Zhao, Y.; Chang, X.-H.; Liu, G.-Z.; Ma, L.-F.; Wang, L.-Y. Five Mn(II) Coordination Polymers Based on 2,3',5,5'-Biphenyl Tetracarboxylic Acid: Syntheses, Structures, and Magnetic Properties. *Cryst. Growth Des.* **2015**, *15*, 966–974.
43. Malenov, D.P.; Ninković, D.B.; Sredojević, D.N.; Zarić, S.D. Stacking of benzene with metal chelates: Calculated CCSD(T)/CBS interaction energies and potential-energy curves. *Chemphyschem* **2014**, *15*, 2458–2461.
44. Malenov, D.P.; Ninković, D.B.; Zarić, S.D. Stacking of metal chelates with benzene: Can dispersion-corrected DFT be used to calculate organic-inorganic stacking? *Chemphyschem* **2015**, *16*, 761–768.
45. Karabıyık, H.; Karabıyık, H.; Ocak İskeleli, N. Hydrogen-bridged chelate ring-assisted π-stacking interactions. *Acta Cryst. B* **2012**, *68*, 71–79.
46. Blagojević, J.P.; Zarić, S.D. Stacking interactions of hydrogen-bridged rings—Stronger than the stacking of benzene molecules. *Chem. Commun.* **2015**, *51*, 12989–12991.

47. Yeo, C.I.; Halim, S.N.A.; Ng, S.W.; Tan, S.L.; Zukerman-Schpector, J.; Ferreira, M.A.B.; Tiekink, E.R.T. Investigation of putative arene-C-H···π(quasi-chelate ring) interactions in copper(I) crystal structures. *Chem. Commun.* **2014**, *50*, 5984–5986.
48. Sobczyk, L.; Grabowski, S.J.; Krygowski, T.M. Interrelation between H-bond and Pi-electron delocalization. *Chem. Rev.* **2005**, *105*, 3513–3560.
49. Lyssenko, K.A.; Antipin, M.Y. The nature and energy characteristics of intramolecular hydrogen bonds in crystals. *Russ. Chem. Bull.* **2006**, *55*, 1–15.
50. Sanz, P.; Mó, O.; Yañez, M.; Elguero, J. Resonance-assisted hydrogen bonds: A critical examination. Structure and stability of the enols of beta-diketones and beta-enaminones. *J. Phys. Chem. A* **2007**, *111*, 3585–3591.
51. Bruno, I.J.; Cole, J.C.; Edgington, P.R.; Kessler, M.; Macrae, C.F.; McCabe, P.; Pearson, J.; Taylor, R. New software for searching the Cambridge Structural Database and visualizing crystal structures. *Acta Cryst. Sect. B—Struct. Sci.* **2002**, *58*, 389–397.
52. Møller, C.; Plesset, M.S. Note on an Approximation Treatment for Many-Electron Systems. *Phys. Rev.* **1934**, *46*, 618–622.
53. Kendall, R.A.; Dunning, T.H., Jr.; Harrison, R.J. Electron affinities of the first-row atoms revisited. Systematic basis sets and wave functions. *J. Chem. Phys.* **1992**, *96*, 6796–6806.
54. Raghavachari, K.; Trucks, G.W.; Pople, J.A.; Head-Gordon, M. A fifth-order perturbation comparison of electron correlation theories. *Chem. Phys. Lett.* **1989**, *157*, 479–483.
55. Boys, S.F.; Bernardi, F. The calculation of small molecular interactions by the differences of separate total energies. Some procedures with reduced errors. *Mol. Phys.* **1970**, *19*, 553–566.
56. Frisch, M.J.; Trucks, G.W.; Schlegel, H.B.; Scuseria, G.E.; Robb, M.A.; Cheeseman, J.R.; Scalmani, G.; Barone, V.; Mennucci, B.; Petersson, G.A.; *et al. Gaussian 09 (Revision D.01)*; Gaussian, Inc.: Wallingford, CT, USA, 2013.
57. Bulat, F.A.; (Fable Theory & Computation LLC, Washington, DC, USA); Toro-Labbe, A.; (Pontificia Universidad Católica de Chile, Santiago, Chile). Personal communication, 2013.
58. Bulat, F.A.; Toro-Labbé, A.; Brinck, T.; Murray, J.S.; Politzer, P. Quantitative analysis of molecular surfaces: Areas, volumes, electrostatic potentials and average local ionization energies. *J. Mol. Model.* **2010**, *16*, 1679–1691.

Comparisons between Crystallography Data and Theoretical Parameters and the Formation of Intramolecular Hydrogen Bonds: Benznidazole

Boaz G. Oliveira, Edilson B. Alencar Filho and Mário L. A. A. Vasconcellos

Abstract: The conformational preferences of benznidazole were examined through the application of DFT, PCM and QTAIM calculations, whose results were compared with crystallography data. The geometries were fully optimized with minimum potential energy surface by means of the Relaxed Potential Energy Surface Scan (RPESS) at AM1, followed by the B3LYP/6-311++G(d,p) theoretical level. As a result, the *s-cis* conformation (**1C**) was shown to be more stable (4.78 kcal·mol^{-1}) than *s-trans* (**1T**). The Quantum Theory Atoms in Molecules (QTAIM) was applied in order to characterize the (N–H···O=N) and (C–H···=N) intramolecular hydrogen bonds. The simulation of solvent effect performed by means of the implicit Polarized Continuum Model (PCM) revealed great results, such as, for instance, that the conformation **1W** is more stable (23.17 kcal·mol^{-1}) in comparison to **1C**. Our main goal was stressed in the topological description of intramolecular hydrogen bonds in light of the QTAIM approach, as well as in the solvent simulation to accurately obtain an important conformation of benznidazole.

> Reprinted from *Crystals*. Cite as: Oliveira, B.G.; Filho, E.B.A.; Vasconcellos, M.L.A.A. Comparisons between Crystallography Data and Theoretical Parameters and the Formation of Intramolecular Hydrogen Bonds: Benznidazole. *Crystals* **2016**, *6*, 56.

1. Introduction

In medicinal chemistry, the studies of new compounds with high biological activities and less toxic effects have attracted the attention of many researchers around the world [1]. This scientific interest concerns the potential of several compounds in treating serious diseases, ranging from "neglected" diseases such as *Schistosoma mansoni* [2] to disorders such as cancer [3]. In the context of neglected diseases, for which big pharmaceutical companies are not investing in the search for therapeutic alternatives, Chagas disease, also called *American trypanosomiasis*, is considered the most important parasitic infection of Latin America [4]. Chagas disease is manifested in humans and domestic animals living in extreme poverty and rural areas [5]. In practice, the infection is caused by the *Trypanosoma cruzi* protozoan although it is transmitted by the *Triatoma infestans* insect. In the actual treatment, basically two drugs have been widely used: Benznidazole (Rochagan®,

Basel, Switzerland) and Nifurtimox (Lampit®, Leverkusen, Germany). Due to the possibility to obtain new compounds to be used in the treatment against Chagas disease, a lot of studies have been conducted in order to improve the potentiality of benznidazole.

Some time ago, a theoretical conformational study of benznidazole was performed [6]. The results point to the existence of local minima in the potential energy surface, whose structures were compared with some active compounds against *T. cruzi*: derivatives of tetrahydro-β-carboline. This study was carried out in gas phase by using the AM1 semi empirical Hamiltonian, whose application was conditioned to the modeling of the Potential Energy Surface Scan (PESS). Only then, the optimized geometry in each point of minimum has been determined through the B3LYP functional jointly with the 6-311+G(d) basis set. From an experimental viewpoint, Soares Sobrinho *et al.* [7] have published a crystallography study about the benznidazole structure based on X-ray diffraction, whose results were very different from those in gas phase, including the formation of intermolecular hydrogen bonds on the crystal packaging. In molecular modeling [8], it is well established that the identification of hydrogen bonds is one of the most important criteria in analysis of biological compounds [9]. These studies may be performed by virtual screening [10] or quantum chemical calculations [11,12] in the pursuit of achieving structure-activity relationships. Historically, the formation of intermolecular or intramolecular hydrogen bonds has been discussed by taking into account the van der Waals radii of the electron donor-acceptor. If we consider this statement [5], the distance values for typical hydrogen bonds should be exactly the same as or shorter than 2.6 Å, of course regarding the F, O and N atoms, or even the π cloud as proton-acceptor centers [13–16]. In fact, the characterization of hydrogen bonds is not just dependent on structural parameters, actually, the application of quantum mechanical criteria is feasible in this regard.

The Quantum Theory of Atoms in Molecules (QTAIM) [17] represents a useful tool in studies of molecular stability and strength of chemical bonds [18], but is also widely applied in investigations of hydrogen bonds through the analysis of electronic density and its topological parameters [19]. According to some publications [20], the QTAIM approach has been also useful in studies of biological systems [21], mainly to characterize the intramolecular hydrogen bonds of compounds with biological activity [22]. In this sense, we are convinced that QTAIM can be decisive in our investigation not only regarding the intramolecular hydrogen bonds [23–31], but also to unveil the most stable conformation of benznidazole. Another interesting viewpoint is that the solvent effect may be responsible for drastic changes in several molecular properties, e.g., geometrical deformations; increase in the reaction rate; control of products along the reaction paths, for instance [32,33]. If we consider that intramolecular hydrogen bonds can be formed in the minimum structure of benznidazole, the specialized literature informs us that the application of the Polarized Continuum Model (PCM) [34] to evaluate the solvent effect is recommended [35–39]. By taking into account the importance of drug action in human organisms, that are predominantly aqueous, this article also presents a theoretical investigation of the solvent effect on the benznidazole structure through the application of PCM protocol, demonstrating the importance of solvation in the conformational study of this bioactive molecule [40]. In the context of molecular stabilization, a topological description of intramolecular hydrogen bonds using the QTAIM approach is another important contribution of this work.

2. Computational Procedure

The procedure of Relaxed Potential Energy Surface Scan (RPESS) was performed by taking particular care with the dihedral angles (θ_1, θ_2, θ_3, and θ_4), which are illustrated in Figure 1. These calculations were executed by using the AM1 semiempirical level with angle variations from 0° to 360°, in intervals of 10°. After that, the minimum conformation was fully optimized by the B3LYP/6-311++G(d,p) level of calculation without any symmetry constraint and with all geometries modeled at a minimum of potential energy because no imaginary frequencies were found. As a result, the conformation recognized as **1C** was generated.

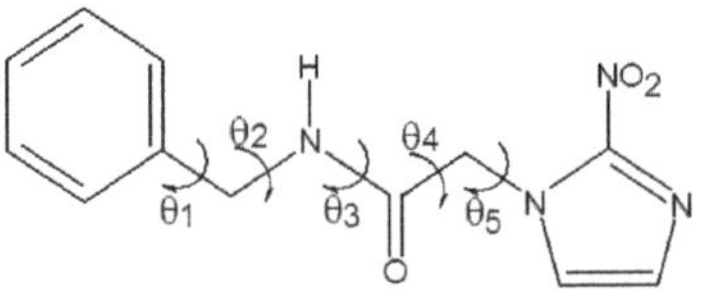

Figure 1. Structure of benznidazole and dihedral angles examined in RPESS procedure.

The crystal structure (**1K**) was optimized using the same level of calculation presented above. The conformation **1T** was obtained from an arbitrary variation in the geometry of **1C**, corresponding to a previously reported geometry [6], which remains as a minimum conformation (less stable than **1C**) after B3LYP/6-311++G(d,p) optimization. The solvent effect was evaluated on both **1K** and **1C** through the arguments of the Polarized Continuum Model (PCM) both providing **1W**. All these calculations were carried out by GAUSSIAN 03W quantum suite of codes [41]. Both the nature of electronic density and the characterization of intramolecular hydrogen bonds were investigated through the QTAIM formalism, whose calculations were performed by the AIM 2000 1.0 software package [42].

3. Results and Discussion

3.1. Minimum Structure, Bond Lenghts and Vibration Modes

Particularly, the study of dihedral θ_3 (Figure 1) is important due to the high energetic barrier of amide group (HN–C=O) [43]. In line with this, it is worthy to assume that two conformations (*s-trans* and *s-cis*) must coexist (Figure 2) when the energetic interconversion between them is about 20–22 kcal·mol^{-1}. The terminology of *s-cis* and *s-trans* (s means to conformation into N–C single bond) was used to describe the relative positioning between the R (RNH) and C=O groups when they are in the same direction (*s-cis*) or even in an opposite alignment (*s-trans*), respectively. From RPESS protocol, the lowest energy conformation was selected, and after undergoing an optimization procedure at the B3LYP/6-311++G(d,p) level of theory, the conformation symbolized as **1C** was obtained (Figure 3). As such, **1C** shows both RN and C=O groups in *s-cis* position. Moreover, the conformation **1T** (Figure 4) was also obtained from the geometry described previously [6] through the optimization at B3LYP/6-311++G(d,p) level of theory. Note that the groups (R–N and C=O) are in *s-trans* position in **1T**.

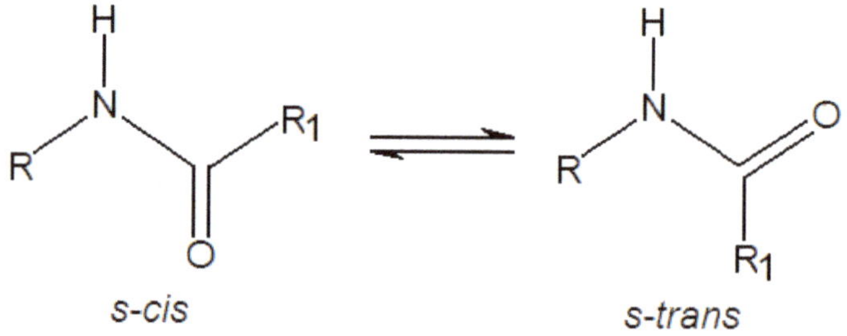

Figure 2. Conformations *s-cis* and *s-trans*.

By analyzing Figure 3, it can be seen that **1C** leads to the formation of two intramolecular hydrogen bonds (N–H^a···O^b=N) and (C–H^c···O^b=N) with respective distance values of 2.125 Å and 2.287 Å, which contributes to a substantial stabilization of this conformation. In **1T**, however, the C–H^d···O^b=N and C–H^e···O^b=N hydrogen bonds were considered, but their lengths of 2.347 Å and 2.562 Å are longer than those results of **1C**. From a structural viewpoint, this discovery provides solid evidence about the preferential conformation of **1C**, *s-cis*, as has already been demonstrated in many similar works [21,44–47]. Regarding **1K** and **1W** (see Figures 3 and 4), B3LYP/6-311++G(d,p) calculations revealed that **1K** presents the same intramolecular C–H^c···O^b=N hydrogen bond of **1C**, once the distance values of 2.340 Å (**1K**) differ slightly from the value of 2.287 Å (**1C**). At last, **1W** presents only one hydrogen bond, C–H^c···O^b=N, whose distance value of 2.468 Å is much longer than C–H^d···O^b=N and N=O^b···C–H^e. Although it should be expected that the most stable conformation must be the solvated structure [48,49], geometrically the strength of the intramolecular N=O^b···H^c–C hydrogen bond is not sufficient evidence for that. The strongest statement regarding the hydrogen bonds' formation, either intra or intermolecular, actually has its basis in the deformations of the bond lengths that compose it.

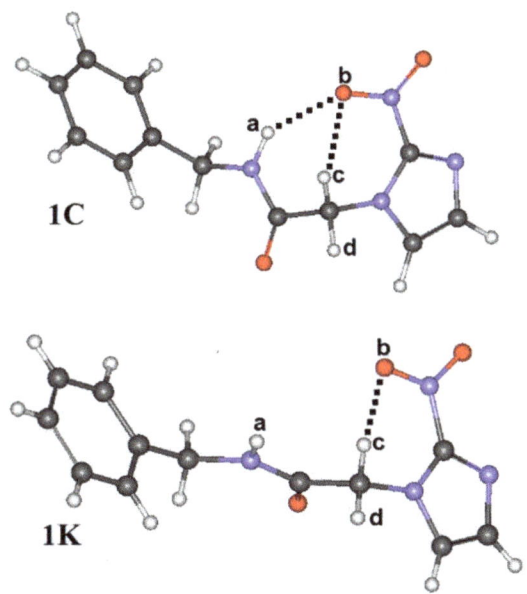

Figure 3. Conformations (1C and 1K) of benznidazole.

Besides the intramolecular hydrogen bonds, Table 1 also enumerates the bond length results of the donors X–H (N–H^a, C–H^c, C–H^e and C–H^d) and acceptors Y (O^b=N) of protons. Although a precise measurement of the bond length variation is not allowed because the monomer configuration of the proton donors cannot be attained, it can be seen that the HBond distances are not in agreement with the bond length variations of X–H as well as of Y, whose values are quite similar. Regarding the characterization of the infrared vibration modes as one of the preconditions to recognize the most stable conformation, the values of stretch frequencies and absorption intensities are invariable, even though a slight difference of 97.3 (1C) and 107.6 km·mol^{-1} (1K) for IN–H^a has been observed. Nevertheless, the new vibration modes could not be unveiled, although by taking into account the long values of the HBond distances beyond 2.000 Å, their stretch frequencies and absorption intensities must be active in an infrared region lower than 50 cm^{-1}.

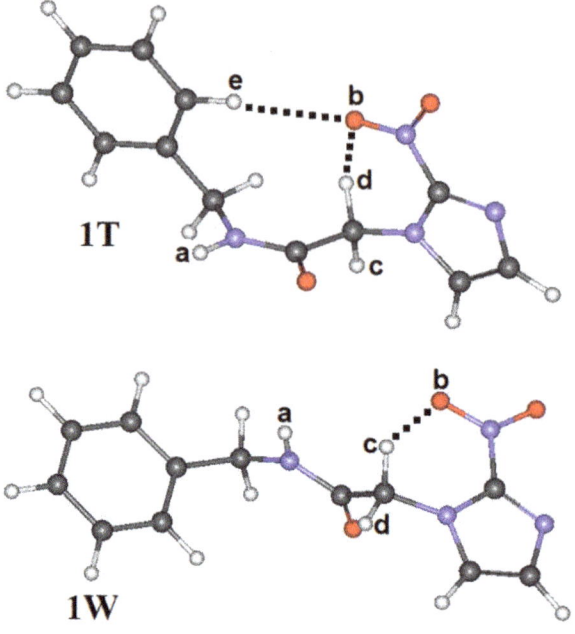

Figure 4. Conformations (**1T** and **1W**) of benznidazole.

Table 1. Values of HBond distances and bond lengths of donors and acceptors of protons in the **1C**, **1T**, **1K** and **1W** conformations of benznidazole.

HBonds and Bonds *	Conformations			
	1C	1T	1K	1W
RN–H^a···O^b=N	2.125	-	-	-
RC–H^c···O^b=N	2.287	-	2.340	2.468
RC–H^d···O^b=N	-	2.347	-	-
RC–H^e···O^b=N	-	2.562	-	-
rN–H^a	1.012	-	1.013	-
rC–H^c	1.085	-	1.086	1.085
rC–H^e	-	1.084	-	-
rC–H^d	-	1.085	-	-
rO^b=N	1.243	1.239	1.243	1.242
υN–H^a	3570.1	-	3562.7	-
IN–H^a	97.3	-	107.6	-
υC–H^c	3172.3	-	3172.4	3172.3
IC–H^c	0.70	-	0.20	0.40
υC–H^e	-	3183.0	-	-
IC–H^e	-	8.0	-	-
υC–H^d	-	3160.0	-	-
IC–H^d	-	2.4	-	-

* Values of R (Intramolecular hydrogen bonds) and r (bond lengths) are given in Å; Values of υ (stretch frequencies) and I (absorption intensity) are given in cm^{-1} and $km\cdot mol^{-1}$.

3.2. PCM Calculations

The stabilization of **1C**, **1T** and **1W** can be discussed on the basis of thermodynamic properties, whose results organized in Table 2 represent the sum of electronic and zero-point energies (ε_0 + ZPE), sum of electronic and thermal free energies (ε_0 + G_{corr}) and ΔG, obtained from the difference between (e_0 + G_{corr}) for conformation and the corresponding value for **1C**. The PCM calculations were performed on **1C**, although this procedure has been based on the X-Ray crystal structure (**1K**), which both culminated with the development of the **1W** conformation. Regarding Figure 3, there is less tendency to form an intramolecular hydrogen bond, which can be explained by the solvent effect in the stabilization of benznidazole, although the structural interaction strength recommended by the RC–H^c···O^b=N contact points out that **1K** is slightly more strongly bonded than **1C**. The value of -23.17 kcal·mol^{-1} may mean that **1W** is the most stable structure. However, this HBond presents a distance of 2.468 Å, which, being one of the longest interactions, which seems to conflict with the stabilization statement presented above. It is important to point out that the solvent effect should reveal electronic energy values that may differ from a hypersurface investigated in the gas phase. Thus, the displayed geometry is consistent with the weakening of the intramolecular interaction favoring the influence of the external environment (aqueous). Really, the solvent effect should be carefully used to explain the additional molecular stabilization, depending on the type of system studied [50], including drug molecules that act in predominantly aqueous environments.

Table 2. Thermochemical parameters calculated at B3LYP/6-311++G(d,p) level (298.15 K and 1 atm) obtained from gas phase and PCM solvent model [a].

Conformations	Parameters		
	(ε_0 + ZPE)	(ε_0 + G_{corr})	ΔG
1C	−909.100529	−909.149975	-
1T	−909.093925	−909.142360	4.78
1W	−909.138240	−909.186911	−23.17

[a] **1C** is the reference minimum.

3.3. QTAIM Parameters and Intramolecular Hydrogen Bonds

Bond pathways, Bond Critical Points (BCP), and values of ρ followed by $\nabla^2\rho$ were characterized in terms of the QTAIM topological integrations of the electronic density. We can see, in Figures 5 and 6 the presence of a Bond Critical Point (BCP) between (N–H^a···O^b=N), indicating the formation of intramolecular hydrogen bonds, corroborating therefore the formulated hypothesis in this work. The same conformation shows a BCP between (C–H^c···O^b=N) groups, forming a ring of six members [9]. The participation of the C–H group in intramolecular hydrogen bonds has been established [51]. In our current case, the simultaneous positions of the α-carbonyl and α-nitrogen-hydrogen can contribute to a peculiar acidity of C–H and also lead to the formation of an intramolecular hydrogen bond. The values of ρ and $\nabla^2\rho$ that characterize these interactions can be visualized in Table 3. By taking into account the *virial* theorem [17], G and U idealize the kinetic (always positive) and potential (always negative) energy density functions. Then, positive values of $\nabla^2\rho$ indicate a depletion of electronic potential energy density at BCP in favor of the kinetic energy because it outweighs the potential one, showing a concentration of electronic charge in separated nuclei [51–53].

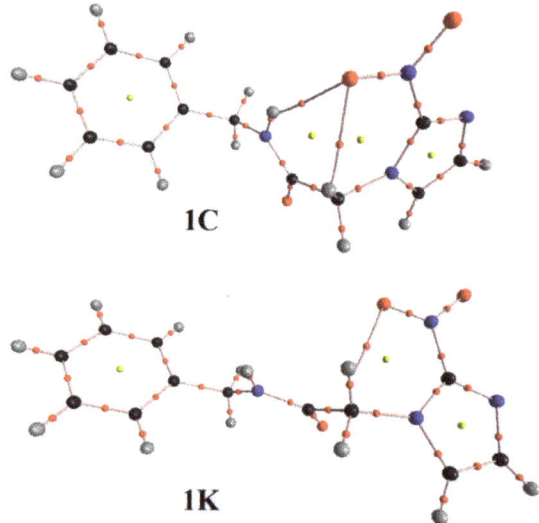

Figure 5. BP and BCP for conformations (**1C** and **1K**) of benznidazole.

These values also indicate intramolecular hydrogen bonds with a significant covalent character, as argued by Gilli *et al.* [54] and Siskos *et al.* [55]. All these considerations can also be visualized in terms of the Resonance Assisted Hydrogen Bond phenomenon (RAHB) [56,57], in which the π electrons of the pirazole system are in resonance with the NO$_2$ group, increasing the electron density at the BCP and decreasing the bond lengths, that are considerably smaller than sum of van der Walls radii. On the other hand, it is important to be highlighted that the values of $\nabla^2\rho$, ρ presented here are much smaller than the ones obtained at BCP for common covalent interactions.

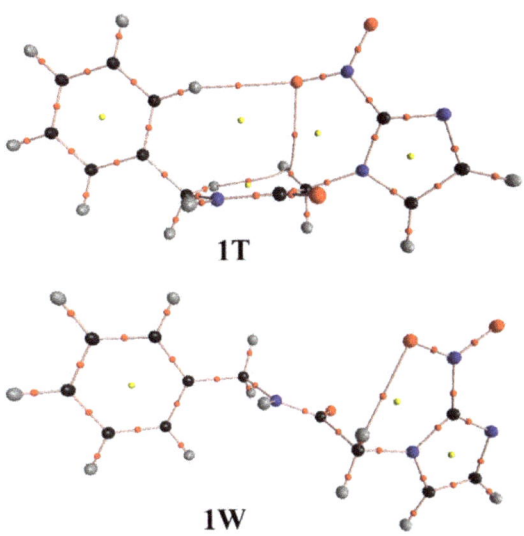

Figure 6. BP and BCP for conformations (**1T** and **1W**) of benznidazole.

Table 3. Topological parameters obtained from QTAIM calculations.

QTAIM Parameters *	Conformations			
	1C	1T	1K	1W
N–H^a···O^b=N				
$\rho\ (\nabla^2\rho)$	0.0169 (0.0634)	-	-	-
U (G)	−0.0114 (0.0136)	-	-	-
−G/U	1.192	-	-	-
C–H^c···O^b=N				
$\rho\ (\nabla^2\rho)$	-	-	0.0164 (0.0645)	0.0147 (0.0561)
U (G)	-	-	−0.0191 (0.0140)	−0.0168 (0.0124)
−G/U	-	-	0.733	0.738
C–H^e···O^b=N				
$\rho\ (\nabla^2\rho)$	-	0.0059 (0.0213)	-	-
U (G)	-	−0.0036 (0.0044)	-	-
−G/U	-	1.222	-	-
C–H^d···O^b=N				
$\rho\ (\nabla^2\rho)$	-	0.0168 (0.0648)	-	-
U (G)	-	−0.0121 (0.0141)	-	-
−G/U	-	1.165	-	-
C–H^d···H^f–C				
$\rho\ (\nabla^2\rho)$	-	0.0130 (0.0570)	-	-
U (G)	-	−0.0074 (0.0096)	-	-
−G/U	-	1.297	-	-

* All values are given in electronic units (e.u.).

Regarding the interaction strength, the −G/U ratio is a qualitative manner to express the appearing of covalent effect [32]. According to values of 0.733 and 0.738 gathered in Table 3, the C–H^c···O^b=N intramolecular hydrogen bonds behave with a trend of covalence in **1K** and **1W**, although the electronic density values are not the largest, and the bond lengths are not the shortedones, as demonstrated in Figure 7. This can be stated once these results are smaller than 1.0 [32,58]. In regards to the remaining hydrogen bonds, all of them are non-covalent. However, besides the hydrogen bonds, **1T** also reveals the formation of a dihydrogen bond [59–61], namely C–H^d···H^f–C, wherein by means of the values of ρ and ∇^2ρ as well as −G/U, this interaction is also non-covalent.

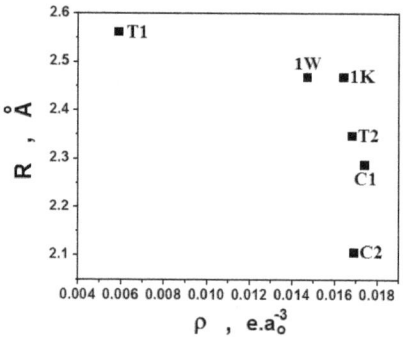

Figure 7. Relationship between the distance values of intramolecular hydrogen bonds and QTAIM electronic density amounts.

4. Conclusions

In this work, we observed that intramolecular hydrogen bonds can control the conformations of benznidazole. The structural parameters embodied as bond lengths and intermolecular distances revealed unsystematic tendencies in comparison with the stabilization criteria. Also, the new vibration modes of the intramolecular hydrogen bonds could not be unveiled, and in addition, the difficulties in identifying the frequency shifts of the proton donors prevent us from correlating this parameter with the stabilization criteria. The QTAIM calculations were able to characterize classical and non-classical intramolecular hydrogen bonds by means of the values of electronic density and Laplacian. Also, the appearance of a partial covalent character was observed in the C–H^c···O^b=N links of **1K** and **1W**. Approaches considering solvent or gas phase effect can lead to deviations of the crystal structure, but they are all *s-cys* and not *s-trans*. This must be taken into account in studies of the structure-activity relationships of benznidazole.

Acknowledgments: The authors thank "Coordenação de Aperfeiçoamento de Pessoal de Nível Superior" (CAPES) and "Conselho Nacional de Desenvolvimento Científico e Tecnológico" (CNPq) for financial support.

Author Contributions: Mário L. A. A. Vasconcellos and Edilson B. Alencar Filho conceived and designed the experiments; Boaz G. Oliveira, Mário L. A. A. Vasconcellos and Edilson B. Alencar Filho performed the experiments; Boaz G. Oliveira, Mário L. A. A. Vasconcellos and Edilson B. Alencar Filho analyzed the data; Boaz G. Oliveira, Mário L. A. A. Vasconcellos and Edilson B. Alencar Filho contributed reagents/materials/analysis tools; Boaz G. Oliveira, Mário L. A. A. Vasconcellos and Edilson B. Alencar Filho wrote the paper. Authorship must be limited to those who have contributed substantially to the work reported.

Conflicts of Interest: The authors declare no conflict of interest.

References

1. Thomas, G. *Medicinal Chemistry: An Introduction*; John Wiley and Sons: Chichester, UK, 2000.
2. Pitta, M.G.R.; Silva, A.C.A.; Neves, J.K.A.L.; Silva, P.G.; Irmão, J.I.; Malagueño, E.; Santana, J.V.; Lima, M.C.A.; Galdino, S.L.; Pitta, I.R.; *et al*. New imidazolidinic bioisosters: Potencial candidates for antischistosomal drugs. *Mem. Inst. Oswaldo Cruz* **2006**, *101*, 313–316.
3. Goodell, J.R.; Ougolkov, A.V.; Hiasa, H.; Kaur, H.; Remmel, R.; Billadeau, D.D.; Ferguson, D.M. Acridine-based agents with *Topoisomerase* II activity inhibit pancreatic cancer cell proliferation and induce apoptosis. *J. Med. Chem.* **2008**, *51*, 179–182.
4. World Health Organization. Available online: http://www.who.int/whr/2004/en/ (accessed on 14 September 2009).
5. Tarleton, R.L.; Reithinger, R.; Urbina, J.A.; Kitron, U.; Gurtler, R.E. The challenges of Chagas disease—Grim outlook or glimmer of hope? *PLoS Med.* **2007**, *4*, 1852–1957.
6. Tonin, L.T.D.; Barbosa, V.A.; Bocca, C.C.; Ramos, E.R.F.; Nakamura, C.V.; Costa, W.F.; Basso, E.A.; Nakamura, T.U.; Sarragiotto, M.H. Comparative study of the trypanocidal activity of the methyl 1-nitrophenyl-1,2,3,4-9H-tetrahydro-β-carboline-3-carboxylate derivatives and benznidazole using theoretical calculations and cyclic voltammetry. *Eur. J. Med. Chem.* **2008**, *44*, 1745–1750.
7. Soares-Sobrinho, J.L.; Cunha-Filho, M.S.S.; Rolim-Neto, P.J.; Torres-Labandeira, J.J.; Dacunha-Marinho, B. Benznidazole. *Acta Crystallogr. E* **2008**, *64*, o634.
8. Höltje, H.-H.; Sippl, W.; Rognan, D.; Folkers, G. *Molecular Modeling: Basic Principles and Applications*; Wiley-VHC: Weinhelm, Germany, 2003.
9. Filho, E.B.A.; Ventura, E.; do Monte, A.S.; Oliveira, B.G.; Junior, C.G.L.; Rocha, G.B.; Vasconcellos, M.L.A.A. Synthesis and conformational study of a new class of highly bioactive compounds. *Chem. Phys. Lett.* **2007**, *449*, 336–340.
10. Santos Filho, J.M.; Leite, A.C.L.; Oliveira, B.G.; Moreira, D.R.M.; Lima, M.S.; Soares, M.B.P.; Leite, L.F.C.C. Design, synthesis and *cruzain* docking of 3-(4-substituted-aryl)-1,2,4-oxadiazole-*N*-acylhydrazones as anti-*Trypanosoma cruzi* agents. *Bioorg. Med. Chem.* **2009**, *17*, 6682–6691.

11. Oliveira, B.G.; Araujo, R.C.M.U.; Carvalho, A.B.; Ramos, M.N. DFT calculations on the cyclic ethers hydrogen-bonded complexes: Molecular parameters and the non-linearity of the hydrogen bond. *Spectrochim. Acta A* **2007**, *68*, 626–631.
12. Oliveira, B.G.; Leite, L.F.C.C. A quantum chemical study of red-shift and blue-shift hydrogen bonds in bimolecular and trimolecular methylhydrazine-hydrate complexes. *J. Mol. Struct. THEOCHEM* **2009**, *915*, 38–42.
13. Oliveira, B.G.; Araujo, R.C.M.U.; Carvalho, A.B.; Ramos, M.N. A chemometrical study of intermolecular properties of hydrogen-bonded complexes formed by $C_2H_4O\cdots HX$ and $C_2H_5N\cdots HX$ with X = F, CN, NC, and CCH. *J. Mol. Model.* **2009**, *15*, 421–432.
14. Oliveira, B.G.; Araujo, R.C.M.U. Theoretical aspects of binary and ternary complexes of aziridine$\cdots$ammonia ruled by hydrogen bond strength. *J. Mol. Model.* **2012**, *18*, 2845–2854.
15. Oliveira, B.G.; Araújo, R.C.M.U.; Ramos, M.N. Evidence for blue-shifting and red-shifting effects in the $C_2H_4\cdots HCF_3$, $C_2H_3(CH_3)\cdots HCF_3$ and $C_2H_2(CH_3)_2\cdots HCF_3$ complexes: π and improper-π hydrogen bonds. *J. Mol. Struct. THEOCHEM* **2010**, *944*, 168–172.
16. Oliveira, B.G.; Zabardasti, A.; Goudarziafshar, H.; Salehnassaj, M. The electronic mechanism ruling the dihydrogen bonds and halogen bonds in weakly bound systems of $H_3SiH\cdots HOX$ and $H_3SiH\cdots XOH$ (X = F, Cl, and Br). *J. Mol. Model.* **2015**, *21*, 77–87.
17. Bader, R.F.W. *Atoms in Molecules: A Quantum Theory*; Oxford University Press: Oxford, UK, 1990.
18. Cortés-Guzmán, F.; Bader, R.F.W. Complementarity of QTAIM and MO theory in the study of bonding in donor-acceptor complexes. *Coord. Chem. Rev.* **2005**, *249*, 633–662.
19. Bader, R.F.W. A quantum theory of molecular structure and its applications. *Chem. Rev.* **1991**, *91*, 893–928.
20. Šafář, P.; Žúžiová, J.; Marchalín, Š.; Prónayová, N.; Švorc, Ľ.; Vrábel, V.; Šesták, S.; Rendić, D.; Tognetti, V.; Joubert, L.; et al. Combined chemical, biological and theoretical DFT-QTAIM study of potent glycosidase inhibitors based on quaternary indolizinium salts. *Eur. J. Org. Chem.* **2012**, *2012*, 5498–5514.
21. LaPointe, S.M.; Farrag, S.; Bohrquez, H.J.; Boyd, R.J. QTAIM study of an alpha-helix hydrogen bond network. *J. Phys. Chem. B* **2009**, *113*, 10957–10964.
22. Oliveira, B.G.; Lima, M.C.A.; Pitta, I.R.; Galdino, S.L.; Hernandes, M.Z. A theoretical study of red-shifting and blue-shifting hydrogen bonds occurring between imidazolidine derivatives and PEG/PVP polymers. *J. Mol. Model.* **2010**, *16*, 119–127.
23. Rozas, I.; Alkorta, I.; Elguero, J. Intramolecular hydrogen bonds in ortho-substituted hydroxybenzenes and in 8-susbtituted 1-hydroxynaphthalenes: Can a methyl group be an acceptor of hydrogen bonds? *J. Phys. Chem. A* **2001**, *105*, 10462–10467.
24. Zevallos, J.; Toro-Labbé, A.; Mó, O.; Yáñez, M. The role of intramolecular hydrogen bonds *versus* other weak interactions on the conformation of hyponitrous acid and its mono- and dithio-derivatives. *Struct. Chem.* **2005**, *16*, 295–303.
25. Grabowski, S.J.; Małecka, M. Intramolecular H-Bonds: DFT and QTAIM studies on 3-(aaminomethylene)pyran-2,4-dione and its derivatives. *J. Phys. Chem. A* **2006**, *110*, 11847–11854.

26. Deshmukh, M.M.; Gadre, S.R.; Bartolotti, L.J. Estimation of intramolecular hydrogen bond energy via molecular tailoring approach. *J. Phys. Chem. A* **2006**, *110*, 12519–12523.
27. Deshmukh, M.M.; Suresh, C.H.; Gadre, S.R. Intramolecular hydrogen bond energy in polyhydroxy systems: A critical comparison of molecular tailoring and isodesmic approaches. *J. Phys. Chem. A* **2007**, *111*, 6472–6480.
28. Sørensen, J.; Clausen, H.F.; Poulsen, R.D.; Overgaard, J.; Schiøtt, B. Short strong hydrogen bonds in 2-acetyl-1,8-dihydroxy-3,6-dimethylnaphthalene: An outlier to current hydrogen bonding theory? *J. Phys. Chem. A* **2007**, *111*, 345–351.
29. Deshmukh, M.M.; Gadre, S.R. Estimation of $N-H\cdots O=C$ intramolecular hydrogen bond energy in polypeptides. *J. Phys. Chem. A* **2009**, *113*, 7927–7932.
30. Fuster, F.; Grabowski, S.J. Intramolecular hydrogen bonds: The QTAIM and ELF characteristics. *J. Phys. Chem. A* **2011**, *115*, 10078–10086.
31. Deshmukh, M.M.; Bartolotti, L.J.; Gadre, S.R. Intramolecular hydrogen bond energy and cooperative interactions in α-, β-, and γ-cyclodextrin conformers. *J. Comput. Chem.* **2011**, *32*, 2996–3004.
32. Oliveira, B.G. Structure, energy, vibrational spectrum, and Bader's analysis of $\pi\cdots H$ hydrogen bonds and $H^{-\delta}\cdots H^{+\delta}$ dihydrogen bonds. *Phys. Chem. Chem. Phys.* **2013**, *15*, 37–79.
33. Reichardt, C. *Solvents and Solvent Effects in Organic Chemistry*; Wiley-VCH: Weinheim, Germany, 1988.
34. Tomasi, J.; Persico, M. Molecular interactions in solution: An overview of methods based on continuous distributions of the solvent. *Chem. Rev.* **1994**, *94*, 2027–2094.
35. Yasuda, T.; Ikawa, S.-I. On the dielectric continuum solvent model for theoretical estimates of the conformational equilibrium of molecules with an intramolecular hydrogen bond. *Chem. Phys.* **1998**, *238*, 173–178.
36. Abkowicz-Bieńko, A.; Biczysko, M.; Latajka, Z. Solvent effect on hydrogen bonded ammonia-hydrogen halide complexes: Continuum medium *versus* cluster models. *Comput. Chem.* **2000**, *24*, 303–309.
37. Oliveira, B.G.; Araújo, R.C.M.U.; Carvalho, A.B.; Ramos, M.N.; Hernandes, M.Z.; Cavalcante, K.R. A theoretical study of the solvent effects in ethylene oxide: Hydrofluoric acid complex using continuum and new discrete models. *J. Mol. Struct. THEOCHEM* **2007**, *802*, 91–97.
38. López-Vallejo, F.; Medina-Franco, J.L.; Hernández-Campos, A.; Rodríguez-Morales, S.; Yépez, L.; Cedillo, R.; Castillo, R. Molecular modeling of some 1H-benzimidazole derivatives with biological activity against Entamoeba histolytica: A comparative molecular field analysis study. *Bioorg. Med. Chem.* **2007**, *15*, 1117–1126.
39. Lee, C.T.; Yang, W.T.; Parr, R.G. Development of the Colle-Salvetti correlation-energy formula into a functional of the electron density. *Phys. Rev. B* **1988**, *37*, 785–789.
40. Becke, A.D. A new mixing of Hartree-Fock and local density-functional theories. *J. Chem. Phys.* **1933**, *98*, 1372–1377.

41. Frisch, M.J.; Trucks, G.W.; Schlegel, H.B.; Scuseria, G.E.; Robb, M.A.; Cheeseman, J.R.; Montgomery, J.A., Jr.; Vreven, T.; Kudin, K.N.; Burant, J.C.; et al. *Gaussian 03 Revision D.02*; Gaussian Inc.: Wallingford, CT, USA, 2004.
42. Biegler-König, F. *AIM 2000 1.0*; University of Applied Sciences: Bielefeld, Germany, 2000.
43. Wang, W.; Pu, X.; Zheng, W.; Wong, N.-B.; Tian, A. Hyperconjugation *versus* intramolecular hydrogen bond: Origin of the conformational preference of gaseous glycine. *Chem. Phys. Lett.* **2003**, *370*, 147–153.
44. Shagidullin, R.R.; Chernova, A.V.; Shagidullin, R.R. Intramolecular hydrogen bonds and conformations of the 1,4-butanediol molecule. *Rus. Chem. Bull.* **1993**, *42*, 1505–1510.
45. Lee, H.-J.; Jung, H.-J.; Kim, J.K.; Park, H.-M.; Lee, K.-B. Conformational preference of azaglycine-containing dipeptides studied by PCM and IPCM methods. *Chem. Phys.* **2003**, *294*, 201–210.
46. Hopkins, W.S.; Hasan, M.; Burt, M.; Marta, R.A.; Fillion, E.; McMahon, T.B. Persistent intramolecular C–H···X (X = O or S) hydrogen-bonding in benzyl meldrum's acid derivatives. *J. Phys. Chem. A* **2014**, *118*, 3795–3803.
47. Alcântara, A.F.C.; Teixeira, A.F.; Silva, I.F.; Almeida, W.B.; Piló-Veloso, D. NMR Investigation and theoretical calculations of the effect of solvent on the conformational analysis of 4′,7-di-hydroxy-8-prenylflavan. *Quim. Nova* **2004**, *27*, 371–377.
48. Tsuzuki, S.; Honda, K.; Uchimaru, T.; Mikami, M.; Tanabe, K. The magnitude of the CH/π interaction between benzene and some model hydrocarbons. *J. Am. Chem. Soc.* **2000**, *122*, 3746–3753.
49. Takahashi, O.; Kohno, Y.; Gondoh, Y.; Saito, K.; Nishio, M. General preference for alkyl/phenyl folded conformations. Relevance of the CH/pi and CH/O interactions to stereochemistry as evidenced by *ab Initio* MO calculations. *Bull. Chem. Soc. Jpn.* **2003**, *76*, 369–374.
50. Lithoxoidou, A.T.; Bakalbassis, E.G. PCM study of the solvent and substituent effects on the conformers, intramolecular hydrogen bonds and bond dissociation enthalpies of 2-substituted phenols. *J. Phys. Chem. A* **2005**, *109*, 366–377.
51. Oliveira, B.G.; Araújo, R.C.M.U. Bonding topology, hydrogen bond strength, and vibrational chemical shifts on hetero-ring hydrogen-bonded complexes—Theoretical insights revisited. *Can. J. Chem.* **2012**, *90*, 368–375.
52. Oliveira, B.G.; Araújo, R.C.M.U.; Carvalho, A.B.; Ramos, M.N. The molecular properties of heterocyclic and homocyclic hydrogen-bonded complexes evaluated by DFT calculations and AIM densities. *J. Mol. Model.* **2009**, *15*, 123–131.
53. Oliveira, B.G.; Araújo, R.C.M.U.; Chagas, F.F.; Carvalho, A.B.; Ramos, M.N. The electronic structure of the C_2H_4O···2HF tri-molecular heterocyclic hydrogen-bonded complex: A theoretical study. *J. Mol. Model.* **2008**, *14*, 949–955.
54. Gilli, P.; Bertolasi, V.; Ferretti, V.; Gilli, G. Evidence for resonance-assisted hydrogen bonding. 4. Covalent nature of the strong homonuclear hydrogen bond. Study of the O–H–O system by crystal structure correlation methods. *J. Am. Chem. Soc.* **1994**, *116*, 909–915.

55. Małecka, M. Intramolecular N–H⋯O resonance-assisted hydrogen bonds in crystal structures of oxaphosphinanes and chromones—DFT calculations and AIM analysis. *J. Mol. Struct.* **2007**, *831*, 135–143.
56. Sanz, P.; Mó, O.; Yáñez, M.; Elguero, J. Bonding in tropolone, 2-aminotropone, and aminotroponimine: No evidence of resonance-assisted hydrogen-bond effects. *Chemistry* **2008**, *14*, 4225–4232.
57. Trujillo, C.; Sánchez-Sanz, G.; Alkorta, I.; Elguero, J.; Mó, O.; Yáñez, M. Resonance assisted hydrogen bonds in open-chain and cyclic structures of malonaldehyde enol: A theoretical study. *J. Mol. Struct.* **2013**, *1048*, 138–151.
58. Grabowski, S.J. What is the covalency of hydrogen bonding? *Chem. Rev.* **2011**, *111*, 2597–2625.
59. Oliveira, B.G.; Araújo, R.C.M.U.; Silva, J.J.; Ramos, M.N. A theoretical study of three and four proton donors on linear HX⋯BeH$_2$⋯HX and bifurcate BeH$_2$⋯2HX trimolecular dihydrogen-bonded complexes with X = CN and NC. *Struct. Chem.* **2010**, *21*, 221–228.
60. Oliveira, B.G.; Vasconcellos, M.L.A.A.A. B3LYP and QTAIM study of a new proton donor for dihydrogen bonds: The case of the $C_2H_5^+$⋯nBeH$_2$ complexes (n = 1 or 2). *Struct. Chem.* **2010**, *20*, 897–902.
61. Oliveira, B.G. Interplay between dihydrogen and alkali–halogen bonds: Is there some covalency upon complexation of ternary systems? *Comput. Theor. Chem.* **2012**, *998*, 173–182.

$H_2XP:OH_2$ Complexes: Hydrogen vs. Pnicogen Bonds

Ibon Alkorta, Janet E. Del Bene and Jose Elguero

Abstract: A search of the Cambridge Structural Database (CSD) was carried out for phosphine-water and arsine-water complexes in which water is either the proton donor in hydrogen-bonded complexes, or the electron-pair donor in pnicogen-bonded complexes. The range of experimental P-O distances in the phosphine complexes is consistent with the results of *ab initio* MP2/aug'-cc-pVTZ calculations carried out on complexes $H_2XP:OH_2$, for X = NC, F, Cl, CN, OH, CCH, H, and CH_3. Only hydrogen-bonded complexes are found on the $H_2(CH_3)P:HOH$ and $H_3P:HOH$ potential surfaces, while only pnicogen-bonded complexes exist on $H_2(NC)P:OH_2$, $H_2FP:OH_2$, $H_2(CN)P:OH_2$, and $H_2(OH)P:OH_2$ surfaces. Both hydrogen-bonded and pnicogen-bonded complexes are found on the $H_2ClP:OH_2$ and $H_2(CCH)P:OH_2$ surfaces, with the pnicogen-bonded complexes more stable than the corresponding hydrogen-bonded complexes. The more electronegative substituents prefer to form pnicogen-bonded complexes, while the more electropositive substituents form hydrogen-bonded complexes. The $H_2XP:OH_2$ complexes are characterized in terms of their structures, binding energies, charge-transfer energies, and spin-spin coupling constants $^{2h}J(O-P)$, $^{1h}J(H-P)$, and $^1J(O-H)$ across hydrogen bonds, and $^{1p}J(P-O)$ across pnicogen bonds.

Reprinted from *Crystals*. Cite as: Alkorta, I.; Del Bene, J.E..; Elguero, J. $H_2XP:OH_2$ Complexes: Hydrogen vs. Pnicogen Bonds. *Crystals* **2016**, *6*, 19.

---

1. Introduction

Chloroform, dichloromethane, and water have been observed as solvent molecules in X-ray structures of crystals [1–6]. Such structures have long been used as a tool for identifying and confirming the presence of weak intermolecular interactions. The most prevalent intermolecular interaction in the Cambridge Structural Database (CSD) is the X-H ··· Y hydrogen bond, which has been at the forefront of intermolecular interactions since Pimentel's book "The Hydrogen Bond" [7]. The hydrogen bond is defined as an attractive interaction between a hydrogen atom from a molecule or a molecular fragment X–H in which X is more electronegative than H, and an atom or a group of atoms in the same or a different molecule, in which there is evidence of bond formation [8,9]. Of particular interest are the hydrogen bonds in the X-ray structures of organic hydrates. Hydrogen bonds involving water molecules interacting with different chemical groups have been identified and classified [3–6].

A relatively new intermolecular interaction, the pnicogen bond, was initially detected in the crystal structures of 1,2-dicarba-*closo*-dodecaboranes [10,11] and aminoalkyl-ferrocenylphosphanes [12]. The structures of two of these complexes are illustrated in Figure 1. A large number of intermolecular and intramolecular pnicogen interactions have also been observed in the solid phase [13–15]. Pnicogen bonds were first described theoretically for model complexes [16,17], and subsequent studies confirmed the stabilizing nature of pnicogen interactions [18–20]. The pnicogen bond is a Lewis acid-Lewis base interaction in which the Lewis acid is a group 15 element (N, P, As, or Sb) acting as an electron-pair acceptor.

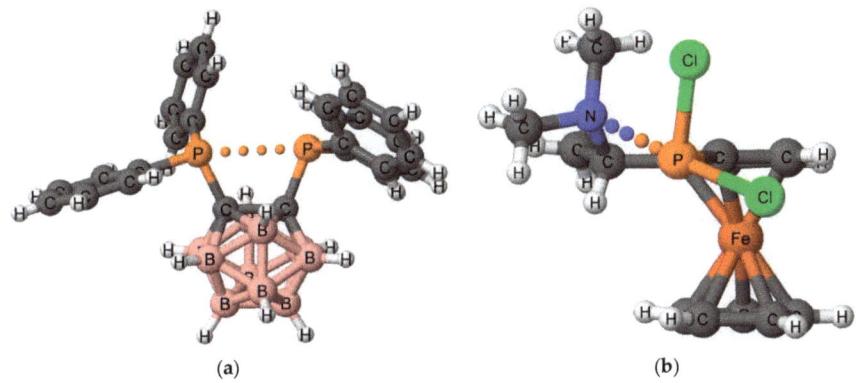

Figure 1. X-ray structure of Cambridge Structural Database (CSD) Refcodes (**a**) XEBBEM01 and (**b**) QEZDOP. The pnicogen bond interaction is indicated with dots.

In the present article, we present the results of our search of the CSD for phosphine-water and arsine-water complexes in which water is either the proton donor in hydrogen-bonded complexes, or the electron-pair donor in pnicogen-bonded complexes. We also report the results of *ab initio* calculations on a series of complexes $H_2XP:OH_2$, for X = NC, F, Cl, CN, OH, CCH, H, and CH_3, stabilized by either hydrogen bonds or pnicogen bonds. We present and discuss the structures, binding energies, and charge-transfer energies of these complexes, as well as equation-of-motion coupled cluster singles and doubles (EOM-CCSD) spin–spin coupling constants across hydrogen bonds and pnicogen bonds.

2. Methods

2.1. Cambridge Structural Database Search

The Cambridge Structural Database [21] version 5.36 with updates from November 2014, February 2015, and May 2015 was searched for complexes that contain P(III) and As(III) with water molecules. Included structures have a distance of 2.0 to 4.0 Å between the pnicogen atom and the oxygen atom of water.

2.2. Ab Initio *Calculations*

The structures of the isolated monomers and the binary complexes $H_2XP:OH_2$ were optimized at second-order Møller-Plesset perturbation theory (MP2) [22–25] with the aug'-cc-pVTZ basis set [26]. This basis set is derived from the Dunning aug-cc-pVTZ basis set [27,28] by removing diffuse functions from H atoms. Frequencies were computed to establish that the optimized structures correspond to equilibrium structures on their potential surfaces. Optimization and frequency calculations were performed using the Gaussian 09 program [29]. The binding energies (ΔE) of all complexes have been calculated as the total energy of the complex minus the sum of the total energies of the corresponding isolated monomers.

The electron densities of complexes have been analyzed using the Atoms in Molecules (AIM) methodology [30–33] employing the AIMAll [34] program. The topological analysis of the electron density produces the molecular graph of each complex. This graph identifies the location of electron density features of interest, including the electron density (ρ) maxima associated with the various nuclei, and saddle points which correspond to bond critical points (BCPs). The zero gradient line which connects a BCP with two nuclei is the bond path.

The Natural Bond Orbital (NBO) [35] method has been used to analyze the stabilizing charge-transfer interactions using the NBO6 program [36]. Since MP2 orbitals are nonexistent, the charge-transfer interactions have been computed using the B3LYP functional [37,38] with the aug'-cc-pVTZ basis set at the MP2/aug'-cc-pVTZ complex geometries, so that at least some electron correlation effects could be included.

Spin-spin coupling constants were evaluated using the EOM-CCSD method in the CI (configuration interaction)-like approximation [39,40], with all electrons correlated. For these calculations, the Ahlrichs [41] qzp basis set was placed on ^{13}C, ^{15}N, ^{17}O, and ^{19}F, and the qz2p basis set on ^{31}P, ^{35}Cl, and hydrogen-bonded ^{1}H atoms. The Dunning cc-pVDZ basis set was placed on all other ^{1}H atoms. All terms, namely, the paramagnetic spin-orbit (PSO), diamagnetic spin orbit (DSO), Fermi contact (FC), and spin dipole (SD), have been evaluated. The EOM-CCSD calculations were performed using ACES II [42] on the IBM Cluster 1350 (Glenn) at the Ohio Supercomputer Center.

3. Results and Discussion

3.1. CSD Search

The CSD search found only three water-phosphine complexes and seven water-arsine complexes that have a distance of 2.0 to 4.0 Å between the pnicogen atom and the oxygen atom of water. Two of the water–phosphine complexes are hydrogen bonded (CSD Refcodes: AGAHIB and BEZTOR) with P···H distances of 2.48 and 2.72 Å, and P···O distances of 3.315 and 3.465 Å, respectively. The third structure (NOPYEX) corresponds to a pnicogen-bonded complex between a triphenylphosphine derivative and water with a longer P···O distance of 3.76 Å. Four of the water-arsine complexes (TELFAR, NOPYAT, FUTDUU, and IBAKIH) are pnicogen-bonded, and three (NIVWAQ, FUTDUU, and HAVKEW) are hydrogen-bonded. The pnicogen-bonded complex TELFAR which is illustrated in Figure 2 has a short As-O pnicogen bond distance of 2.56 Å. The O-As-O angle is 171°, which allows for the interaction of the O of water with the σ-hole of As. The short distance for this bond suggests that it has significant covalent character. The As···O distances in the other pnicogen-bonded complexes range between 3.61 and 3.78 Å. The hydrogen-bonded complexes have As···H distances between 2.81 and 3.22 Å, and As···O distances between 3.70 and 3.96 Å.

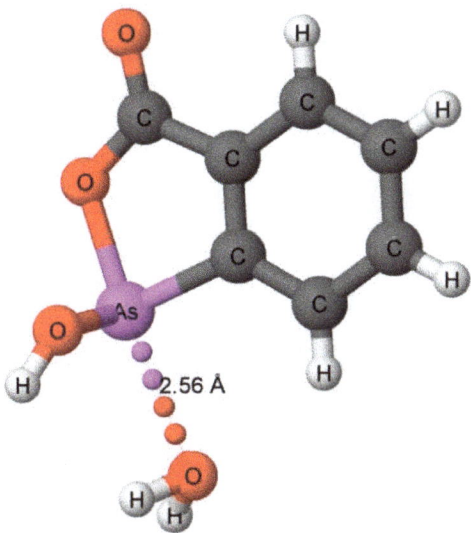

Figure 2. The X-ray structure of CSD Refcode: TELFAR showing the As···O pnicogen bond interaction.

3.2. Computational Results

We have attempted to optimize eight complexes H_2XP:HOH with O-H $\cdots$ P hydrogen bonds, and eight complexes with P $\cdots$ O pnicogen bonds, with X = NC, F, Cl, CN, OH, CCH, H, and CH_3. However, only four hydrogen-bonded (HB) and six pnicogen-bonded (ZB) equilibrium complexes have been found on the potential surfaces. The structures, total energies, and molecular graphs of these complexes are reported in Table S1 of the Supporting Information, and their binding energies are given in Table 1. ZB complexes have binding energies which vary from −11.3 to −21.1 kJ·mol^{-1}, while HB complexes have binding energies between −9.4 and −15.4 kJ·mol^{-1}. The absolute values of the binding energies of the ZB complexes decrease in the order

$$NC > F > Cl > CN > OH > CCH$$

while those of the HB complexes decrease in the reverse order

$$CH_3 > H > CCH > Cl.$$

It is apparent from Table 1 that the more electronegative substituents prefer to form pnicogen-bonded complexes, while the more electropositive substituents form hydrogen-bonded complexes. This trend follows the general trend of the Molecular Electrostatic Potential (MEP) values around the phosphorous atom [43]. H_2ClP and H(CCH)P form both hydrogen-bonded and pnicogen-bonded complexes with H_2O, with the latter more stable by 7.9 and 0.5 kJ·mol^{-1}, respectively.

Table 1. Binding energies of equilibrium pnicogen-bonded (ZB) and hydrogen-bonded (HB) complexes of H_2XP with H_2O.

H_2XP, X =	Binding Energies (ΔE, kJ·mol^{-1})	
	ZB	HB
NC	−21.1	
F	−19.5	
Cl	−17.3	−9.4
CN	−16.9	
OH	−12.8	
CCH	−11.3	−10.8
H		−11.1
CH_3		−15.4

3.2.1. Hydrogen-Bonded Complexes

Only four equilibrium hydrogen-bonded complexes have been found on the H_2XP:HOH surfaces; namely, H_2ClP:HOH, H_2(CCH)P:HOH, H_3P:HOH, and

$H_2(CH_3)P:HOH$. The binding energies of these are given in Table 1, and selected structural parameters are reported in Table 2. The shortest intermolecular $O \cdots P$ and $H \cdots P$ distances are 3.360 and 2.536 Å, respectively, in $H_2(CH_3)P:HOH$. The remaining three complexes have $O \cdots P$ distances between 3.55 and 3.59 Å, and $H \cdots P$ distances between 2.62 and 2.66 Å. These distances are consistent with the experimental distances from the CSD. As evident from Table 2, the hydrogen bonds in $H_2(CCH)P:HOH$ and $H_3P:HOH$ are close to linear, with H-O-P angles of 7°, while these bonds in $H_2ClP:HOH$ and $H_2(CH_3)P:HOH$ are nonlinear with values of 19 and 27°, respectively. Moreover, the H_2O molecule is positioned similarly in $H_2ClP:HOH$ and $H_2(CCH)P:HOH$, but has a different orientation in $H_3P:HOH$ and $H_2(CH_3)P:HOH$, as illustrated in Figure 3. In these two complexes, but particularly $H_2(CH_3)P:HOH$, there appears to be an attractive interaction between the water oxygen and the hydrogens of the substituent. All of these differences lead to a lack of correlation between the binding energies of these complexes and the $O \cdots P$ distances. However, the electrostatic minimum associated with the lone pair of the phosphorous atom [43] and the $H \cdots P$ distances do correlate. Although not included in Tables 1 and 2 there is an equilibrium hydrogen-bonded complex formed between $H_2(OH)P$ and H_2O with a binding energy of -28.5 kJ·mol^{-1}. However, it is stabilized by an $O-H \cdots O$ hydrogen bond with the substituent O-H as the proton donor to H_2O.

Table 2. $O \cdots P$, $H \cdots P$, and O-H distances, and H-O-P angles for hydrogen-bonded complexes $H_2XP:HOH$ with C_s symmetry.

H_2XP, X =	Distance (R, Å)			Angles (<, °)
	R(O$\cdots$P)	R(H$\cdots$P)	R(O-H) [a]	<H-O-P
Cl	3.553	2.656	0.965	19
CCH	3.585	2.629	0.966	7
H	3.575	2.620	0.966	7
CH$_3$	3.360	2.536	0.967	27

(a) The O-H distance in isolated H_2O is 0.961 Å.

Table 3 presents the stabilizing NBO $P_{lp} \rightarrow \sigma^*$H-O charge-transfer energies for the four hydrogen-bonded complexes. These energies range from 9 kJ·mol^{-1} for $H_2ClP:HOH$ to between 14 and 15 kJ·mol^{-1} for the remaining complexes. The charge-transfer energies do not correlate with either the $O \cdots P$ or the $H \cdots P$ distances.

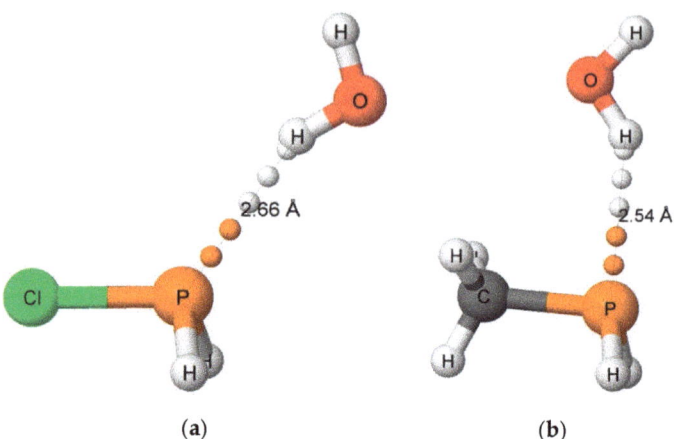

Figure 3. The hydrogen-bonded complexes $H_2ClP:HOH$ (a) and $H_2(CH_3)P:HOH$ (b). The orientation of the H_2O molecule in $H_2(CCH)P:HOH$ is similar to that in $H_2ClP:HOH$, while the orientation of H_2O in $H_3P:HOH$ is similar to that in $H_2(CH_3)P:HOH$.

Table 3. Charge-transfer energies ($P_{lp} \rightarrow \sigma^*H-O$) and coupling constants $^{2h}J(O-P)$, $^{1h}J(H-P)$, and $^1J(O-H)$ for hydrogen-bonded complexes $H_2XP:HOH$.

H_2XP, X =	Charge-Transfer Energies (kJ·mol^{-1})	Coupling Constants (Hz)		
	$P_{lp} \rightarrow \sigma^*H-O$ [a]	$^{2h}J(O-P)$	$^{1h}J(H-P)$	$^1J(O-H)$ [b]
Cl	9.0	−18.2	−12.9	−78.0
CCH	13.8	−14.4	−13.0	−77.9
H	14.0	−14.0	−13.5	−78.1
CH$_3$	15.0	−24.1	−15.0	−78.4

[a] The $O_{lp} \rightarrow \sigma^*H-O$ charge-transfer energy in the complex of $H_2(OH)P$ with H_2O that has the substituent O-H as the proton donor is 41.0 kJ·mol^{-1}; [b] $^1J(O-H)$ in isolated H_2O is −77.0 Hz.

The two-bond coupling constant $^{2h}J(O-P)$ and the one-bond coupling constant $^{1h}J(H-P)$ across the hydrogen bonds, and the one-bond coupling constant $^1J(O-H)$ for the hydrogen-bonded O-H group are also reported in Table 3. The components of these coupling constants are reported in Table S2 of the Supporting Information. Values of $^{2h}J(O-P)$ vary from −14 to −24 Hz, while values of $^{1h}J(H-P)$ lie between −13 and −15 Hz. The dependence of these two coupling constants on the corresponding O···P and H···P distances is shown graphically in Figure 4. Since there are only 4 points in each set and at least two of them have similar values of the coupling constant and the corresponding distance, only linear trendlines were used to illustrate the distance dependence. The correlation coefficients of these trendlines are 0.915 for $^{2h}J(O-P)$ and 0.973 for $^{1h}J(H-P)$. The third coupling constant, $^1J(O-H)$ has a value of −77.0 Hz in the isolated H_2O molecule, and increases in absolute value only slightly upon complex formation to between −77.9 and −78.4 Hz. The O-H distance of 0.961 Å in the monomer also increases only slightly upon complexation, with values between 0.965 and 0.967 Å. The one-bond coupling constant $^1J(O-H)$ for the non-hydrogen-bonded O-H decreases to about −62 Hz.

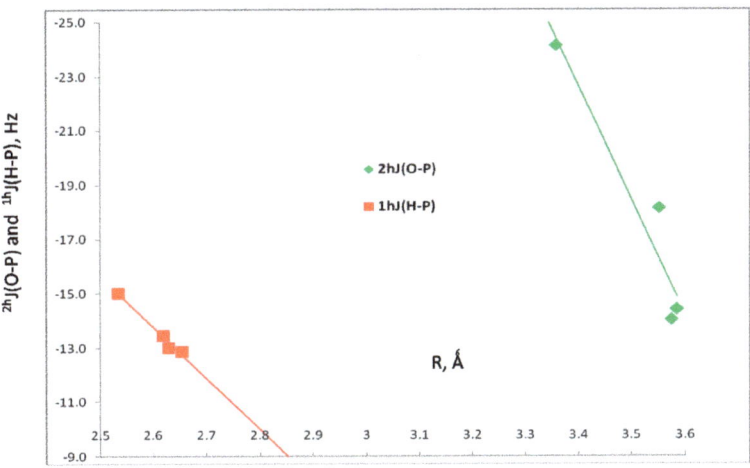

Figure 4. $^{2h}J(O-P)$ *versus* the O···P distance, and $^{1h}J(H-P)$ *versus* the H···P distance for HB complexes $H_2XP:HOH$.

3.2.2. Pnicogen-Bonded Complexes

Three of the six equilibrium pnicogen-bonded $H_2XP:OH_2$ complexes have C_s symmetry with the H_2O molecule in the symmetry plane, and three have C_1 symmetry. For reasons of computational efficiency, particularly for the coupling constant calculations, we have re-optimized the C_1 structures under the constraint of C_s symmetry with an in-plane H_2O molecule. The binding energies of C_1 and C_s structures are compared in Table 4. The C_s structures of $H_2FP:OH_2$, $H_2ClP:OH_2$, and $H_2(OH)P:OH_2$ with H_2O in the C_s symmetry plane are only 0.1 to 0.4 kJ·mol^{-1} less stable than the C_1 equilibrium structures. Moreover, the ordering of complexes according to decreasing binding energy is the same for the equilibrium structures and those with C_s symmetry, and the structures of C_1 and corresponding C_s complexes are very similar. To ensure that there are no other pnicogen-bonded complexes with P···O pnicogen bonds, we also optimized a set of these complexes with C_s symmetry in which the H_2O molecule does not lie in the symmetry plane. All of these complexes have one imaginary frequency, and smaller absolute values of the binding energies than the complexes with in-plane H_2O molecules, as evident from Table 4. The two types of C_s complexes are illustrated in Figure 5.

The structures, total energies, and molecular graphs of the three pnicogen-bonded ZB complexes $H_2XP:OH_2$ with C_s symmetry, one imaginary frequency, and in-plane H_2O molecules are reported in Table S3 of the Supporting Information. This table also provides these data for the six less-stable ZB complexes with C_s symmetry, one imaginary frequency, and out-of-plane H_2O molecules. Table 5 presents selected data for the more stable C_s complexes with in-plane H_2O molecules. The P$\cdots$O distances in these complexes range from 2.755 Å in $H_2FP:OH_2$ to 3.036 Å in $H_2(CCH)P:OH_2$. Their binding energies do not correlate well with the P$\cdots$O distances, as indicated by correlation coefficients of 0.7 for linear, quadratic, and exponential trendlines. The O-P-A angles, with A the atom of X directly bonded to P, are also reported in Table 5. These angles vary between 161 and 168°, indicating that the O-P-A arrangement approaches linearity. These values are consistent with the values of the P-P-A and N-P-A angles in the pnicogen-bonded complexes $(H_2XP)_2$ [44] and $H_2XP:NXH_2$ [45].

Table 4. Binding energies of pnicogen-bonded complexes with C_1 and C_s symmetries.

H_2XP, X =	Equilibrium Symmetry	ΔE (kJ·mol^{-1}) for Equilibrium Structures	ΔE (kJ·mol^{-1}) for C_s Structures with H_2O in-Plane	ΔE (kJ·mol^{-1}) for C_s Structures with H_2O out-of-Plane
NC	C_s	−21.1	−21.1	−19.9[b]
F	C_1	−19.5	−19.2[a]	−18.5[b]
Cl	C_1	−17.3	−17.2[a]	−16.3[b]
CN	C_s	−16.9	−16.9	−15.7[b]
OH	C_1	−12.8	−12.4[a]	−12.1[b]
CCH	C_s	−11.3	−11.3	−10.1[b]

[a] These C_s structures with water in-plane have one imaginary frequency; [b] These complexes with out-of-plane H_2O molecules have one imaginary frequency.

(a)   (b)

Figure 5. The pnicogen-bonded complexes $H_2ClP:OH_2$ (a) with the H_2O molecule in the C_s symmetry plane; and $H_2FP:OH_2$ (b) with C_s symmetry and an out-of-plane H_2O molecule.

Table 5 also presents the NBO $O_{lp} \rightarrow \sigma^*$P-A charge-transfer energies in these pnicogen-bonded complexes. Charge-transfer energies vary from 8 kJ·mol^{-1} in $H_2(CCH)P:OH_2$ to 20 kJ·mol^{-1} in $H_2(NC)P:OH_2$. They exhibit an exponential dependence on the P$\cdots$O distance, with a correlation coefficient of 0.949.

Table 5. P⋯O distances, O-P-A angles, charge-transfer energies, and ^1pJ(P-O) coupling constants for pnicogen-bonded complexes $H_2XP:OH_2$ with C_s symmetry and in-plane H_2O molecules.

H_2XP, X =	Distance (R, Å)	Angles (<, °)	Charge-Transfer Energies (kJ·mol^{-1})	Coupling Constants (Hz)
	R(P⋯O)	<O-P-A[a]	$O_{lp} \rightarrow \sigma^*$P-A	^1pJ(P-O)
NC[b]	2.800	165	20.3	−62.5
F	2.755	167	19.5	−69.8
Cl	2.835	166	17.4	−61.2
CN[b]	2.944	161	12.3	−41.9
OH	2.919	166	11.6	−46.7
CCH	3.036	167	8.0	−36.7

[a] A is the atom of X directly bonded to P. [b] The atom written first is directly bonded to P.

Table 5 also reports the spin–spin coupling constants ^1pJ(P-O) across the pnicogen bonds. The components of these coupling constants are reported in Table S4 of the Supporting Information. ^1pJ(P-O) values are dominated by the Fermi-contact terms, and vary from −37 Hz in H_2(CCH)P:OH$_2$ to −70 Hz in H_2FP:OH$_2$. Figure 6 illustrates the second-order dependence of ^1pJ(P-O) on the P⋯O distance, with a correlation coefficient of 0.979.

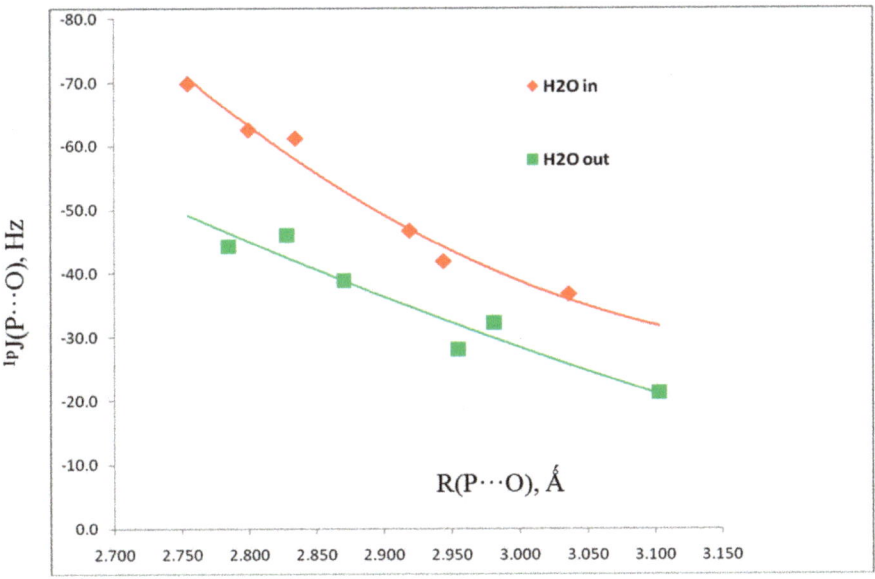

Figure 6. ^1pJ(P-O) *versus* the P⋯O distance for pnicogen-bonded complexes $H_2XP:OH_2$ with C_s symmetry and in-plane and out-of-plane H_2O molecules.

A second reason for optimizing the set of complexes with C_s symmetry and out-of-plane H_2O molecules is to examine the coupling constants of these structures, the components of which are reported in Table S5. Since $^1pJ(P-O)$ values are also dominated by the Fermi-contact terms, it is expected that the s electron densities at O and at P interacting with O will be very different in the ground state and the excited states which couple to the ground state for these two orientations of H_2O molecules. It is apparent from Figure 6 that such is the case, since at the same P···O distances, the points for structures with out-of-plane H_2O molecules lie below those for in-plane H_2O molecules.

Finally, a plot of $^{2h}J(O-P)$ for hydrogen-bonded complexes and $^1pJ(P-O)$ for pnicogen-bonded complexes with C_s symmetry and in-plane H_2O molecules *versus* the P···O distance is reported as Figure S1 of the Supporting Information. At the shorter P···O distances, the absolute values of $^{2h}J(O-P)$ are greater than the values of $^1pJ(P-O)$ at longer distances, but a single second-order trendline with a correlation coefficient of 0.981 describes the distance dependence of both coupling constants.

4. Conclusions

Crystal structures have long been used as a tool for identifying and confirming the presence of weak intermolecular interactions, including hydrogen bonds and pnicogen bonds. A search of the CSD for complexes of water with phosphine and arsine identified two water–phosphine complexes stabilized by hydrogen bonds and one stabilized by a pnicogen bond, as well as three water–arsine complexes with hydrogen bonds and four with pnicogen bonds. The range of P···O distances in the phosphine complexes is consistent with the results of *ab initio* MP2/aug'-cc-pVTZ calculations carried out on complexes $H_2XP:OH_2$, for X = NC, F, Cl, CN, OH, CCH, H, and CH_3. Only hydrogen-bonded complexes are found on the $H_2(CH_3)P:OH_2$ and $H_3P:OH_2$ potential surfaces, while only pnicogen-bonded complexes exist on the $H_2(NC)P:OH_2$, $H_2FP:OH_2$, $H_2(CN)P:OH_2$, and $H_2(OH)P:OH_2$ surfaces. Both hydrogen-bonded and pnicogen-bonded complexes are found on the $H_2ClP:OH_2$ and $H_2(CCH)P:OH_2$ surfaces, with the pnicogen-bonded complexes more stable than the corresponding hydrogen-bonded complexes. It is apparent that the more electronegative substituents prefer to form pnicogen-bonded complexes, while the more electropositive substituents form hydrogen-bonded complexes. The binding energies of pnicogen-bonded complexes range from −11 to −21 kJ·mol^{-1}, while the pnicogen-bonded complexes have binding energies from −9 to −15 kJ·mol^{-1}. The hydrogen-bonded complexes are stabilized by charge transfer from the lone pair on P to the antibonding σ^* H-O orbital, while pnicogen-bonded complexes are stabilized by charge transfer from the lone pair on O to the antibonding σ^* P-A orbital, with A the atom of X directly bonded to P.

Spin-spin coupling constants ^{2h}(O-P) and ^{1h}J(H-P) correlate with the O$\cdots$P and H$\cdots$P distances, respectively, while ^{1}J(O-H) for the hydrogen-bonded O-H group increases in absolute value only slightly upon complex formation. ^{1p}J(P-O) coupling constants were computed for two sets of pnicogen-bonded complexes, one with C_s symmetry and the H_2O molecule in the symmetry plane, and the other also with C_s symmetry but with an out-of-plane H_2O molecule. ^{1p}J(P-O) for both sets are quadratically dependent on the P$\cdots$O distance. The different orientations of the H_2O molecule in these two sets alter the s electron densities at O and P, and lead to greater absolute values of ^{1p}J(P-O) for complexes with in-plane H_2O molecules compared to those with out-of-plane H_2O molecules.

Supplementary Materials: Supplementary materials can be found at http://www.mdpi.com/2073-4352/6/2/19/s1.

Acknowledgments: This work was carried out with financial support from the Ministerio de Economía y Competitividad (Project No. CTQ2012-35513-C02-02) and Comunidad Autónoma de Madrid (Project FOTOCARBON, ref. S2013/MIT-2841). Thanks are also given to the Ohio Supercomputer Center and CTI (CSIC) for their continued computational support.

Author Contributions: Ibon Alkorta and Janet E. Del Bene carried out the calculations and analyzed the data. All three authors contributed to the manuscript.

Conflicts of Interest: The authors declare no conflict of interest.

References

1. Allen, F.H.; Wood, P.A.; Galek, P.T.A. Role of chloroform and dichloromethane solvent molecules in crystal packing: An interaction propensity study. *Acta Cryst.* **2013**, *B69*, 379–388.
2. Van de Streek, J. All series of multiple solvates (including hydrates) from the Cambridge Structural Database. *CrystEngComm* **2007**, *9*, 350–352.
3. Infantes, L.; Chisholm, J.; Motherwell, S. Extended motifs from water and chemical functional groups in organic molecular crystals. *CrystEngComm* **2003**, *5*, 480–486.
4. Infantes, L.; Motherwell, S. Water clusters in organic molecular crystals. *CrystEngComm* **2002**, *4*, 454–461.
5. Gillon, A.L.; Feeder, N.; Davey, R.J.; Storey, R. Hydration in Molecular Crystals-A Cambridge Structural Database Analysis. *Cryst. Growth Des.* **2003**, *3*, 663–673.
6. Infantes, L.; Fabian, L.; Motherwell, S. Organic crystal hydrates: What are the important factors for formation. *CrystEngComm* **2007**, *9*, 65–71.
7. Pimentel, G.C.; McClellan, A.L. *The Hydrogen Bond*; W.H. Freeman: San Francisco, CA, USA, 1960.
8. Arunan, E.; Desiraju Gautam, R.; Klein Roger, A.; Sadlej, J.; Scheiner, S.; Alkorta, I.; Clary David, C.; Crabtree Robert, H.; Dannenberg Joseph, J.; Hobza, P.; *et al.* Defining the hydrogen bond: An account (IUPAC Technical Report). *Pure Appl. Chem.* **2011**, *83*, 1619–1636.

9. Arunan, E.; Desiraju Gautam, R.; Klein, R.-A.; Sadlej, J.; Scheiner, S.; Alkorta, I.; Clary, D.C.; Crabtree, R.H.; Dannenberg, J.J.; Hobza, P.; et al. Definition of the hydrogen bond (IUPAC Recommendations 2011). *Pure Appl. Chem.* **2011**, *83*, 1637–1641.
10. Sundberg, M.R.; Uggla, R.; Viñas, C.; Teixidor, F.; Paavola, S.; Kivekäs, R. Nature of intramolecular interactions in hypercoordinate C-substituted 1,2-dicarba-closo-dodecaboranes with short P···P distances. *Inorg. Chem. Commun.* **2007**, *10*, 713–716.
11. Bauer, S.; Tschirschwitz, S.; Lönnecke, P.; Frank, R.; Kirchner, B.; Clarke, M.L.; Hey-Hawkins, E. Enantiomerically pure bis(phosphanyl)carbaborane(12) compounds. *Eur. J. Inorg. Chem.* **2009**, *2009*, 2776–2788.
12. Tschirschwitz, S.; Lonnecke, P.; Hey-Hawkins, E. Aminoalkylferrocenyldichlorophosphanes: Facile synthesis of versatile chiral starting materials. *Dalton Trans.* **2007**, 1377–1382.
13. Politzer, P.; Murray, J.; Janjić, G.; Zarić, S. σ-Hole interactions of covalently-bonded nitrogen, phosphorus and arsenic: A survey of crystal structures. *Crystals* **2014**, *4*, 12–31.
14. Sánchez-Sanz, G.; Alkorta, I.; Trujillo, C.; Elguero, J. Intramolecular pnicogen interactions in PHF-$(CH_2)_n$-PHF (n = 2–6) systems. *ChemPhysChem* **2013**, *14*, 1656–1665.
15. Sánchez-Sanz, G.; Trujillo, C.; Alkorta, I.; Elguero, J. Intramolecular pnicogen interactions in phosphorus and arsenic analogues of proton sponges. *Phys. Chem. Chem. Phys.* **2014**, *16*, 15900–15909.
16. Scheiner, S. A new noncovalent force: Comparison of P···N interaction with hydrogen and halogen bonds. *J. Chem. Phys.* **2011**, *134*, 094315.
17. Zahn, S.; Frank, R.; Hey-Hawkins, E.; Kirchner, B. Pnicogen bonds: A new molecular linker? *Chem. Eur. J.* **2011**, *17*, 6034–6038.
18. Scheiner, S. The pnicogen bond: Its relation to hydrogen, halogen, and other noncovalent bonds. *Acc. Chem. Res.* **2012**, *46*, 280–288.
19. Scheiner, S. Detailed comparison of the pnicogen bond with chalcogen, halogen, and hydrogen bonds. *Int. J. Quantum Chem.* **2013**, *113*, 1609–1620.
20. Del Bene, J.E.; Alkorta, I.; Elguero, J. The pnicogen bond in review: Structures, binding energies, bonding properties, and spin-spin coupling constants of complexes stabilized by pnicogen bonds. In *Noncovalent Forces*; Challenges and Advances in Computational Chemistry and Physics Series; Scheiner, S., Ed.; Springer: Gewerbestrasse, Switzerland, 2015; Volume 19, pp. 191–263.
21. Allen, F. The Cambridge Structural Database: A quarter of a million crystal structures and rising. *Acta Crystallogr. Sect. B* **2002**, *58*, 380–388.
22. Pople, J.A.; Binkley, J.S.; Seeger, R. Theoretical models incorporating electron correlation. *Int. J. Quantum Chem. Quantum Chem. Symp.* **1976**, *10*, 1–19.
23. Krishnan, R.; Pople, J.A. Approximate fourth-order perturbation theory of the electron correlation energy. *Int. J. Quantum Chem.* **1978**, *14*, 91–100.
24. Bartlett, R.J.; Silver, D.M. Many-body perturbation theory applied to electron pair correlation energies. I. Closed-shell first-row diatomic hydrides. *J. Chem. Phys.* **1975**, *62*, 3258–3268.

25. Bartlett, R.J.; Purvis, G.D. Many-body perturbation theory, coupled-pair many-electron theory, and the importance of quadruple excitations for the correlation problem. *Int. J. Quantum Chem.* **1978**, *14*, 561–581.
26. Del Bene, J.E. Proton affinities of ammonia, water, and hydrogen fluoride and their anions: A quest for the basis-set limit using the dunning augmented correlation-consistent basis sets. *J. Phys. Chem.* **1993**, *97*, 107–110.
27. Dunning, T.H. Gaussian Basis Sets for Use in Correlated Molecular Calculations. I. The Atoms Boron through Neon and Hydrogen. *J. Chem. Phys.* **1989**, *90*, 1007–1023.
28. Woon, D.E.; Dunning, T.H. Gaussian Basis Sets for Use in Correlated Molecular Calculations. V. Core-Valence Basis Sets for Boron Through Neon. *J. Chem. Phys.* **1995**, *103*, 4572–4585.
29. Frisch, M.J.; Trucks, G.W.; Schlegel, H.B.; Scuseria, G.E.; Robb, M.A.; Cheeseman, J.R.; Scalmani, G.; Barone, V.; Mennucci, B.; Petersson, G.A.; *et al. Gaussian–09, Revision D.01*; Gaussian, Inc.: Wallingford, CT, USA, 2009.
30. Bader, R.F.W. A quantum theory of molecular structure and its applications. *Chem. Rev.* **1991**, *91*, 893–928.
31. Bader, R.F.W. *Atoms in Molecules, A Quantum Theory*; Oxford University Press: Oxford, UK, 1990.
32. Popelier, P.L.A. *Atoms in Molecules. An Introduction*; Prentice Hall: Harlow, UK, 2000.
33. Matta, C.F.; Boyd, R.J. *The Quantum Theory of Atoms in Molecules: From Solid State to DNA and Drug Design*; Wiley-VCH: Weinheim, Germany, 2007.
34. Keith, T.A. *AIMALL*; Version 15.09.27. TK Gristmill Software: Overland Park, KS, USA, 2011. Available online: http://aim.tkgristmill.com (accessed on 31 December 2015).
35. Reed, A.E.; Curtiss, L.A.; Weinhold, F. Intermolecular interactions from a Natural Bond Orbital, donor-acceptor viewpoint. *Chem. Rev.* **1988**, *88*, 899–926.
36. Glendening, E.D.; Badenhoop, J.K.; Reed, A.E.; Carpenter, J.E.; Bohmann, J.A.; Morales, C.M.; Landis, C.R.; Weinhold, F. *NBO 6.0*; University of Wisconsin: Madison, WI, USA, 2013.
37. Becke, A.D. Density functional thermochemistry. III. The role of exact exchange. *J. Chem. Phys.* **1993**, *98*, 5648–5652.
38. Lee, C.; Yang, W.; Parr, R.G. Development of the Colle-Salvetti correlation-energy formula into a functional of the electron density. *Phys. Rev. B* **1988**, *37*, 785789.
39. Perera, S.A.; Nooijen, M.; Bartlett, R.J. Electron correlation effects on the theoretical calculation of Nuclear Magnetic Resonance spin–spin coupling constants. *J. Chem. Phys.* **1996**, *104*, 3290–3305.
40. Perera, S.A.; Sekino, H.; Bartlett, R.J. Coupled–cluster calculations of indirect nuclear coupling constants: The importance of non–Fermi-contact contributions. *J. Chem. Phys.* **1994**, *101*, 2186–2196.
41. Schäfer, A.; Horn, H.; Ahlrichs, R. Fully optimized contracted Gaussian basis sets for atoms Li to Kr. *J. Chem. Phys.* **1992**, *97*, 2571–2577.

42. Stanton, J.F.; Gauss, J.; Watts, J.D.; Nooijen, M.; Oliphant, N.; Perera, S.A.; Szalay, P.S.; Lauderdale, W.J.; Gwaltney, S.R.; Beck, S.; *et al.* *ACES II*; University of Florida: Gainesville, FL, USA, 2014.
43. Del Bene, J.E.; Alkorta, I.; Elguero, J. Characterizing Complexes with Pnicogen Bonds Involving sp2 Hybridized Phosphorus Atoms: $(H_2C=PX)_2$ with X = F, Cl, OH, CN, NC,CCH, H, CH_3, and BH_2. *J. Phys. Chem. A* **2013**, *117*, 6893–6903.
44. Del Bene, J.E.; Alkorta, I.; Sanchez-Sanz, G.; Elguero, J. ^{31}P-^{31}P Spin-spin coupling constants for Pnicogen Homodimers. *Chem. Phys. Lett.* **2011**, *512*, 184–187.
45. Del Bene, J.E.; Alkorta, I.; Sanchez-Sanz, G.; Elguero, J. Structures, Energies, Bonding, and NMR Properties of Pnicogen Complexes $H_2XP:NXH_2$ (X = H, CH_3, NH_2, OH, F, Cl). *J. Phys. Chem. A* **2011**, *115*, 13724–13731.

Theoretical Studies on Hydrogen Bonds in Anions Encapsulated by an Azamacrocyclic Receptor

Jing Wang, Jiande Gu, Md. Alamgir Hossain and Jerzy Leszczynski

Abstract: Hydrogen bonds in two halides encapsulated by an azamacrocyclic receptor were studied in detail by the density functional theory (DFT) approaches at B3LYP/6-311++G(d,p) and M06-2X/6-311++G(d,p) levels. The atoms in molecules (AIM) theory and the electron density difference maps were applied for characterizing the hydrogen bond patterns. The results suggest that the fluoride complex has a unique binding pattern which shows a hydrogen bond augmented with ionic bond characteristics.

Reprinted from *Crystals*. Cite as: Wang, J.; Gu, J.; Hossain, M.A.; Leszczynski, J. Theoretical Studies on Hydrogen Bonds in Anions Encapsulated by an Azamacrocyclic Receptor. *Crystals* **2016**, *6*, 31.

1. Introduction

The specific interactions between anions and their supramolecular receptors have been widely studied [1–4]. Among the various anions, halides play an important role in environments and life [5]. Interactions between halides and synthetic receptors have been broadly investigated both experimentally and theoretically [2,6–12]. Studies indicate that these interactions are mainly attributed to hydrogen bonding and electrostatic attractions. The binding patterns and selectivity vary with different spacers and the linking amine groups [13–16].

A simple monocyclic polyamine incorporated with N-methyl-2,2'-diaminodiethylamine may selectively encapsulate a sulfate anion through multiple hydrogen bonds [17]. Studies reported that this receptor was able to encapsulate two chlorides in its macrocyclic cavity as well [18]. The X-ray crystallography structure of the charge-assisted two chlorides encapsulated by a hexaprotonated azamacrocycle ($[H_6L]^{6+}$) demonstrates that two chlorides are located on the opposite side of the macrocycle through trigonal recognition by hydrogen-bonding interactions [19]. Hence, the interactions for this receptor and halides were further investigated by experiments and theoretical methods [20]. The binding between halide and a *p*-xylyl-based azamacrocycle was studied experimentally by ^{1}H-NMR titrations and single crystal X-ray diffraction analysis. Meanwhile, the binding properties and the hydrogen bonds in the complexes were studied by density functional theory (DFT) at the M06-2X/6-311G(d,p) level. Both

1:1 and 1:2 complexes of $[H_6L(X)_2]^{4+}$ with halides were investigated. The 1:2 binding mode was found to be energetically favorable [19]. The structural analysis reveals that each halide forms three hydrogen bonds with the neighboring protonated amine nitrogen for each 1:2 halides binding complex. However, the predicted binding energies for complexes of $[H_6L(F)_2]^{4+}$ (−158.5 kcal/mol) are too strong for characterizing six hydrogen bonds. Considering that there are six H-F bonds in this complex and each H-F bonding energy amounts to 26.4 kcal/mol [19], it is clear that these H-F bonds should not be classified as normal H bonds. To understand the selectivity of the azamacrocycle it is necessary to explore the nature of these halide-amine interactions.

To further understand and shed a light on such strong interactions between fluoride anions and the binding center (amine), their binding patterns have been analyzed based on the electron density formation approach. This approach is proved to be able to reveal the binding patterns at the very basic electron density level of theory. For comparison, the atom in molecule (AIM) approach has also been applied to compensate the quantitating the binding patterns [21,22]. To avoid the bias of the specific DFT functional, both B3LYP/6-311++G(d,p) and M06-2X/6-311++G(d,p) methods were applied for all the structure optimizations and the electron density calculations and analysis.

2. Computational Methods

The density functional theory (DFT)-based calculations were carried out for the anion-bound azamacrocycle compounds. The DFT approach through Becke's three parameter (B3) exchange functional along with the Lee-Yang-Parr (LYP) nonlocal correlation functional B3LYP [20,23–25] with an optimized weight of the exact HF exchange was applied in the present study. The Minnesota density functional M06-2X [26–28] level calculations were also performed. The pseudopotential basis set LanL2DZ was applied for iodine. The standard valence triple zeta basis set, augmented with d-type polarization and diffuse functions for heavy elements, and p-type polarization and diffuse functions for H, 6-311++G(d,p) was used for all the other elements [29]. The Barone-Tomasi polarizable continuum model (PCM) [30] with the standard dielectric constant of water (ε = 78.39) was applied to simulate the solvated environment of an aqueous solution. The force constants were determined analytically in the analysis of harmonic vibrational frequencies for all of the complexes. The GAUSSIAN 09 system of DFT programs [31] was used for all computations.

To analyze the H-bonding pattern in the anion-bonded azamacrocycle systems, the atoms-in-molecules (AIM) theory of Bader [21,22] was applied, and the analysis was performed by the AIM2000 program [32]. The AIM analysis is based upon the density obtained at the B3LYP/6-311++G(d,p) level. Hydrogen bonding can also be characterized by the change of the electron density for the bonded moiety. The electron density around the proton and the proton acceptor decreases, while the density between the proton and its acceptor increases as the results of the formation of a hydrogen bond [33].

3. Results and Discussions

The present studied models of $[H_6L(X)_2]^{4+}$ have multiple charges. The interactions in the gas phase are mainly caused by the strong static electronic interactions between $[H_6L]^{6+}$ and the anions. In order to eliminate the influences of the static electronic interactions and reveal, more accurately, the hydrogen bonding, all the models were fully optimized including the solvent effects through the PCM model in which water is applied as the solvent. All the discussions will be focused on the results with solvent obtained using PCM models. The optimized structural parameters indicate that there is no significant difference between the structures predicted by the B3LYP method and those predicted by the M06-2X method (the largest deviations in the hydrogen bonding atomic distances are less than 0.02 Å, except for HB2 of $[H_6L(F)_2]^{4+}$, which is 0.09 Å longer in the prediction with B3LYP than that with M06-2X)). Therefore, the following discussion is mainly based on the results of the B3LYP/6-311++G(d,p) level of theory.

The azamacrocycle ring shows an ellipsoid structure, of which two methyl groups are trans-orientated. As shown in Figure 1, all six nitrogen atoms (from N1 to N6) are fully protonated. Studies show that two halide anions can be encapsulated in the azamacrocyclic cavity, forming stable 1:2 (ligand:anion) complexes [13]. Here we use the width and height values to describe the approximate size of the cavity. The atomic distance between N2 and N5 is used to represent the width of the azamacrocycle. The two aromatic benzene planes are parallel. The distance between the two parallel planes is used to describe the height of the cavity. The optimized cavity of the azamacrocycle has the width of 9.966 Å and height of 6.838 Å. The fully-optimized structures of the halide-encapsulated azamacrocyclic complexes display symmetric geometry (C_{2h}). Six related hydrogen bonds are formed between the halides and the six neighboring protonated amine nitrogen for the complex, HB1, HB2, HB3, HB4, HB5, and HB6. Due to the structural symmetry, hydrogen bonds (HB4, HB5, HB6) have the same properties as hydrogen bonds of HB1 HB2, and HB3 (Figure 1). Therefore, in the following discussions, only HB1, HB2, and HB3 will be discussed in detail to avoid redundancy.

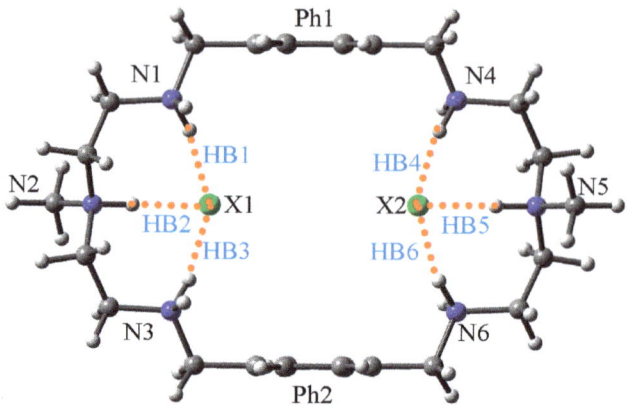

Figure 1. The scheme of the $[H_6L(X)_2]^{4+}$ structure. (X = F$^-$, Cl$^-$, Br$^-$, I$^-$).

3.1. Two Chlorides Encapsulated in $[H_6L]^{6+}$

The crystallography data of the chloride azamacrocyclic compound reveals that the host's cavity keeps the two chlorides inside with six hydrogen bonds [16]. Figure 2a shows the optimized structure resulted from the computations, which is highly consistent with that obtained from the X-ray data (see Supporting Information). In the theoretical structure, the width and the height of the cavity are calculated to be 10.768 Å and 7.927 Å, compared to the width and the height of 10.338 Å and 7.842 Å in the crystal structure (Table 1). The atomic distance of the chlorides is predicted to be around 5.628 Å, while in X-ray data the two chlorides are apart by about 4.433 Å. The atomic distance between N1 and Cl1 is predicted to be 3.172 Å (3.146 Å in crystal structure). In addition, the atomic distance between N2 and Cl1 is calculated to be 3.117 Å while the X-ray data reveals that the distance amounts to 3.067 Å. The binding energy for the chlorides with the azamacrocycle cavity is predicted to be −52.2 kcal/mol.

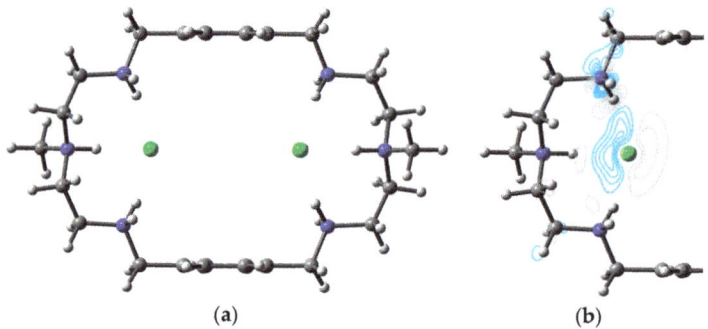

(a)   (b)

Figure 2. *Cont.*

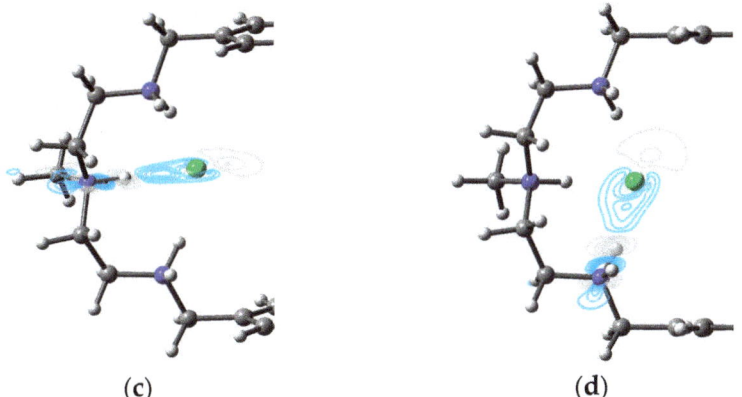

(c) (d)

Figure 2. Two chlorides encapsulated by the azamacrocyclic compound $(H_6L)^{6+}$. (a) Optimized $[H_6LCl_2]^{4+}$ structure; (b) electron density differences ($\Delta\rho$) maps of HB1 in the plane of N1-H-Cl1; (c) electron density differences ($\Delta\rho$) maps of HB2 in the plane of N2-H-Cl1; and (d) electron density differences ($\Delta\rho$) maps of HB3 in the plane of N3-H-Cl1. Contours in the deformation density map are shown at ± 0.001 au. Grey lines indicate deficiency of density, and blue lines indicate increasing density. ($\Delta\rho$) = $\rho[H_6LCl_2]^{4+} - \rho[H_6L]^{6+} - \rho(Cl_2)^{2-}$.

Table 1. Parameters for the hydrogen bonds at the B3LYP/6-311++G(d,p) level and AIM analysis (results at M06-2X/6-311++G(d,p) level are listed in parenthesis).

	Hydrogen Bonds	NH...X (Å)	ρ(BCPs) (au)	$\nabla^2\rho$ (au)	X...X (Å)	N2...N2' (Å)	H (Å)	ΔE (kcal/mol)	logK₂ (Reference [19])
[H6L]F₂	HB1	1.612 (1.627)	0.0514	0.0414	6.563 (6.502)	11.314 (11.106)	7.119 (7.172)	−95.9 (−108.9)	2.82
	HB2	1.510 (1.413)	0.0691	0.0485					
	HB3	1.612 (1.627)	0.0514	0.0414					
[H6L]Cl₂	HB1	2.147 (2.130)	0.0296	0.0171	5.628 (5.628)	10.768 (10.678)	7.927 (7.736)	−52.2 (−66.5)	2.70
	HB2	2.068 (2.049)	0.0368	0.0181					
	HB3	2.147 (2.130)	0.0296	0.0171					
[H6L]Br₂	HB1	2.311 (2.312)	0.0261	0.0135	5.683 (4.831)	10.705 (10.344)	8.054 (7.952)	−45.5 (−60.4)	2.28
	HB2	2.233 (2.233)	0.0319	0.0143					
	HB3	2.311 (2.312)	0.0261	0.0135					
[H6L]I₂	HB1	2.571 (2.548)	0.0209	0.0128	5.435 (4.821)	10.502 (10.252)	8.347 (8.102)	−45.7 (−63.8)	2.20
	HB2	2.509 (2.493)	0.0245	0.0134					
	HB3	2.571 (2.548)	0.0209	0.0128					

Notes: ρ(BCPs): density of the bond critical points. $\nabla^2\rho$: Laplacian of the density at the bond critical points. H: the distance between the two benzene planes. ΔE: binding energies of the complex $[H_6L(X)_2]^{4+}$, ΔE = $E([H_6L(X)_2]^{4+}) - E([H6L]^{6+}) - 2E(X)^-$. logK₂: binding data of the ligand for halides in D2O at 298K. (Reference [19]).

The atomic distance of N1H ... Cl1 and N3H ... Cl1 for HB1 and HB3 is both calculated to be 2.147 Å. Their density of the BCP is 0.0296 au while the Laplacian of the density is about 0.0171 au. HB2 reveals a shorter atomic distance of 2.068 Å. The density of the BCP and the Laplacian of the density are larger than those of HB1 and HB3, which read as 0.0368 au and 0.0181 au, respectively. These values are typical for the descriptions of hydrogen bonding [34,35]. HB2 is characterized by a stronger hydrogen bond than HB1 and HB3, which is also confirmed by X-ray data where the Cl ion is closer to N2 than to N1 and N3.

The deformation electron density maps for the three H bonds (Figure 2b–d) clearly demonstrate the typical characteristics of the H bonds. The electron density increases between the donating protons and the chloride anion. The electron density increase along HB2 is larger than that of the HB1 and HB3, indicating that hydrogen bond HB2 is strong than the other two. This observation is consistent to the shorter H ... Cl atomic distance in HB2 than that in HB1 and HB3. The electron density difference map of $[H_6LCl_2]^{4+}$ also shows a decrease in density at the proton position H and at the lone electron pair of N for all three hydrogen bonds. The binding energy in $[H_6LCl_2]^{4+}$ suggests that the average H bonding amounts to 8~9 kcal/mol.

3.2. Two Bromides Encapsulated in $[H_6L]^6$

It should be noted here that the crystallography results are available only for $[H_4L(Br)2]^{2+}$ and $[H_4LI2]^{2+}$ where N2 and N5 are not protonated. The experimental data reveal that bromides and iodides are not encapsulated in the azamacrocyclic cavity (see supplementary materials). For comparison and consistency purposes, in the present study, we consider the six protonated azamacrocyclic cavity for all the models as those in reference [19].

As depicted in Figure 3a, the optimized structure of compound $[H_6LBr_2]^{4+}$ has a C_{2h} symmetric geometry. The size of the azamacrocycle cavity in this compound is measured by width of 10.705 Å and height of 8.054 Å, which show that the paralleled aromatic benzene planes are parted farther, and N2 and N5 becomes closer with a shorter atomic distance when comparing to those of $[H_6LCl_2]^{4+}$ complexes. The two bromides are separated by about 5.683 Å. The atomic distance of N1H ... Br and N3H ... Br for HB1 and HB3 are calculated to be 2.311 Å and 2.309 Å, respectively. Both of the corresponding densities of BCPs is ~0.026 au and the Laplacian of density through AIM analysis is around 0.013 au. Comparison with HB1 and HB3, HB2 demonstrates a shorter atomic distance for N2H ... Br (2.233 Å) and a larger density of BCP (0.0319 au) and a larger Laplacian of density (0.0143). This implies that hydrogen bond HB2 is stronger than hydrogen bonds of HB1 and HB3. The electron density differences maps also reveal a stronger HB2, where the electron density increases along N2H ... Br of HB2 (Figure 4c) more than those along N1H ... Br of HB1 and N3H ... Br of HB3 (Figure 3b,d).

The binding energies for the two bromides in $[H_6LBr_2]^{4+}$ is calculated to be −45.5 kcal/mol in the solvent, which is about 7 kcal/mol weaker than the binding of $[H_6LCl_2]^{4+}$.

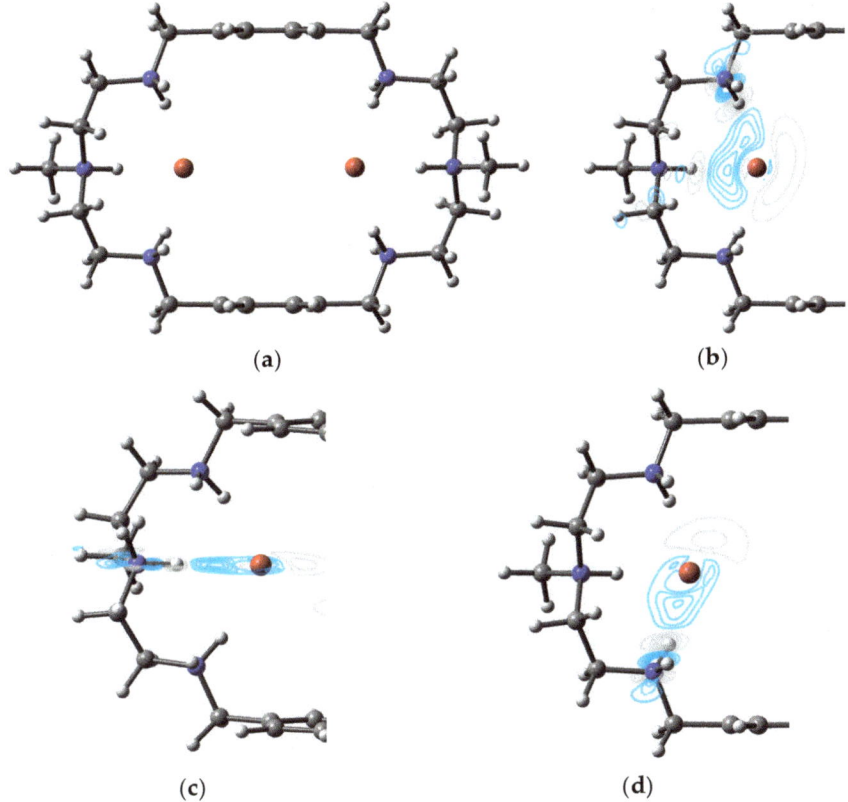

Figure 3. Two bromides encapsulated by the azamacrocyclic compound. (a) Optimized [H$_6$LBr$_2$]$^{4+}$ structure; (b) electron density differences ($\Delta\rho$) maps of HB1 in the plane of N1–H–Br1; (c) electron density differences ($\Delta\rho$) maps of HB2 in the plane of N2-H-Br1; and (d) electron density differences ($\Delta\rho$) maps of HB3 in the plane of N3-H-Br1. Contours in the deformation density map are shown at ± 0.001 au. Grey lines indicate deficiency of density, and blue lines indicate increasing density. ($\Delta\rho$) = ρ[H$_6$LBr$_2$]$^{4+}$ − ρ[H$_6$L]$^{6+}$ − ρ(Br$_2$)$^{2-}$.

3.3. Two Iodides Encapsulated in [H$_6$L]^6

The azamacrocycle cavity which encapsulates two iodide anions has a size characterized by a width of 10.502 Å and a height of 8.347 Å. The atomic distance between the two iodides is 5.435 Å. The bond distances for NH . . . I are predicted to be 2.573 Å, 2.509 Å, and 2.571 Å for HB1, HB2, and HB3, respectively. This shows that HB2 is stronger than HB1 and HB2. The same feature can also be seen in the related electron density difference maps (Figure 4b–d). The density of BCP for HB2 is 0.0245 au, which is about 0.004 au larger than those of HB1 and HB3, while the

Laplacian of density for HB2 is 0.0134 au, which is slightly larger than those of HB1 and HB2.

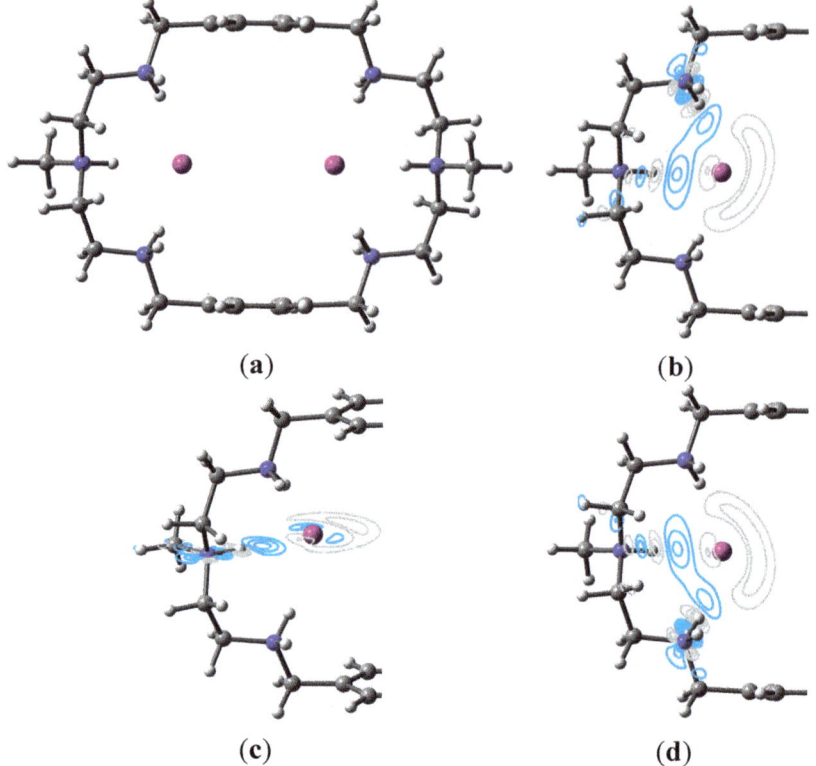

Figure 4. Two iodides encapsulated by the azamacrocyclic compound. (a) Optimized $[H_6LI_2]^{4+}$ structure; (b) electron density differences ($\Delta\rho$) maps of HB1 in the plane of N1-H-I1; (c) electron density differences ($\Delta\rho$) maps of HB2 in the plane of N2–H–I1; and (d) electron density differences ($\Delta\rho$) maps of HB3 in the plane of N3–H–I1. Contours in the deformation density map are shown at ± 0.001 au. Grey lines indicate deficiency of density, and blue lines indicate increasing density. ($\Delta\rho$) = $\rho[H_6LI_2]^{4+} - \rho[H_6L]^{6+} - \rho(I_2)^{2-}$.

The binding energy of −45.7 kcal/mol characterizes complexes between iodides and the azamacrocycle. This shows similar binding strength as that predicted for analogous complex with bromide anions, while it has weaker binding characteristics than those of chloride anions.

3.4. Two Fluorides Encapsulated in $[H_6L]^6$

The PCM model optimized structure of model $[H_6LF_2]^{4+}$ complex is shown in Figure 5a. For this compound, the width of the azamacrocycle cavity is calculated to be 11.314 Å. The height is predicted to be 7.119 Å. The overall shape is elongated for the $[H_6LF_2]^{4+}$ complex, as compared to the other complexes. This elongation is partly due to the stronger charge repulsion between the F anions. This can be further confirmed by the significant separation of the two F anions in the complex. The two fluoride anions are encapsulated inside the cavity at the two corners. The atomic distance between the two anions is estimated to be 6.563 Å. For the comparison, the other halide anions in the cavity are separated by less than 5.7 Å. As depicted in Figure 5, each fluoride anion is bounded within the cavity through three hydrogens of the nearby protonated amine nitrogens. Compared to the other halide complexes, the HBs in $[H_6LF_2]^{4+}$ complex are significantly shorter than those found in other halides complexes. For hydrogen bonds 1 and 3 (HB1 and HB3), the atomic distances of N1H ... F1 is 1.612 Å.

Based upon the AIM analysis, the density of the bond critical point (BCP) for HB1 (or HB3) is about 0.0514 au and the corresponding Laplacian of the density is predicted to be 0.0414 au. This suggests that HB1 is a strong hydrogen bond. The atomic distance of N2H ... F1 of HB2 is measured to be 1.510 Å, which is about 0.1 Å shorter than that of HB1. The density of BCP and the Laplacian of the density of HB2 are calculated to be 0.0691 au and 0.0485 au, respectively (Table 1). Here one can see that the density of BCP of HB2 is about 0.018 au larger than that of HB1 or HB3. This large BCP density suggests that the corresponding bond should not be attributed to the H bond only.

To explore the nature of this extraordinary N–H–F interaction the analysis of the electron density differences of this complex has been performed. The deformation density maps for the three hydrogen bonds (HB1, HB2, and HB3) are depicted in Figure 5. The electronic structures of the H-bond reveal themselves in the deformation density map of $[H_6LF_2]^{4+}$. For HB1 and HB3 (Figure 5b,d), it can be seen that the concentration of electron density arises between the proton H and the proton acceptor (fluoride) with a corresponding electron density deficiency at the positions of the proton H and the acceptor lone pair of nitrogen. For HB2, the electron density deformation map (Figure 5c) shows an inescapable different as compared to those of HB1 and HB3. The electron density increase on the N–H–F plane is not just limited to the space between H and F atomic centers. The electron density increase is surrounding the fluoride center in this HB2 bond plane. Moreover, the electron density increase along H-bond HB2 is much larger compared to that of the H bonds of HB1 and HB3. This implies that HB2 is a much stronger hydrogen bond than HB1 and HB3. These differences are well correlated with the notable short atomic distance of 1.510 Å for HB2. The density of BCP (around 0.07) also shows a bigger density than the regular hydrogen bonds. Considering these factors, one can describe HB2 as a hybrid hydrogen bond combined with some character of ionic bonding between the proton and the fluoride. Therefore, HB2 contributes to the binding more than HB1 and HB3.

The binding energies for the two fluorides anions with $[H_6L]^{6+}$ are calculated to be about -95.9 kcal/mol in polarizable environments, suggesting a very strong binding for the fluorides and the azamacrocycle cavity. This provides further proof that HB2 is not just a pure hydrogen bond, but a hybrid hydrogen bond augmented with ionic binding which shows a much stronger binding than those of HB1 and HB3.

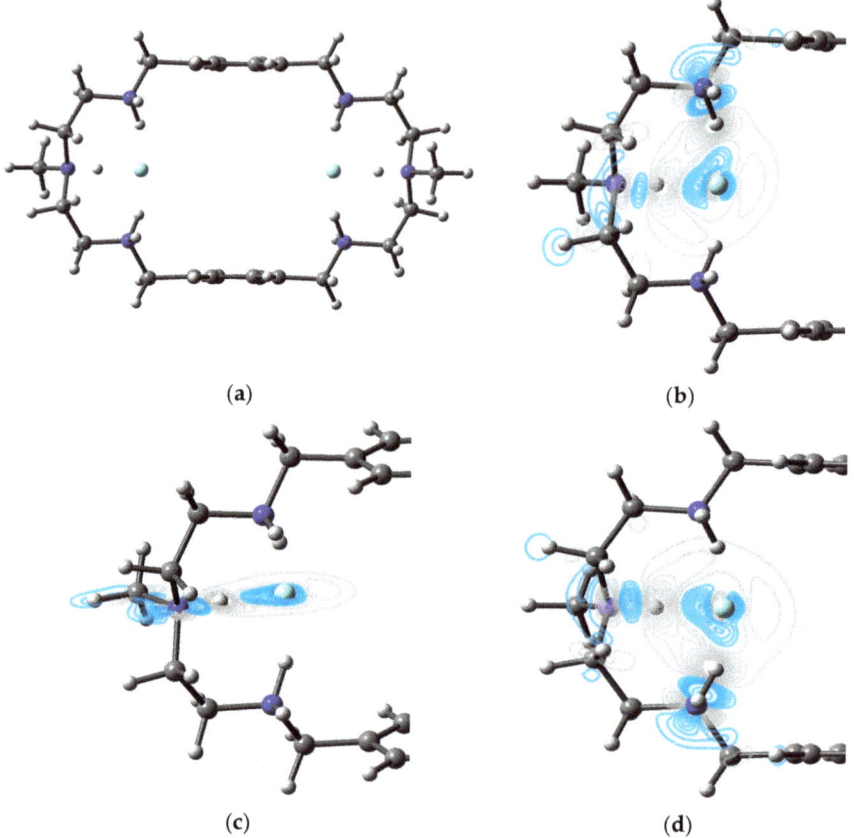

Figure 5. Two fluorides encapsulated by the azamacrocyclic compound. (a) Optimized $[H_6LF_2]^{4+}$ structure; (b) electron density differences ($\Delta\rho$) maps of HB1 in the plane of N1–H–F1; (c) electron density differences ($\Delta\rho$) maps of HB2 in the plane of N2–H–F1; and (d) electron density differences ($\Delta\rho$) maps of HB3 in the plane of N3–H–F1. Contours in the deformation density map are shown at $\pm\,0.001$ au. Grey lines indicate deficiency of density, and blue lines indicate increasing density. $\Delta\rho = \rho([H_6LF_2]^{4+} - \rho[H_6L]^{6+} - \rho(F_2)^{2-}$.

As shown in the binding ligand–halides data for the 1:2 binding complexes at 298K in D2O (Table 1), the calculated binding strength of the studies species has the same trend as observed in experiments. The fluorides have the strongest binding of all the considered species and the chlorides have the less pronounced binding potential.

There is significant influence of the complexation of halide ions on the structure of azamacrocycle. It is noted that the cavity of azamacrocycle ring becomes larger in order to accommodate the halides. The fluorides complex is characterized by the longest width and the shortest height of the cavity, while in the iodides complex has the narrowest width and the longest height. Due to the strong charge repulsion, the two fluoride ions are located at the largest distance apart, about 1 Å farther than the other three halides species. The binding energy suggests that the fluorides have the strongest binding affinity to the receptor's cavity.

4. Conclusions

The structural and bonding characteristics of halide ion complexes with an azamacrocycle have been investigated using a computational, DFT-based approach. We observed that the cavity of an azamacrocycle ring expands in order to accommodate the halides. The fluoride complex has the longest width and the shortest height of the cavity while the iodide complex is characterized by the narrowest width and the longest height. Due to the strong charge repulsion, the two fluorides are located at the largest distance apart, about 1 Å farther than the other three halides species. The values of binding energy suggest that the fluorides have the strongest interactions with the azamacrocycle. As investigated by the electron density analysis, the interaction between NH–F is not limited to traditional H bonding, but it could be described as a hybrid hydrogen bond mixed with contributions of ionic bonding between the proton and the fluoride. On the other hand, the binding strength of the chloride, bromide, and iodide complexes is only half (or less) of that of the fluorides. The electron density analysis indicates that the halides in these complexes are stabilized through the normal hydrogen bonds.

Supplementary Materials: The supplementary materials are available online at http://www.mdpi.com/ 2073-4352/6/3/31/s1.

Acknowledgments: The authors thank for support of the NSF CREST Interdisciplinary Nanotoxicity Center NSF-CREST—Grant # HRD-0833178; NSF-EPSCoR Award Number: 362492-190200-01-0903787. The computation time was provided by the Extreme Science and Engineering Discovery Environment (XSEDE) by National Science Foundation Grant Number OCI-1053575 and XSEDE award allocation Number DMR110088 and by the Mississippi Center for Supercomputer Research.

Author Contributions: Jing Wang and Jerzy Leszczynski conceived and designed the theoretical study; Md. Alamgir Hossain performed the experiments; Jing Wang performed the calculations; Jing Wang and Jiande Gu analyzed the data; Jing Wang wrote the paper.

Conflicts of Interest: The authors declare no conflict of interest.

References

1. Wenzel, M.; Hiscock, J.R.; Gale, P.A. Anion receptor chemistry: Highlights from 2010. *Chem. Soc. Rev.* **2012**, *41*, 480–520.
2. Bianchi, A.; Bowman-James, K.; Garcia-Espana, E. *Supramolecular Chemistry of Anions*; Wiley: New York, NY, USA, 1997.
3. Amendola, V.; Esteban-Gomez, D.; Fabbrizzi, L.; Licchelli, M. What anions do to N–H-Containing Receptors. *Acc. Chem. Res.* **2006**, *39*, 343–353.
4. Wichmann, K.; Antonioli, B.; Sohnel, T.; Wenzel, M.; Gloe, K.; Gloe, K.; Price, J.R.; Lindoy, L.F.; Blake, A.J.; Schroder, M. Pylymine-based anion receptors: Extraction and structural studies. *Coord. Chem. Rev.* **2006**, *250*, 2987–3003.
5. Hossain, M.A. Inclusion complexes of halide anions with macrocyclic receptors. *Curr. Org. Chem.* **2008**, *12*, 1231–1256.
6. Révész, Á.; Schröder, D.; Svec, J.; Wimmerová, M.; Sindelar, V. Anion binding by bambus[6]uril probed in the gas phase and in solution. *J. Phys. Chem. A* **2011**, *115*, 11378–11386.
7. Pramanik, A.; Powell, D.R.; Wong, B.M.; Hossain, M.A. Spectroscopic, structural, and theoretical studies of halide complexes with a urea-based tripodal receptor. *Inorg. Chem.* **2012**, *51*, 4274–4284.
8. Turner, D.R.; Paterson, M.J.; Steed, J.W. A conformationally flexible, urea-based tripodal anion receptor: solid-state, solution, and theoretical studies. *J. Org. Chem.* **2006**, *71*, 1598–1608.
9. Chen, Y. Theoretical study of interactions between halogen-substituted s-triazine and halide anions. *J. Phys. Chem. A* **2013**, *117*, 8081–8090.
10. Chaumont, A.; Wipff, G. Macrotricyclic quaternary tetraammonium receptors: Halide anion recognition and interfacial activity at an aqueous interface. A molecular dynamics investigation. *J. Comput. Chem.* **2002**, *23*, 1532–1543.
11. Wang, D.X.; Zheng, Q.Y.; Wang, Q.Q.; Wang, M.X. Halide recognition by tetroxacalix[2]arene[2]triazine receptors: Concurrent nocovalent halide-pi and lone-pair-pi interactions in host-halide-water ternary complexes. *Angew. Chem. Int. Ed. Engl.* **2008**, *47*, 7485–7488.
12. Liu, Y.-Z.; Yuan, K.; Lv, L.-L.; Zhu, Y.-C.; Yuan, Z. Designation and exploration of halide-anion recognition based on cooperative noncovalent interactions including hydrogen bonds and anion-π. *J. Phys. Chem. A* **2015**, *119*, 5842–5852.
13. Rhaman, M.M.; Ahmed, L.; Wang, J.; Powell, D.R.; Leszczynski, J.; Hossain, M.A. Encapsulation and selectivity of sulfate with a furan-based hexaazamacrocyclic receptor in water. *Organ. Biomol. Chem.* **2014**, *12*, 2045–2048.
14. Valencia, L.; Bastida, R.; García-España, E.; de Julián-Ortiz, J.V.; Llinares, J.M.; Macías, A.; Pérez Lourido, P. Nitrate encapsulation within the cavity of polyazapyridinophane. Considerations on nitrate-pyridine interactions. *Cryst. Growth Des.* **2010**, *10*, 3418–3423.

15. Chauhan, S.M.S.; Garg, B.; Bisht, T. Synthesis and anion binding of 2-Arylalazo-meso-octamethylcalix[4] pyrroles. *Supramol. Chem.* **2009**, *21*, 394–400.
16. Juwarker, H.; Lenhardt, J.M.; Castillo, J.C.; Zhao, E.; Krishnamurthy, S.; Jamiolkowski, R.M.; Kim, K.-H.; Graig, S.L. Anion binding of short, flexible aryl triazole oligomers. *J. Org. Chem.* **2009**, *74*, 8924–8934.
17. Mendy, J.S.; Pilate, M.L.; Horne, T.; Dey, V.W.; Hossain, M.A. Encapsulation and selective recognition of sulfate anion in an azamacrocycle in water. *Chem. Commun.* **2010**, *46*, 6084–6086.
18. Hossain, M.A.; Saeed, M.A.; Fronczek, F.R.; Wong, B.M.; Deay, K.R.; Mendy, J.S.; Gibson, D. Charge-assisted encapsulation of two chlorides by a hexaprotonated azamacrocycle. *Cryst. Growth Des.* **2010**, *10*, 1478–1481.
19. Ahmed, L.; Rhaman, M.M.; Mendy, J.S.; Wang, J.; Fronczek, F.R.; Powell, D.R.; Leszczynski, J.; Hossain, M. Experimental and theoretical studies on halide binding with a p-Xylyl-based azamacrocycle. *J. Phys. Chem. A* **2015**, *119*, 383–394.
20. Becke, A.D. Density-functional thermochemistry. III. The role of exact exchange. *J. Chem. Phys.* **1993**, *98*, 5648–5652.
21. Bader, R.F.W. *Atoms in Molecules: A Quantum Theory*; Clarendon Press: Oxford, UK, 1990.
22. Bader, R.F.W. A quantum theory of molecular structure and its applications. *Chem. Rev.* **1991**, *91*, 893–928.
23. Lee, C.; Yang, W.; Parr, R.G. Development of the Colle-Salvetti correlation-energy formula into a functional of the electron density. *Phys. Rev. B* **1988**, *37*, 785–789.
24. Stephens, P.J.; Devlin, F.J.; Chabalowski, C.F.; Frisch, M.J. Ab inition calculation of vibrational absorption and circular dichroism spectra using density functional force fields. *J. Phys. Chem.* **1994**, *98*, 11623–11627.
25. Miehlich, B.; Savin, A.; Stoll, H.; Preuss, H. Results obtained with the correlation energy density functionals of becke and Lee, Yang and Parr. *Chem. Phys. Lett.* **1989**, *157*, 200–206.
26. Zhao, Y.; Truhlar, D.G. Applications and validations of the minnesota density functional. *Chem. Phys. Lett.* **2011**, *502*, 1–13.
27. Zhao, Y.; Truhlar, D.G. The M06 suite of density functional for main group thermochemistry, thermochemical kinetics, noncovalent interactions, excited states, and transition elements: two new functional and systematic testing of four M06-class functional and 12 other functional. *Theor. Chem. Acc.* **2008**, *120*, 215–241.
28. Zhao, Y.; Truhlar, D.G. Density functional with broad applicability in chemistry. *Acc. Chem. Res.* **2008**, *41*, 157–167.
29. Hehre, W.J.; Radom, L.; Schleyer, P.R.; Pople, J.A. *Ab initio Molecular Orbital Theory*; Wiley: New York, NY, USA, 1986.
30. Cossi, M.; Barone, V.; Cammi, R.; Tomasi, J. Ab initio study of solvated molecules: A new implementation of the polarizable continuum model. *Chem. Phys. Lett.* **1996**, *255*, 327–335.
31. Frisch, M.J.; Trucks, G.W.; Schlegel, H.B.; Scuseria, G.E.; Robb, M.A.; Cheeseman, J.R.; Scalmani, G.; Barone, V.; Mennucci, B.; Petersson, G.A.; et al. *Gaussian 09, Revision D.01*; Gaussian, Inc.: Wallingford, CT, USA, 2009.

32. Biegler-König, F.; Schönbohm, J.; Bayles, D. AIM2000—A program to analyze and visualize atoms in molecules. *J. Comput. Chem.* **2001**, *22*, 545–559.
33. Vanquickenborne, L.G. Quantum chemistry of hydrogen bonds. In *Intermolecular Forces*; Huyskens, P.L., Luck, W.A.P., ZeegersHuyskens, T., Eds.; Springer: Berlin/Heidelberg, Germany, 1991; p. 41.
34. Grabowski, S.J.; Leszczynski, J. Unrevealing the nature of hydrogen bonds: PI-electron delocalization shapes H-Bond features. Intramolecular and intermolecular resonance-assisted hydrogen bonds in hydrogen bonding new insights. In *Challenges and Advances in Computational Chemistry and Physics*; Leszczynski, J., Ed.; Springer: Dordrecht, The Netherlands, 2006; Volume 3.
35. Lipkowski, P.; Grabowski, S.J.; Leszczynski, J. Properties of the halogen-hydride interaction: An ab initio and "atoms in molecules" analysis. *J. Phys. Chem. A* **2006**, *110*, 10296–10302.

$RCH_3\cdots O$ Interactions in Biological Systems: Are They Trifurcated H-Bonds or Noncovalent Carbon Bonds?

Antonio Bauzá and Antonio Frontera

Abstract: In this manuscript, we combine high-level *ab initio* calculations on some model systems (XCH_3 σ-hole/H-bond donors) and a Protein Data Bank (PDB) survey to distinguish between trifurcated H-bonds and noncovalent carbon bonds in $XCH_3\cdots O$ complexes (X = any atom or group). Recently, it has been demonstrated both experimentally and theoretically the importance of noncovalent carbon bonds in the solid state. When an electron-rich atom interacts with a methyl group, the role of the methyl group is commonly viewed as a weak H-bond donor. However, if the electron-rich atom is located equidistant from the three H atoms, the directionality of each individual H-bond in the trifurcated binding mode is poor. Therefore, the $XCH_3\cdots O$ interaction could be also defined as a tetrel bond ($C\cdots O$ interaction). In this manuscript, we shed light into this matter and demonstrate the importance of $XCH_3\cdots O$ noncovalent carbon bonding interactions in two relevant protein-substrate complexes retrieved from the PDB.

Reprinted from *Crystals*. Cite as: Bauzá, A.; Frontera, A. $RCH_3\cdots O$ Interactions in Biological Systems: Are They Trifurcated H-Bonds or Noncovalent Carbon Bonds? *Crystals* **2016**, *6*, 26.

1. Introduction

Supramolecular chemistry is a multidisciplinary field of research that develops very fast and has a deep impact [1,2] in the scientific community. Undoubtedly, the comprehension of the great deal of noncovalent forces, which are the basis of highly specific recognition, is crucial for the chemists working in this discipline. For instance, interactions between hosts and guests govern the creation of assemblies with high affinities even in highly competitive media [3–6]. For this reason, the correct description and understanding of noncovalent interactions between molecules is essential for being successful in this field of research. In general, strong and highly directional interactions, such as hydrogen bonding and σ-hole bonding [7–17], and less directional forces like ion pairing are used for this purpose.

The allocation of a hydrogen atom between a donor (D) and acceptor (A) moiety, $D-H\cdots A$ hydrogen bonding [18] is a particularly well studied and established supramolecular interaction ranging in strength from very weak (~1 kcal/mol) [19] to very strong (~40 kcal/mol) [20]. The weakest H-bonding interactions are established between weakly polarized D–H bonds (e.g., aliphatic CH) and mildly electron-rich H-acceptor moieties (e.g., alkene π-electrons) [21–23]. While a single interaction is energetically insignificant, when several weak H-bonds coexist they can stabilize protein structures [24–26] and contribute to the binding of proteins to carbohydrates [27–31]. Apart from this weak bonding interaction, recent theoretical explorations have suggested that pnicogen and tetrel atoms in their sp^3 hybridized form can act as electron poor entities suitable to accommodate an electron-rich guest [16,32–37]. Surprisingly few studies have been dedicated to studying the most abundant of these atoms, namely, the carbon atom. As a matter of fact, it has recently been reported that strong complexes are formed between electron-rich entities and 1,1,2,2-tetracyanocyclopropane [38–40]. Moreover, the ability of the carbon atom in a methyl group or in an aliphatic chain (sp^3 hybridized) to participate in σ-hole interactions (as a σ-hole donor, *i.e.* electron acceptor) has been explored by Mani and Arunan [41]. They demonstrated that the carbon atom in fact could act as an electrophilic center which can non-covalently bond with electron-rich entities leading to noncovalent carbon bonding, following a nomenclature analogous to other σ-hole interactions [7–17]. The theoretical predictions were confirmed experimentally by Guru Row's group [42], thus validating the existence of this type of bonding by means of X-ray charge density analysis. Electron density topologies in two prototypical crystal structures with potential carbon bonding motifs (R_3N^+–CH_3···O/Cl) were reported and revealed two distinct features of bond paths. That is, for the X-ray structure with the R_3N^+–CH_3···Cl motif, the bond path revealed a C–H···Cl hydrogen bond and for the other motif the bond path connected the electron-rich oxygen atom with the –CH_3 carbon atom; remarkably, no other bond paths connected the oxygen atom to the C–H hydrogen atoms. More recently, cooperativity effects involving carbon bonding interactions, and other noncovalent interactions have been analyzed in several theoretical studies [43–45].

The presence of methyl groups is abundant in many biologically relevant ligands and electron-rich O atoms are ubiquitous in proteins. Therefore, we wondered if perhaps weak non-covalent bonding with sp^3 hybridized carbon could have some relevance in ligand-protein complexes. Our approach was to conduct a rigorous statistical survey of the PDB together with quantum mechanical computations on the RI-MP2/aug-cc-pVTZ level of theory on some model systems. For the present theoretical study, we considered the sp^3 C-atom in several exemplifying molecules (**1–3**, see Figure 1) and computed their complexes with the electron-rich O atom of two molecules (formaldehyde and HO^-). In addition, using Bader's theory of "atoms in molecules" [46], we analyzed the bond path connecting the O atom with the CH_3 group in several complexes with the purpose of differentiating the H-bonding from the carbon bonding. From the PDB search, we analyzed the impact of the X–CH_3···O interactions in biologically relevant protein-ligand complexes.

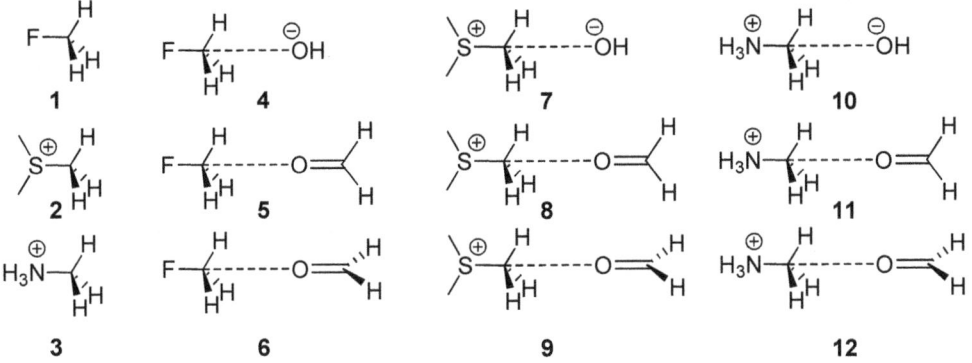

Figure 1. Compounds and complexes **1–12** studied in this work.

2. Results and Discussion

2.1. Preliminary MEP Analysis

We firstly computed the molecular electrostatic potential (MEP) mapped onto the van der Waals surface in very simple compounds (XCH_3) to explore the existence/absence of a σ-hole in the carbon atom (along the extension of the X–C bond) and to compare its electrostatic potential to that measured along the C–H bonds of the methyl group. In particular we computed the MEP surfaces of fluoromethane, dimethylether, dimethylthioether, dimethylformamide (DMF), acetonitrile, and methylamine, and the results are shown in Figure 2. In fluoromethane and dimethylether, an evident σ-hole is observed (see Figure 2a,b) in the C atom with a MEP value that is comparable to that observed in the H atoms. Therefore, either the carbon or the hydrogen bonding interactions should be equally favored, at least electrostatically, in both molecules. For the DMF and acetonitrile molecules, a perfectly defined σ-hole is not observed; however, both molecules present a significantly positive value of MEP at the C atom. Finally, the methylamine and dimethylthioether molecules (see Figure 2d,e) do not present a σ-hole at the C atom, although the electrostatic potential is slightly positive. Consequently, both molecules are better H-bond donors than carbon bond donors in terms of electrostatic effects.

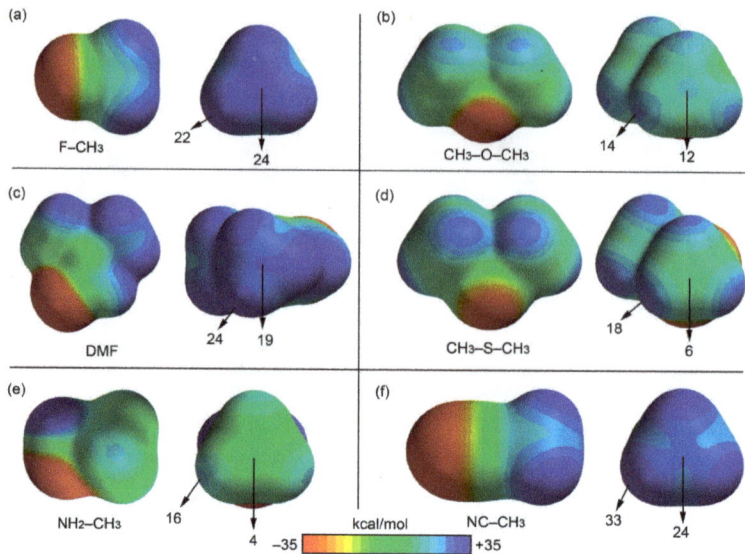

Figure 2. MEP surfaces of fluoromethane (**a**); dimethylether (**b**); DMF (**c**); dimethylthioether (**d**); methylamine (**e**); and acetonitrile (**f**). The MEP values at selected points are indicated.

2.2. Energetic and Geometric Results

Table 1 reports the interaction energies and equilibrium distances of the optimized complexes **4–12** (see Figures 1 and 3) computed at the RI-MP2/aug-cc-pVTZ level of theory. In complexes **7** and **10**, where the OH$^-$ anion interacts with charged carbon bond donors, the noncovalent complex was not found; instead, a nucleophilic S$_N$2 reaction occurs in the optimization. In formaldehyde complexes, we studied the influence of the orientation of the lone pairs of the O atom on the interaction energy and equilibrium distance. At this point, it should be noted that we used two charged carbon bond donors for the following reasons. First, the utilization of trimethylsulfonium was chosen to mimic the SAM cofactor (important in methyl transfer enzymatic processes). Second, we used methanaminium because, at physiological conditions, the amine groups are likely protonated.

From the inspection of the results summarized in Table 1 and Figure 3, several points are worth discussing. First, in the fluoromethane complexes, the interaction energy with the charged OH$^-$ guest is large and negative (−12.9 kcal/mol), modest for the complex with the neutral formaldehyde guest (−1.9 kcal/mol). Interestingly, the carbon bonding complexes (or trifurcated H-bonded) were found to be minima on the potential hypersurface. Moreover, the orientation of the oxygen lone pairs in the formaldehyde in its complexes with fluoromethane does not influence either the interaction energy or the equilibrium distance (complexes **5** and **6**). Second, the trimethylsulfonium and methanaminium complexes with neutral formaldehyde present large interaction energies, being more favorable with the protonated amine. Third, the interaction energies are identical for both orientations of formaldehyde; however, in the case of trimethylsulfonium, the C···O equilibrium distance is longer in complex **8** than in **9** and the S–C···O angle is smaller (172.0° in **8** and 178.6° in **9**, see Figure 3b). This behavior is not observed in methanaminium complexes **11** and **12**, and both present almost identical geometric and energetic features.

Table 1. Interaction energies (BSSE corrected, ΔE_{BSSE}, kcal/mol) and equilibrium distances (R_1 and R_2, Å) for complexes **4-12** at the RI-MP2/aug-cc-pVTZ level of theory.

Complex	ΔE_{BSSE}	R_1 (C···O)	R_2 (C···H)
4	−12.9	2.654	2.521
5	−1.9	2.993	2.829
6	−1.9	2.993	2.830
7	−[1]	−[1]	−[1]
8	−8.5	2.822	2.502
9	−8.5	2.803	2.665
10	−[1]	−[1]	−[1]
11	−9.7	2.746	2.600
12	−9.7	2.746	2.614

[1] S_N2 attack.

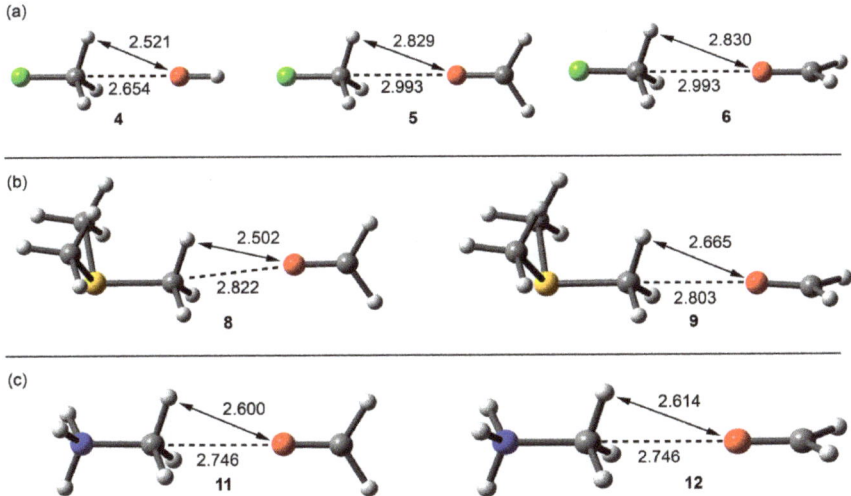

Figure 3. RI-MP2/aug-cc-pVTZ optimized complexes of fluoromethane (**a**); trimethylsulfonium (**b**); and methanaminium (**c**). Distances in Å.

2.3. AIM Analysis

We used Bader's theory of "atoms in molecules" (AIM) to characterize the noncovalent bonds in complexes **4–12** and to differentiate the type of interaction (carbon or hydrogen bonding). A bond critical point (CP) and a bond path connecting two atoms is unambiguous evidence of interaction. The AIM distribution of critical points and bond paths computed for the complexes is shown in Figure 4. In all CH_3F complexes (Figure 4a), the interaction is characterized by the presence of a bond CP that connects the O atom with the carbon atom, thus confirming the carbon bonding nature of this interaction. The value of ρ at the bond CP that emerges upon complexation is larger in complex **4** than in either complex **5** or **6**, in agreement with the interaction energies and equilibrium distances. In trimethylsulfonium complex **9**, where the S–C···O angle is close to 180°, the AIM analysis shows a bond CP and bond path unambiguously connecting the C and O atoms. In complex **8**, where the S–C···O angle is 172°, the bond path presents a different trajectory and, at first sight, seems to connect the O atom of formaldehyde to the C–H bond critical point. However, a closer look reveals that the bond path suddenly deviates when it reaches the C–H bond critical point and finally connects to the C atom. The distribution in methanaminium complexes **11** and **12** (see Figure 4c) clearly shows a bond CP and bond path connecting the C and O atoms, thus confirming the carbon bonding nature of the interaction. In all cases, the Laplacian of ρ at the bond CP that connects the O and C atoms is positive, as is common in closed shell interactions. Interestingly, for the whole series of complexes, the value of ρ strongly correlates with the interaction energies. Therefore, the value of ρ can be used as a measure of bond order in this type of noncovalent bonding.

We also analyzed the effect of the X–C···O angle on the interaction to find the value that causes a change from carbon bonding to hydrogen bonding. For this study, we progressively changed the F–C···O angle (α) starting from 180° and moving the O atom in two opposite directions (see Figure 5a): (i) toward a single H atom ($\alpha < 180°$ values) and (ii) toward the middle of two H atoms ($\alpha > 180°$ values). In each point, we optimized the geometry and simply froze the angle to the desired value. Moreover, we analyzed the distribution of critical points in order to find the value of α where the interaction changes from carbon to hydrogen bonding. If the OH^- moves to a single C–H bond (Figure 5b), the interaction rapidly changes from carbon bonding to hydrogen bonding; therefore, the critical angle is close to 170°. This result agrees with the behavior observed for complex **8** (see Figure 4) that presents an S–C···O angle of 172°, and the bond path trajectory reveals that it is in the borderline between both interactions. When the OH^- moves to the opposite direction (toward the middle of two C–H bonds), the behavior is different: For $\alpha = 190°$ a bond path connects the C and O atoms similarly to fully optimized complex **4**. Curiously, and for $\alpha = 200°$ and 210°, two bond paths connect the C and

O atoms, forming a bifurcated carbon bonding. As a consequence, a ring CP (yellow sphere) is also generated. Finally, at 220°, the carbon bonding changes to a bifurcated H-bonding interaction.

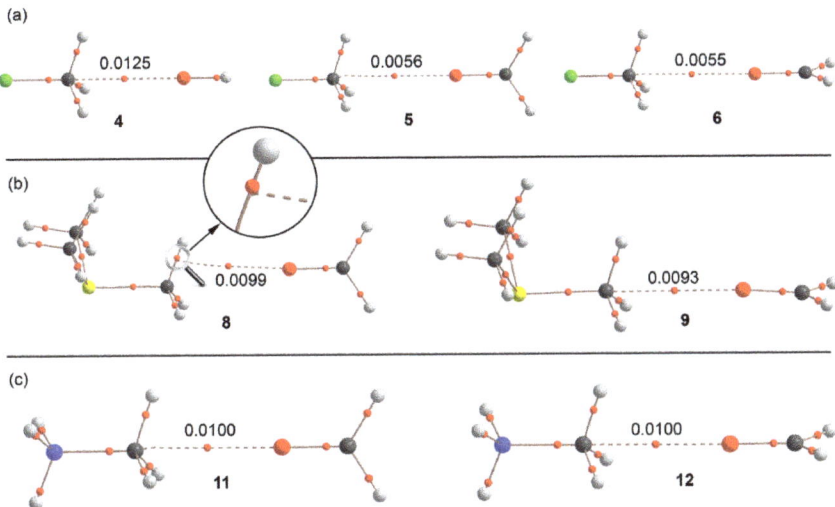

Figure 4. Distribution of critical points (red spheres) and bond paths for complexes of fluoromethane (**a**); trimethylsulfonium (**b**); and methanaminium (**c**) at the RI-MP2/aug-cc-pVTZ level of theory. The value of the charge density (ρ) at the bond critical points that emerge upon complexation are indicated in a.u.

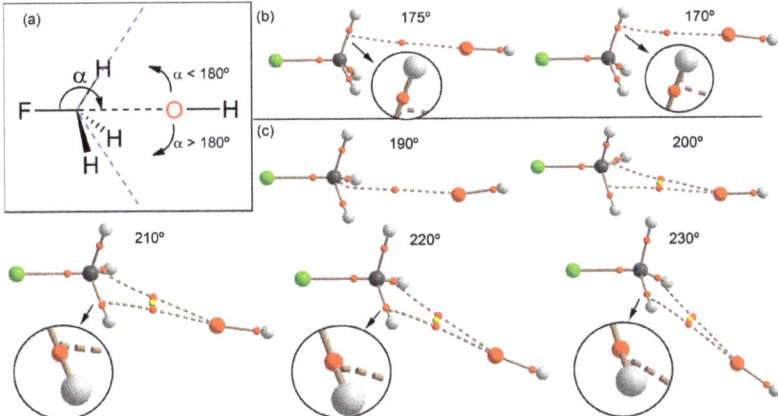

Figure 5. (**a**) Schematic representation of the angle and both directions; (**b**) distribution of critical points (red spheres) and bond paths for complexes of fluoromethane with OH⁻ using F–C···O angles <180° at the RI-MP2/aug-cc-pVTZ level of theory; (**c**) same as (**b**) but using angles α > 180°.

This analysis is useful for defining the search criteria of carbon bonding complexes in X-ray structure databases (CSD, PDB, *etc.*). Since the H atoms bonded to C are usually not experimentally located, there is some uncertainty regarding their real position in the CH_3 group. Therefore, it is recommended that a tight criterion regarding the X–C···O angle (*i.e.*, $\alpha > 170°$) is used when searching for this type of noncovalent carbon bonding X-ray structures.

2.4. PDB Search

We explored the PDB in order to prove the importance of carbon bonding interactions in biologically relevant molecules. The following criteria were used for the search: (i) The CH_3 donor group belongs to the ligand and the O atom is part of the protein; (ii) C···O distance (d) is shorter than 3.2 Å and the X–C···O angle (A) >170°; (iii) only X-ray solid state structures are considered (no NMR resolved); and (iv) a resolution factor <10 Å. The search was performed using the freely available Relibase software [47]. As a result, we found 459 protein-ligand complexes exhibiting this type of bonding (see ESI for the full list of hits and their geometrical features, Table S1). The histograms plots for the angle and distance are represented in Figure 6. The importance of this result should be emphasized since we found a large number of hits, taking into consideration the tight geometric criteria used and that only one electron-rich element was used in the search (O). Therefore, carbon bonding interaction in general has a bright future in this field and will likely become a prominent player for explaining some biological processes, either controlling or fine-tuning the binding of substrates to enzymes.

In the original manuscripts where the protein-ligand complexes were reported, the X–CH_3···O interaction was either overlooked or considered as an H-bonding by the original authors. We selected two examples from this search to further illustrate the importance of this interaction. For the first example [48], we selected a very well resolved X-ray structure (PDB ID: 4NSY, resolution 1.1 Å. see Figure 7a) that is a covalent complex between a trypsin-type serine protease (lysyl endoproteinase, LysC) and its inhibitor N^{α}-p-tosyl-lysyl chloromethylketone (TLCK). Chloromethyl ketones such as TLCK (Figure 7b) are well-known covalent inhibitors of cysteine and serine proteases [49]. Two covalent bonds are formed with serine proteases, each one to the active-site serine and histidine. Similarly, in the LysC enzyme, the TLCK inhibitor is attacked by SER194 and by nucleophilic substitution of the Cl atom by HIS57. This covalently bonded inhibitor interacts with the other subunit by means of a noncovalent carbon bonding interaction (see Figure 7a). We constructed a model to energetically evaluate this interaction. We substituted the SER193 and HIS47 that are covalently bonded to the inhibitor by C–H bonds in order to keep the size of the system computationally approachable. The model also includes the alanine residue and part of the peptide backbone (see Figure 7c). The interaction energy is −1.8 kcal/mol that is similar to that obtained for neutral complexes **5** and **6** (see Table 1). Remarkably, the distribution of CPs shows a bond critical point and a bond path connecting the O atom of the amide group to the carbon atom of the inhibitor, thus confirming the existence of the carbon bonding interaction in 4NSY.

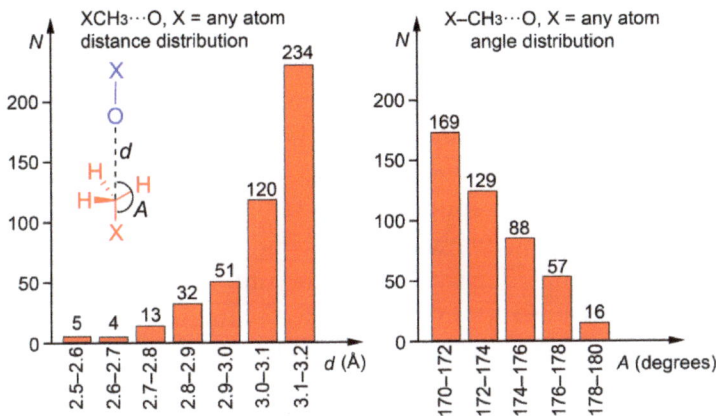

Figure 6. Histograms obtained directly from the Relibase software. Distance distribution is shown on the left and the angle distribution on the right of the figure. Inset figure: definition of *d* and A parameters; the protein is represented in blue and the ligand in red.

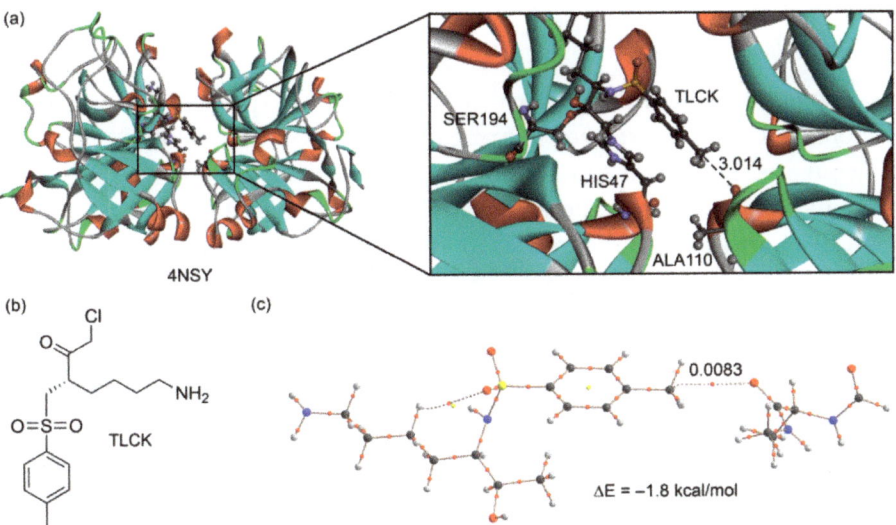

Figure 7. (a) X-ray structure of 4NSY with indication of the carbon bonding interaction (distance in Å); (b) chemical drawing of the TLCK; (c) AIM distribution of critical points of a model derived from the X-ray coordinates.

The second example we selected [50] corresponds to the structure of a M3 muscarinic acetylcholine receptor bound to tiotropium (PDB ID: 4U14, resolution 2.5 Å. see Figure 8a). Tiotropium is a muscarinic receptor antagonist (anticholinergic bronchodilator) used in the management of chronic obstructive pulmonary disease [51]. The presence of a dimethylammonium group in the structure ensures their ability to form electrostatically assisted carbon (or hydrogen) bonding interactions using the R_3N^+–CH_3 groups. As a matter of fact, an aspartate residue of the active site is very close to a carbon atom belonging to one of both methyl groups (2.935 Å) with an almost linear N–C···O angle (177.4°). We used a theoretical model derived from the crystallographic coordinates that includes the antagonist and the aspartate bonded to a fragment of the peptide backbone. The AIM analysis confirms the existence of a carbon bonding interaction (see bond path in Figure 8a). The ASP147 residue is likely deprotonated in the X-ray structure, since both C–O distances are very similar (1.248 and 1.250 Å). In any case, we evaluated the interaction energy considering both possibilities. The interaction energy of the carbon bonding complex using the protonated ASP147 is −9.0 kcal/mol. Moreover, the value of the density at the bond critical point is ρ = 0.0093 a.u. These values strongly agree with those previous ones computed for methanaminium (see Table 1 and Figure 4 (complexes **11** and **12**)). Considering the deprotonated ASP147, the interaction energy is very large (−73.9 kcal/mol) due to the strong electrostatic contribution. Therefore, this charge-assisted carbon bonding interaction has a strong influence on the binding of tiotropium in the muscarinic acetylcholine receptor.

2.5. NBO Analysis

To find out whether orbital effects are important to explain the carbon bonding interactions described above, we performed natural bond orbital (NBO) calculations focusing our attention on the second order perturbation analysis, due to its usefulness for studying donor-acceptor interactions [52]. We carried out the NBO calculations for complex **4** and for the theoretical model used to characterize the interaction in the PDB ID 4U14 (see Figure 8c). The results of the second order perturbation analysis are summarized in Table 2. For complex **4**, we found an important orbital contribution that consists in the interaction of the lone pair orbital (LP) of the donor with the C–F antibonding orbital (BD*) of the acceptor. This interaction is significant (3.5 kcal/mol) since it accounts for approximately 27% of the total interaction energy and further confirms that the interaction in complex **4** is with the σ-hole of the carbon atom. Moreover, the expected LP→BD*(C–H) orbital contribution that is typical of a hydrogen bond is less than 0.1 kcal/mol. Similarly, in the PDB 4U14, we found an interesting LP(O)→BD*(C–N) contribution that is 2.0 kcal/mol and that the LP(O)→BD*(C–H) interaction is less than 0.1 kcal/mol, in sharp agreement with the AIM analysis commented above.

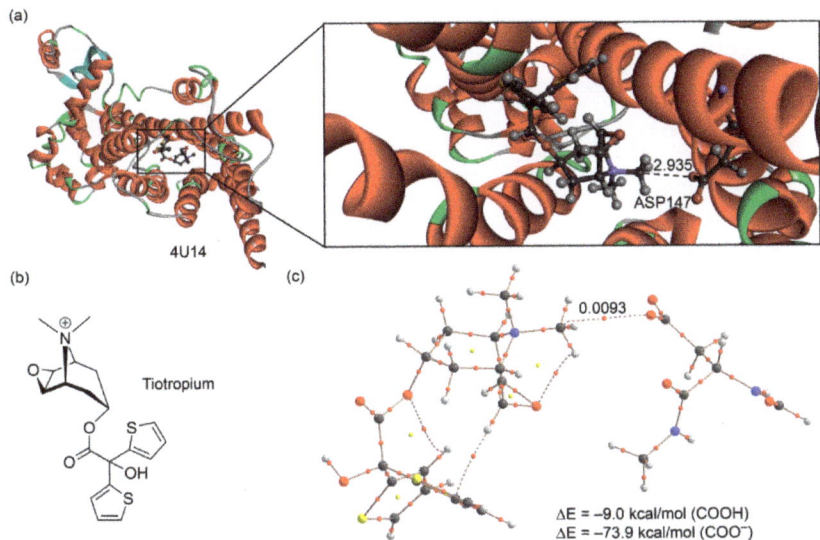

Figure 8. (a) X-ray structure of 4U14 with indication of the carbon bonding interaction (distance in Å); (b) chemical drawing of the tiotropium antagonist; (c) AIM distribution of critical points of a model retrieved from the X-ray coordinates.

Table 2. Donor and acceptor NBOs with indication of the second-order interaction energy $E^{(2)}$ and donor and acceptor orbitals for complexes **4** and PDB ID 4U14. Energy values are in kcal/mol.

Complex	Donor [1]	Acceptor	$E^{(2)}$
4	LP (O)	BD* (C–H)	<0.1
	LP(O)	BD* (C–F)	3.5
4U14	LP (O)	BD* (C–H)	<0.1
	LP(O)	BD* (C–N)	2.0

[1] LP stands for lone pair orbital and BD* for antibonding orbital.

2.6. SAPT Analysis

In Table 3, we summarize the DF-DFT-SAPT energy values relative to some of the carbon bonding complexes showed above in order to show the relative importance of the electrostatic contribution to the total interaction energies, especially those involving charged carbon bonding donors. The total SAPT interaction energies for these three complexes are similar to those obtained using the RI-MP2/aug-cc-pVTZ level of theory (see Table 1), attributing reliability to the partition method and the level of theory used to compute the SAPT. The energetic contributions (Table 3) indicate that the cationic complexes **9** and **12** are clearly dominated by the electrostatic term. Moreover, the contributions of induction and dispersion terms are also important. In the neutral complex **6**, the electrostatic and dispersion terms equally contribute to the total interaction energy.

Table 3. SAPT interaction energies (E_{total}, kcal/mol) and their partitioning into the electrostatic, exchange, induction, dispersion, and contributions (E_{ee}, E_{ex}, E_{ind}, E_{disp}, respectively, kcal/mol) at the RI-DFT/aug-cc-pVTZ level of theory using the DF-DFT-SAPT approach.

Complex	E_{ee}	E_{ex}	E_{ind}	E_{disp}	E_{total}
6	−1.8	2.1	−0.2	−1.6	−1.5
9	−8.3	4.4	−1.3	−2.4	−7.6
12	−9.5	4.8	−1.7	−2.3	−8.7

To further demonstrate the importance of electrostatic forces in the interaction energies of compounds **9** and **12**, we computed the carbon bonding interaction of their equivalent neutral complexes where methylamine and dimethylthioether are used as carbon bond donors. The results are gathered in Figure 9, and it can be observed that the interaction energies are very small (−0.7 kcal/mol for **13** and −1.1 kcal/mol for **14**), in agreement with the MEP surfaces of methylamine and dimethylthioether shown in Figure 2. Moreover, the localization of the minima corresponding to the carbon bonding complexes **13** and **14** (see Figure 9) on the potential surface is worth emphasizing, since the MEP value at the H-atoms is considerably more positive than that at the C atom in both neutral molecules.

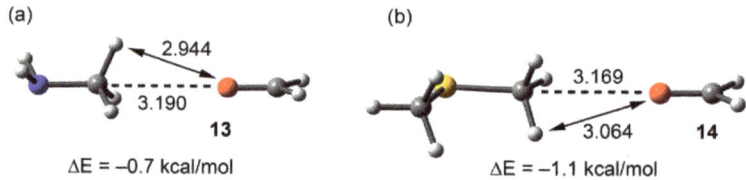

Figure 9. RI-MP2/aug-cc-pVTZ optimized complexes end interaction energies of methylamine (**a**) and dimethylthioether (**b**). Distances in Å.

3. Theoretical Methods

The geometries of the complexes studied herein have been fully optimized at the RI-MP2/aug-cc-pVTZ level of theory. Cartesian coordinates of the optimized complexes are given in the supplementary material file (Tables S2–S8). The calculations have been performed by using the program TURBOMOLE version 7.0 [53]. The interaction energies were calculated with correction for the basis set superposition error (BSSE) by using the Boys–Bernardi counterpoise technique [54]. The C_s symmetry point group has been used in the optimization of the complexes. The minimum nature of the complexes has been confirmed by carrying out frequency calculations. For the theoretical analysis of the noncovalent interactions present in PDB structures 4NSY and 4U14, the BP86-D3/aug-cc-pVTZ level of theory was used; the position of the hydrogen atoms present in these structures was optimized prior to the evaluation of the binding energy values. For the hereoatoms we used the crystallographic coordinates, we used the DFT-D functional with the latest available correction for dispersion (D3) [55]. Bader's "atoms in molecules" theory was used to study the interactions discussed herein by means of the AIMall calculation package [56]. Finally, the partitioning of the interaction energies into the individual electrostatic, induction, dispersion, and exchange-repulsion components was carried out with the symmetry adapted intermolecular perturbation theory approach DFT-SAPT at the BP86/aug-cc-pVTZ level of theory using the aug-cc-pVQZ basis set for the MP2 density fitting [57] by means of the MOLPRO program [58,59].

4. Conclusions

In this manuscript, we analyzed the dual ability of the methyl group (XCH_3) to act as either H-bond or carbon bond donor in complexes with electron-rich oxygen atoms by means of high-level *ab initio* calculations and using Bader's theory of atoms in molecules. For X–C···O complexes exhibiting angles close to linearity, a carbon bonding interaction instead of H-bonding is established as confirmed by AIM and NBO analyses. The importance of electrostatic and dispersion contributions to the interaction energy of the carbon bonding complexes was shown using the SAPT partition scheme. We demonstrated the importance of latter interactions in biological systems by examining the PDB and illustrated this in two selected examples. Since –CH_3 groups are commonplace, non-covalent carbon bonding involving this group might turn out to be as functionally relevant as other σ-hole and hydrogen bonding interactions.

Supplementary Materials: The following are available online at http://www.mdpi.com/2073-4352/6/3/26/s1. Table S1: Results from the PDB search. Distances in Å. Angles in degrees. Table S2: Cartesian coordinates of the complexes **4**. Table S3: Cartesian coordinates of the complexes **5**. Table S4: Cartesian coordinates of the complexes **6**. Table S5: Cartesian coordinates of the complexes **8**. Table S6: Cartesian coordinates of the complexes **9**. Table S7: Cartesian coordinates of the complexes **11**. Table S8: Cartesian coordinates of the complexes **12**.

Acknowledgments: We thank CONSOLIDER-Ingenio 2010 (project CSD2010-0065) and the MICINN of Spain (project CTQ2014-57393-C2-1-P FEDER funds). We thank the CTI for computational facilities.

Author Contributions: Antonio Bauzá and Antonio Frontera conceived and designed the calculations; Antonio Bauzá and Antonio Frontera analyzed the data; Antonio Frontera wrote the paper.

Conflicts of Interest: The authors declare no conflict of interest. The founding sponsors had no role in the design of the study, in the collection, analyses, or interpretation of data, in the writing of the manuscript, or in the decision to publish the results.

Abbreviations

The following abbreviations are used in this manuscript:

AIM	Atoms in molecules
DMF	Dimethylformamide
MEP	Molecular electrostatic potential
MP2	Second order Moller-Plesset
SAM	S-Adenosyl methionine
PDB	Protein Databank
CSD	Cambridge Structural Database
CP	Critical point
LysC	lysyl endoproteinase
TLCK	N^{α}-p-tosyl-lysyl chloromethylketone
ASP	Aspartate
HIS	Histidine
SER	Serine
NBO	Natural Bond Orbital
SAPT	Symmetry adapted perturbation theory

References

1. Schneider, H.J. Binding mechanisms in supramolecular complexes. *Angew. Chem. Int. Ed.* **2009**, *48*, 3924–3977.
2. Schneider, H.J.; Yatsimirski, A. *Principles and Methods in Supramolecular Chemistry*; Wiley: Chichester, UK, 2000.
3. Lehn, J.M. *Supramolecular Chemistry Concepts and Perspectives*; Wiley–VCH: Weinheim, Germany, 1995.
4. Vögtle, F. *Supramolecular Chemistry: An Introduction*; Wiley: New York, NY, USA, 1993.
5. Beer, P.D.; Gale, P.A.; Smith, D.K. *Supramolecular Chemistry*; Oxford University Press: Oxford, UK, 1999.
6. Steed, J.W.; Atwood, J.L. *Supramolecular Chemistry*; Wiley: Chichester, UK, 2000.
7. Destecroix, H.; Renney, H.C.M.; Mooibroek, T.J.; Carter, T.S.; Stewart, P.F.N.; Crump, M.P.; Davis, A.P. Affinity enhancement by dendritic side chains in synthetic carbohydrate receptors. *Angew. Chem. Int. Ed.* **2015**, *54*, 2057–2061.
8. Murray-Rust, P.; Motherwell, W.D.S. Computer retrieval and analysis of molecular geometry. 4. Intermolecular interactions. *J. Am. Chem. Soc.* **1979**, *101*, 4374–4376.

9. Murray-Rust, P.; Stallings, W.C.; Monti, C.T.; Preston, R.K.; Glusker, J.P. Intermolecular interactions of the carbon-fluorine bond: The crystallographic environment of fluorinated carboxylic acids and related structures. *J. Am. Chem. Soc.* **1983**, *105*, 3206–3214.
10. Ramasubbu, N.; Parthasarathy, R.; Murray-Rust, P. Angular preferences of intermolecular forces around halogen centers: Preferred directions of approach of electrophiles and nucleophiles around carbon-halogen bond. *J. Am. Chem. Soc.* **1986**, *108*, 4308–4314.
11. Metrangolo, P.; Neukirch, H.; Pilati, T.; Resnati, G. Halogen bonding based recognition processes: A world parallel to hydrogen bonding. *Acc. Chem. Res.* **2005**, *38*, 386–395.
12. Politzer, P.; Murray, J.S. Halogen bonding: An interim discussion. *ChemPhysChem* **2013**, *14*, 278–294.
13. Bauzá, A.; Mooibroek, T.J.; Frontera, A. The Bright Future of Unconventional σ/π-Hole Interactions. *ChemPhysChem* **2015**, *16*, 2496–2517.
14. Politzer, P.; Murray, J.S.; Clark, T. Halogen bonding: An electrostatically-driven highly directional noncovalent interaction. *Phys. Chem. Chem. Phys.* **2010**, *12*, 7748–7757.
15. Bauzá, A.; Mooibroek, T.J.; Frontera, A. Tetrel Bonding Interactions. *Chem. Rec.* **2016**, *16*, 473–487.
16. Bauzá, A.; Mooibroek, T.J.; Frontera, A. Tetrel-bonding interaction: Rediscovered supramolecular force? *Angew. Chem. Int. Ed.* **2013**, *52*, 12317–12321.
17. Bauzá, A.; Frontera, A. Competition between halogen bonding and π-hole interactions involving various donors: the role of dispersion effects. *ChemPhysChem* **2015**, *16*, 3108–3113.
18. Desiraju, G.R. A bond by any other name. *Angew. Chem. Int. Ed.* **2011**, *50*, 52–59.
19. Desiraju, G.R.; Steiner, T. *The Weak Hydrogen Bond in Structural Chemistry and Biology*; Oxford University Press: Oxford, UK, 1999.
20. Larson, J.W.; McMahon, T.B. Gas-phase bihalide and pseudobihalide ions. An ion cyclotron resonance determination of hydrogen bond energies in XHY- species (X, Y = F, Cl, Br, CN). *Inorg. Chem.* **1984**, *23*, 2029–2033.
21. Nishio, M.; Umezawa, Y.; Fantini, J.; Weiss, M.S.; Chakrabarti, P. CH–π hydrogen bonds in biological macromolecules. *Phys. Chem. Chem. Phys.* **2014**, *16*, 12648–12683.
22. Nishio, M.; Umezawa, Y.; Honda, K.; Tsuboyamad, S.; Suezawae, H. CH/π hydrogen bonds in organic and organometallic chemistry. *CrystEngComm* **2009**, *11*, 1757–1788.
23. Nishio, M.; Hirota, M.; Umezawa, Y. *The C–H/π Interaction: Evidence, Nature, Consequences*; Wiley: New York, NY, USA, 1998.
24. Samanta, U.; Pal, D.; Chakrabarti, P. Environment of tryptophan side chains in proteins. *Proteins* **2000**, *38*, 288–300.
25. Plevin, M.J.; Bryce, D.L.; Boisbouvier, J. Direct detection of CH/π interactions in proteins. *Nat. Chem.* **2010**, *2*, 466–471.
26. Brandl, M.; Weiss, M.M.S.; Jabs, A.; Sühnel, J.; Hilgenfeld, R. C-H...pi-interactions in proteins. *J. Mol. Biol.* **2001**, *307*, 357–377.
27. Quiocho, F.A.; Vyas, N.K. Novel stereospecificity of the L-arabinose-binding protein. *Nature* **1984**, *310*, 381–386.

28. Vyas, N.K.; Vyas, M.N.; Quiocho, F.A. Sugar and signal-transducer binding sites of the *Escherichia coli* galactose chemoreceptor protein. *Science* **1988**, *242*, 1290–1295.
29. Nishio, M. The CH/π hydrogen bond in chemistry. Conformation, supramolecules, optical resolution and interactions involving carbohydrates. *Phys. Chem. Chem. Phys.* **2011**, *13*, 13873–13900.
30. Laughrey, Z.R.; Kiehna, S.E.; Riemen, A.J.; Waters, M.L. Carbohydrate−π Interactions: what are they worth? *J. Am. Chem. Soc.* **2008**, *130*, 14625–14633.
31. Barwell, N.P.; Davis, A.P. Substituent effects in synthetic lectins—Exploring the role of CH−π interactions in carbohydrate recognition. *J. Org. Chem.* **2011**, *76*, 6548–6557.
32. Politzer, P.; Murray, J.S.; Clark, T. Halogen bonding and other σ-hole interactions: A perspective. *Phys. Chem. Chem. Phys.* **2013**, *15*, 11178–11189.
33. Murray, J.S.; Riley, K.E.; Politzer, P.; Clark, T. Directional weak intermolecular interactions: σ-hole bonding. *Aust. J. Chem.* **2010**, *63*, 1598–1607.
34. Clark, T. σ-Holes. *Wiley Interdiscip. Rev. Comput. Mol. Sci.* **2013**, *3*, 13–20.
35. Bauzá, A.; Mooibroek, T.J.; Frontera, A. σ-Hole opposite to a lone pair: unconventional pnicogen bonding interactions between ZF_3 (Z = N, P, As and Sb). *ChemPhysChem* **2016**, *17*.
36. Bundhun, A.; Ramasami, P.; Murray, J.S.; Politzer, P. Trends in σ-hole strengths and interactions of F_3MX molecules (M = C, Si, Ge and X = F, Cl, Br, I). *J. Mol. Model.* **2013**, *19*, 2739–2746.
37. Grabowski, S.J. Tetrel bond–σ-hole bond as a preliminary stage of the SN_2 reaction. *Phys. Chem. Chem. Phys.* **2014**, *16*, 1824–1834.
38. Bauzá, A.; Mooibroek, T.J.; Frontera, A. Small cycloalkane $(CN)_2C–C(CN)_2$ Structures Are Highly Directional Non-covalent Carbon-Bond Donors. *Chem. Eur. J.* **2014**, *20*, 10245–10248.
39. Bauzá, A.; Mooibroek, T.J.; Frontera, A. 1,1,2,2-Tetracyanocyclopropane (TCCP) as supramolecular synthon. *Phys. Chem. Chem. Phys.* **2016**, *18*, 1693–1698.
40. Escudero-Adán, E.C.; Bauzá, A.; Frontera, A.; Ballester, P. Nature of noncovalent carbon-bonding interactions derived from experimental charge-density analysis. *ChemPhysChem* **2015**, *16*, 2530–2533.
41. Mani, D.; Arunan, E. The X–C···Y (X = O/F, Y = O/S/F/Cl/Br/N/P) 'carbon bond' and hydrophobic interactions. *Phys. Chem. Chem. Phys.* **2013**, *15*, 14377–14383.
42. Thomas, S.P.; Pavan, M.S.; Row, T.N.G. Experimental evidence for 'carbon bonding' in the solid state from charge density analysis. *Chem. Commun.* **2014**, *50*, 49–51.
43. Solimannejad, M.; Orojloo, M.; Amani, S. Effect of cooperativity in lithium bonding on the strength of halogen bonding and tetrel bonding: $(LiCN)_n \cdots ClYF_3$ and $(LiCN)_n \cdots YF_3Cl$ (Y = C, Si and n = 1–5) complexes as a working model. *J. Mol. Model.* **2015**, *21*, 183.
44. Marín-Luna, M.; Alkorta, I.; Elguero, J. Cooperativity in Tetrel Bonds. *J. Phys. Chem. A* **2016**, *120*, 648–656.
45. Guo, X.; Liu, Y.-W.; Li, Q.-Z.; Li, W.-Z.; Cheng, J.-B. Competition and cooperativity between tetrel bond and chalcogen bond in complexes involving F2CX (X = Se and Te). *Chem. Phys. Lett.* **2015**, *620*, 7–12.

46. Bader, R.F.W. A quantum theory of molecular structure and its applications. *Chem. Rev.* **1991**, *91*, 893–928.
47. Relibase. Available online: https://www.ccdc.cam.ac.uk/Community/freeservices/Relibase/ (accessed on 25 January 2016).
48. Asztalos, P.; Müller, A.; Hölke, W.; Sobek, H.; Rudolph, M.G. Atomic resolution structure of a lysine-specific endoproteinase from Lysobacter enzymogenes suggests a hydroxyl group bound to the oxyanion hole. *Acta Crystallogr. Sect. D—Biol. Crystallogr.* **2014**, *70*, 1832–1843.
49. Drenth, J. Binding of chloromethyl ketone substrate analogues to crystalline papain. *Biochemistry* **1976**, *15*, 3731–3738.
50. Thorsen, T.S.; Matt, R.; Weis, W.I.; Kobilka, B.K. Modified T4 Lysozyme Fusion Proteins Facilitate G Protein-Coupled Receptor Crystallogenesis. *Structure* **2014**, *22*, 1657–1664.
51. Kato, M.; Komamura, K.; Kitakaze, M. Tiotropium, a novel muscarinic M3 receptor antagonist, improved symptoms of chronic obstructive pulmonary disease complicated by chronic heart failure. *Circ. J.* **2006**, *70*, 1658–1660.
52. Weinhold, F.; Landis, C.R. *Valency and Bonding: A Natural Bond Orbital Donor-Acceptor Perspective*; Cambridge University Press: Cambridge, UK, 2005.
53. Ahlrichs, R.; Bär, M.; Hacer, M.; Horn, H.; Kömel, C. Electronic structure calculations on workstation computers: the program system TURBOMOLE. *Chem. Phys. Lett.* **1989**, *162*, 165–169.
54. Boys, S.B.; Bernardi, F. The calculation of small molecular interactions by the differences of separate total energies. Some procedures with reduced errors. *Mol. Phys.* **1970**, *19*, 553–566.
55. Grimme, S.; Antony, J.; Ehrlich, S.; Krieg, H.A. Consistent and accurate *ab initio* parametrization of density functional dispersion correction (DFT-D) for the 94 elements H-Pu. *J. Chem. Phys.* **2010**, *132*, 154104–154119.
56. Todd, A.; Keith, T.K. *AIMAll*; Version 13.05.06; Gristmill Software: Overland Park, KS, USA, 2013.
57. Heßelmann, A.; Jansen, G. The helium dimer potential from a combined density functional theory and symmetry-adapted perturbation theory approach using an exact exchange–correlation potential. *Phys. Chem. Chem. Phys.* **2003**, *5*, 5010–5014.
58. Werner, H.-J.; Knowles, P.J.; Knizia, G.; Manby, F.R.; Schütz, M. Molpro: A general-purpose quantum chemistry program package. *WIREs Comput. Mol. Sci.* **2012**, *2*, 242–253.
59. Werner, H.-J.; Knowles, P.J.; Knizia, G.; Manby, F.R.; Schütz, M.; Celani, P.; Korona, T.; Lindh, R.; Mitrushenkov, A.; Rauhut, G.; *et al.* MOLPRO, Version 2012.1; a Package of ab Initio Programs. Available online: https://www.molpro.net/info/authors (accessed on 15 March 2016).

Effect of Intra- and Intermolecular Interactions on the Properties of *para*-Substituted Nitrobenzene Derivatives

Halina Szatylowicz, Olga A. Stasyuk, Célia Fonseca Guerra and Tadeusz M. Krygowski

Abstract: To study the influence of intra- and intermolecular interactions on properties of the nitro group in *para*-substituted nitrobenzene derivatives, two sources of data were used: (i) Cambridge Structural Database and (ii) quantum chemistry modeling. In the latter case, "pure" intramolecular interactions were simulated by gradual rotation of the nitro group in *para*-nitroaniline, whereas H-bond formation at the amino group allowed the intermolecular interactions to be accounted for. BLYP functional with dispersion correction and TZ2P basis set (ADF program) were used to perform all calculations. It was found that properties of the nitro group dramatically depend on both its orientation with respect to the benzene ring as well as on the substituent in the *para*-position. The nitro group lies in the plane of the benzene ring for only a small number of molecules, whereas the mean value of the twist angle is 7.3 deg, mostly due to intermolecular interactions in the crystals. This distortion from planarity and the nature of *para*-substituent influence the aromaticity of the ring (described by HOMA index) and properties of the nitro group due to electronic effects. The results obtained by QM calculations fully coincide with observations found for the data set of crystal structures.

> Reprinted from *Crystals*. Cite as: Szatylowicz, H.; Stasyuk, O.A.; Guerra, C.F.; Krygowski, T.M. Effect of Intra- and Intermolecular Interactions on the Properties of *para*-Substituted Nitrobenzene Derivatives. *Crystals* **2016**, *6*, 29.

1. Introduction

The nitro group is one of the most frequently encountered substituents in organic chemistry [1]. In the Cambridge Structural Database (CSD) [2], 38406 crystal structures of chemical compounds containing the NO_2 group are reported, whereas considering more precise measurements—the numbers are 36030 and 20662 with *R*-factor equal to or less than 0.10 and 0.05, respectively. This particular interest in nitro-compounds is due to their very important properties. The nitro group is one of the most typical electron-accepting substituents with the substituent constant $\sigma_p = 0.78$ [3]. Recently, we analyzed the effect of the nitro group with Kohn-Sham molecular orbital theory to separate inductive and resonance effects [4]. Its electron-accepting power (EAP) originates via two mechanisms.

There is a strong inductive activity due to a combined effect of three electronegative atoms (the group electronegativity χ_{NO_2} is equal to 4.00 for coplanar and 4.19 for perpendicular orientation with respect to the benzene ring [5]). Furthermore, the two electronegative oxygen atoms cause an electron deficiency on the nitrogen atom and hence the nitro group may also act as a strong π-electron acceptor; the resonance power of the nitro group becomes significant when the electron-donating substituent is present in the *para*-position, as documented by the σ_p^- constants being equal to 1.27 [3,6]. It is important to note that rotation of this group around the C-N bond with respect to the benzene ring decreases the efficiency of the through-resonance interactions and the substituent constant σ_p^- reduces by almost a half, from 1.27 for planar conformation to 0.72 for the perpendicular one [7].

Understanding of structural aspects of this group is very important because nitro-compounds belong to very important chemicals applied in medicinal and pharmaceutical chemistry [8], used as explosives [9,10] and fertilizers [11] as well as in other fields of chemistry [1].

Deformations in geometric and electronic structure of the nitro group as well as the deviation from coplanarity with the aromatic moiety are of great importance, since these effects may change significantly both the nature of the group as well as the effect on the substituted system. To study these effects we have chosen *para*-substituted nitrobenzene derivatives, since one can expect in these systems a substantial variability of aforementioned properties [12]. To this end, we have undertaken detailed studies on: (i) deformations of para-substituted nitrobenzene derivatives found in the solid state; (ii) controlled deformations carried out by rotation of the nitro group in *para*-nitroaniline (PNA), Scheme 1; (iii) the influence of intermolecular hydrogen (H-) bonding of the amino group on properties of the nitro group in PNA.

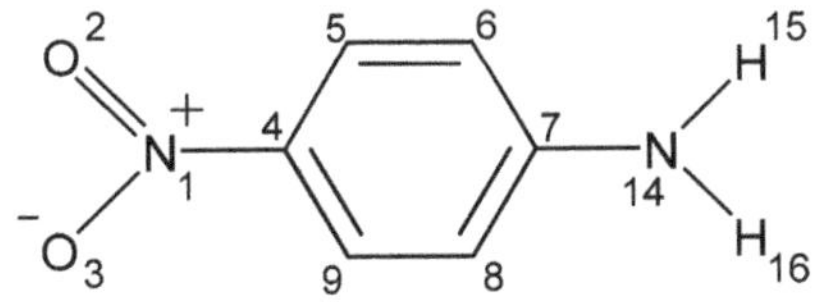

Scheme 1. Labeling of atoms in the *p*-nitroaniline.

2. Experimental and Computational Section

Geometries of the studied derivatives were retrieved from CSD [2] with the following restrictions:

(1) The search was performed for organic compounds containing *para*-substituted nitrobenzene fragment(s).

(2) The search was restricted to structure measurements with the reported mean estimated standard deviation (esd) for the CC bond $\leqslant 0.005$ Å, with 3D coordinates determined, not disordered, without errors, polymeric, ions and powder structures, and with $R \leqslant 0.05$. Sometimes solvent molecules were also present in the crystal latice.

For the analysis of the retrieved data Vista v2.1 [13] was used.

All quantum chemical calculations were carried out using the Amsterdam Density Functional (ADF) program [14,15]. Geometries and energies were calculated using the generalized gradient approximation (GGA) with BLYP functional [16,17] and DFT-D3 correction [18] as the best correction for non-covalent interactions [19]. The MOs were expanded in uncontracted sets of Slater type orbitals (STOs) containing diffuse functions with two sets of polarization functions [20] (TZ2P).

To investigate the effect of the intermolecular H-bond on the NO_2 group properties, two types of H-bonded complexes were studied: (i) with neutral and (ii) charge-assisted hydrogen bonds (Scheme 2). In the first case, $N \cdots HF$ and $H \cdots FH$ interactions were considered. In the second one, $N^- \cdots HF$ and $NH \cdots F^-$ interactions were taken into account.

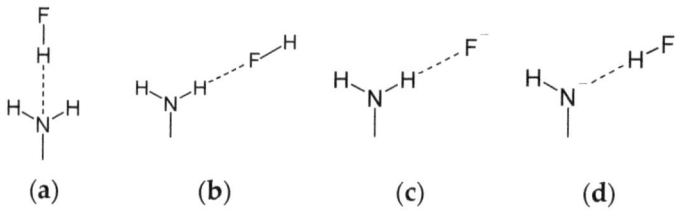

(a)   (b)   (c)   (d)

Scheme 2. Schematic presentation of the studied H-bonded interactions of the amino group.

The total energy of the H-bond consists of two major components: the deformation (ΔE_{def}) and interaction (ΔE_{int}) energies.

$$\Delta E_{tot} = \Delta E_{def} + \Delta E_{int} \qquad (1)$$

The first term represents the amount of energy required to deform the equilibrium geometries (E') of two fragments into their geometries in the complex (E).

$$\Delta E_{def} = (E_A - E_A') + (E_B - E_B') \qquad (2)$$

The second term of Equation (1) corresponds to the actual energy change when the two distorted fragments are combined in the final structure. According to Ziegler-Rauk method [21–23], ΔE_{int} can be decomposed into a number of chemically meaningful components representing different steps toward the formation of a complex from two selected fragments:

$$\Delta E_{int} = \Delta V_{elstat} + \Delta E_{Pauli} + \Delta E_{oi} + \Delta E_{disp} \qquad (3)$$

The term ΔV_{elstat} corresponds to the classical electrostatic interaction between the unperturbed charge distributions of the distorted fragments and is usually attractive. The Pauli repulsion energy, ΔE_{Pauli}, comprises the destabilizing interactions between occupied orbitals and is responsible for the steric repulsion. The orbital interaction energy, ΔE_{oi}, represents the donor–acceptor interactions between the occupied molecular orbitals on one fragment with the unoccupied molecular orbitals of the other fragment, as well as the mixing of occupied and virtual orbitals within the same fragment (intrafragment polarization). The term ΔE_{disp} accounts for the dispersion corrections as introduced by Grimme and co-workers [18,24,25].

The changes in the H-bond strength were achieved by elongation of the distance between N and F atoms from equilibrium to 4.0 Å with step of 0.3 Å. In all H-bonded complexes a linearity of N···H···F atoms was imposed (to maintain the same settings during the modeling of the H-bonded complexes). It is well known that this angle is usually close to 180 deg for strong hydrogen bonds [26,27]. The calculated differences in interaction energies for linear and bent bonding in equilibrium complexes are less than 0.1 kcal/mol. To estimate the effect of *para*-substitution on the benzene ring the same types of H-bonded complexes for aniline and *p*-nitroaniline were studied.

The electron-accepting properties of the nitro group were modified with the rotation of this group from coplanar to perpendicular orientation with respect to the benzene ring. This angle φ ($0 \leq \varphi \leq 90$) was changed with step of 10 and 30 degrees for intra- and intermolecular interactions, respectively (the NH_2 group remains fixed).

The electron density distribution was analyzed using the Voronoi deformation density (VDD) method [28,29], which is an independent basis set and yields chemically meaningful atomic charges.

The charge of the substituent active region, cSAR, [30–32] was used to characterize the substituent. It is defined as the sum of atomic charges of the substituent X and the *ipso* carbon atom:

$$cSAR(X) = q(X) + q(C_{ipso}) \qquad (4)$$

π-Electron delocalization [33] of the phenyl ring was estimated applying the aromaticity index HOMA (Harmonic Oscillator Model of Aromaticity) [34,35] which reads:

$$\text{HOMA} = 1 - \frac{1}{n}\sum_{i=1}^{n} \alpha \left(R_{opt} - R_i\right)^2 \qquad (5)$$

where n is the number of CC bonds taken into account when carrying out the summation, α is a normalization constant ($\alpha_{CC} = 257.7$) fixed to give HOMA = 0 for a model nonaromatic system and HOMA = 1 for the system with all bonds equal to the optimal value $R_{opt,CC} = 1.388$ Å, and R_i denotes the computed bond lengths.

3. Results and Discussion

Three sources of molecular geometry deformations in *para*-substituted nitrobenzene derivatives are discussed: (i) substituents; (ii) intermolecular interactions when the molecule is subjected to chemical reactions, e.g., by protonation process; and (iii) interactions in crystals. The first two cases are relatively simple. The reason for the perturbation is known and its magnitude may be qualitatively estimated by applying the standard rules used for evaluation of substituent effects or by estimation of the rate/equilibrium constants or H-bond energy. In the third case, deformation of every symmetrically independent molecule in the elementary crystal cell may be different even for the same crystal. Consequently, for a data set taken for various *para*-substituted nitrobenzene derivatives in the crystalline state, the source of the structural deformation is the so-called crystal packing forces and hence presents a collection of chaotically acting reasons.

On these grounds our discussion is carried out in two parts. The first one deals with deformations observed for various *para*-substituted nitrobenzene derivatives in the crystalline state taken from CSD as a source of information. Then the application of the quantum chemical modeling (QM) enables us to enlighten the experimental data with some controlled quantitative descriptions.

3.1. Analyses of Molecular Geometry Deformations Based on the Cambridge Structural Database

Two kinds of deformations are considered: (i) changes in geometry of the nitro groups, as ONO angle and NO bond lengths; and (ii) changes in the adjacent parts, as the linking the CN bond length as well as the dihedral angle between the C-NO$_2$ and phenyl ring planes. Figures 1–4 represent the above-mentioned characteristics obtained from CSD (1750 hits with 2071 structural studied fragments) for *para*-substituted nitrobenzene derivatives; phenyl ring aromaticity is shown in Figure 5. All these histograms were prepared using Vista v2.1 [13].

When the ONO angle is considered (Figure 1), the distribution is very symmetrical with a mean value of 123.446 deg. The value of the angle greater than 120 deg, as expected for sp^2 hybridization, may result from repulsion forces due to the interactions between two N(+)O(−) bond dipoles. Low temperature (104K) X-ray determination found this ONO angle equal to 123.2(1) deg [36], whereas 125.3(2) deg from gas phase electron diffraction determination [37] and 124.35(1) deg from microwave determination [37] were determined. It seems probable that the greater angles estimated for free molecules are due to there being more space around the nitro group than in the crystal state.

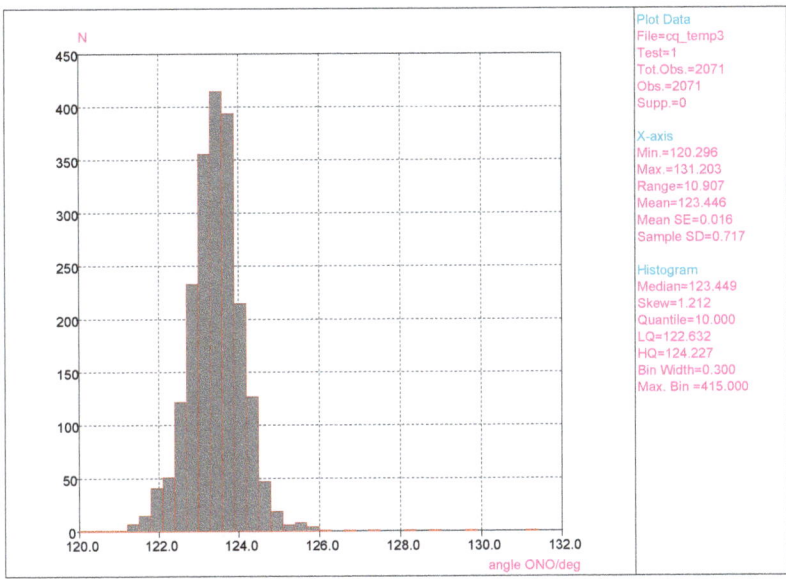

Figure 1. Distribution of ONO angles for *para*-substituted nitrobenzene derivatives.

A similar analysis was carried out for the NO bond lengths; the difference between them, $d(N1O2)-d(N1O3)$ (labeling of atoms in Scheme 1), is presented by histogram (Figure 2). The maximum value of these differences amounts to 0.134 Å, whereas the range in their variability is equal to 0.193 Å for the retrieved data set. The observed changes seem to be mostly associated with intermolecular interactions between the neighboring molecules, the lengths of both NO bonds can be unequal depending on the surroundings (Figure 3).

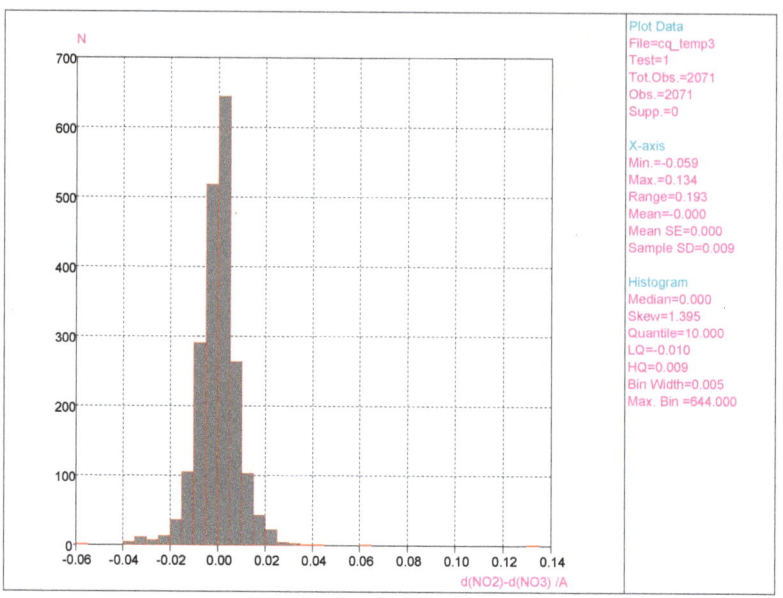

Figure 2. Distribution of differences $d(N1O2)-d(N1O3)$ for *para*-substituted nitrobenzene derivatives (labeling of atoms in Scheme 1).

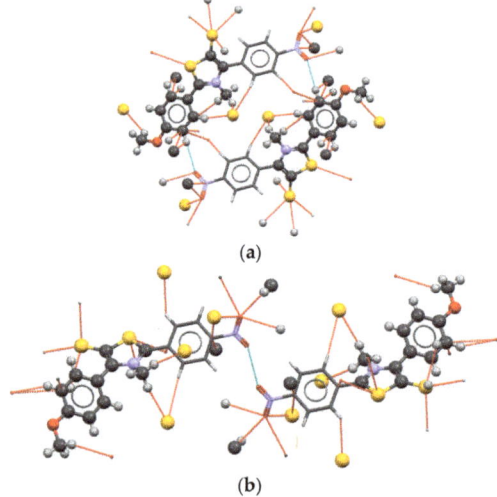

Figure 3. Intermolecular interactions (red or blue dot lines) in crystals of 1,3-thiazolium-5-thiolate derivatives (**a**) MAGTEV, twist angle of the NO_2 group equal to 6.22 deg; and (**b**) MAGTIZ, twist angle = 41.06 deg. The *p*-nitrophenyl part is presented by sticks, red lines show intermolecular interactions while not showing neighboring molecules, whereas bolls (very small and larger) indicate atoms of the molecule.

The distribution of the dihedral angle between the NO_2 group and phenyl ring planes is illustrated in the histogram in Figure 4. The maximum twist (observed for one structure, MAGTIZ) is 41.06 deg and, surprisingly, the smallest twist of the range 0–1 deg was found for ~120 hits. That is only for ~6% of all cases! This means that coplanarity of the nitro group with the aromatic ring can be very easily broken. This is understandable if one takes into account that even in so weakly interacting molecules as those in crystals of nitrobenzene a very weak O⋯H-C bond with a neighboring nitrobenzene molecule causes a dihedral angle of 1.7 deg [36]. Additionally, for both oxygen atoms of the nitro group participating in a similar number of intermolecular interactions, the dihedral angle is close to the one found for the mean value; in the case of *para*-dinitrobenzene it amounts to *ca*. 10 deg [38,39]. However, if due to intermolecular interactions, the two oxygen atoms significantly differ from each other, then the nitro group becomes much more twisted. Good illustrations are 1,3-thiazolium-5-thiolate derivatives: 4-(4-methoxyphenyl)-3-methyl-2-(4-nitrophenyl)-1,3-thiazolium-5-thiolate (MAGTIZ) and 2-(4-methoxyphenyl)-3-methyl-4-(4-nitrophenyl)-1,3-thiazolium-5-thiolate (MAGTEV) [40], see Figure 3. In the first case the maximum twist angle is observed, whereas for MAGTEV the angle amounts to 6.22 deg.

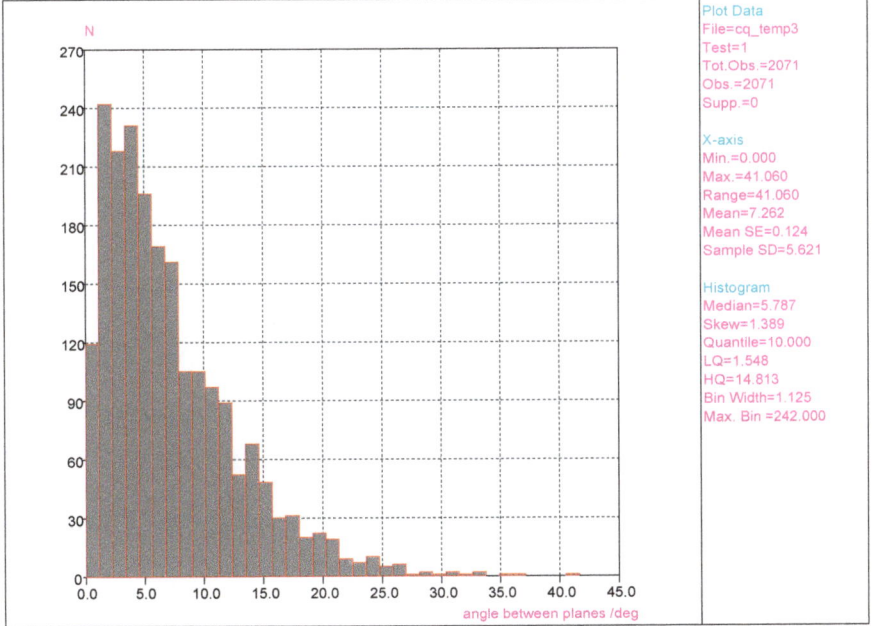

Figure 4. Distribution of the twist angle between NO_2 and phenyl ring planes for *para*-substituted nitrobenzene derivatives.

Changes in CN bond lengths are worthy of deeper analysis because of the large range in their variations, between 1.410 and 1.535 Å. It is well known that the lengths of the exocyclic CN or CC bonds in *para*-substituted benzene derivatives are good estimators of the aromatic character of the ring described by the HOMA index [41]. Shortening of this bond indicates a large contribution of the quinoid structure of the ring and hence a decrease in its aromatic character. Distribution of the CN bond lengths, shown in Figure 5, suggests a significant variability in aromaticity of *para*-substituted nitrobenzene derivatives. The length of the CN bond in nitrobenzene is 1.467 Å [36], which is very close to the mean value (1.469 Å). Thus, one can assume that the great range of CN bond lengths results from the intramolecular interactions (electronic effects) between the electron-accepting NO_2 group and counter-substituents located in the *para*-position. Moreover, it illustrates indirectly the changes in aromaticity due to these influences.

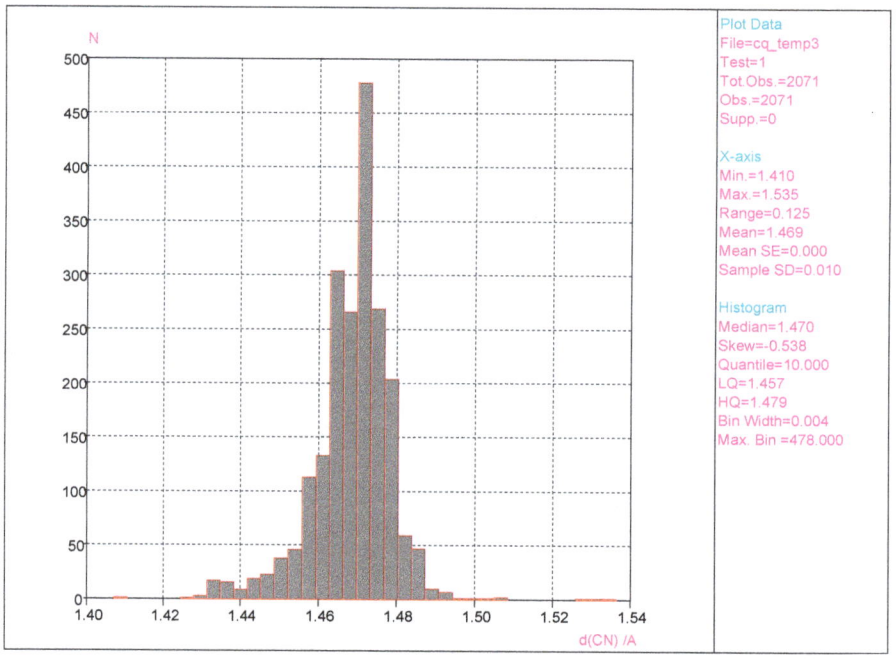

Figure 5. Distribution of CN bond length for *para*-substituted nitrobenzene derivatives.

The obtained results for the retrieved data set (Figure 6) confirm this suggestion, showing the range of HOMA values from 0.826 and in a more regular way between 0.88 and 1.00. A relatively small decrease in the aromatic character of the phenyl ring as an effect of the substituent is in line with more detailed studies based on QM modeling of this problem [42,43].

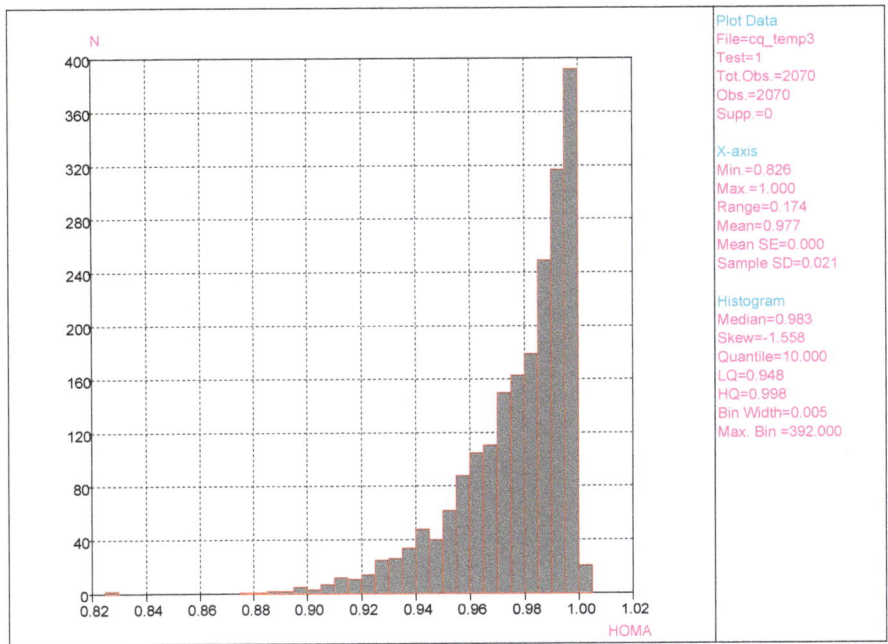

Figure 6. Histogram of the distribution of HOMA data for *para*-substituted nitrobenzene derivatives.

It is important to note that histograms of the data sets for *para*-substituted nitrobenzene derivatives and measurements with $R \leq 0.075$ or 0.1 are characterized by similar shapes and statistical parameters as shown above; only a slightly greater range of variability of the studied parameters and their esds were observed.

Therefore, the aromatic character of the ring in *para*-substituted nitrobenzene derivatives depends on at least two factors: (i) the twist around the CN bond and (ii) the above-mentioned substituent effect. Figure 7 illustrates a direct dependence of the HOMA index on the angle between planes for NO_2 and carbon atoms of the benzene ring. Taking into account that in the HOMA formula (Equation (5)), differences between the d_{opt} and the measured bond lengths are considered, with relatively low accuracy of estimation of the latter, then the obtained HOMA values are biased with a large error. Despite this disadvantage, Figure 7 shows some regularity. The lowest values of the HOMA increase with an increase in the dihedral angle. This means that in this direction the possible contribution of the quinoid form decreases and hence the aromatic character increases.

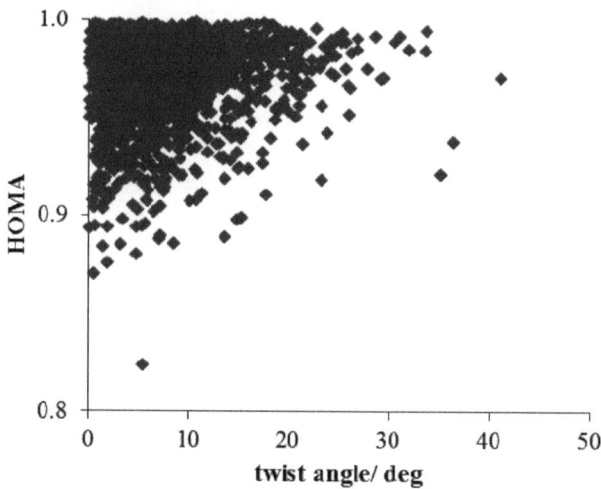

Figure 7. Dependence of HOMA on angle between planes for NO$_2$ and carbon atoms of the benzene ring for *para*-substituted nitrobenzene derivatives.

3.2. Quantum Chemistry Modeling

For clarity, this section is divided into two parts. The first one deals with consequences of "pure" intramolecular interactions where changes are modeled by gradual rotation of the nitro group in *p*-nitroaniline. The second part presents effects of intermolecular interactions that are realized by involving the amino group in H-bond formation. The main object of the study, *p*-nitroaniline (PNA), is presented in Scheme 1 (including numbering of atoms).

3.2.1. Consequences of the Nitro Group Rotation in *p*-Nitroaniline

PNA can be considered as a derivative of aniline, nitrobenzene, and benzene. To visualize how the substitution of the benzene ring affects the distribution of charges, we computed the VDD charges of the atoms in benzene (see Figure 8) for the nitro and amino derivatives, and finally in PNA with the planar and perpendicular conformation (PNA90) of the nitro group. To compare the substituent effects, C$_s$ symmetry for all systems has been imposed (energy differences between planar and non-planar conformations of the amino group in aniline and PNA are equal to 0.75 and 0.14 kcal/mol, respectively).

For nitrobenzene derivatives presented in Figure 8, the charge at the nitrogen atom of the nitro group changes less (0.010 e) than at the oxygen atoms (0.028 e). Similar modifications are observed for the amino group of aniline derivatives, where the change in the nitrogen atom charge is smaller (0.017 e) than in hydrogen atoms (0.020 e). In the perpendicular PNA conformation, the acidity of hydrogens in the amino group is clearly less than in a coplanar one (with charges of 0.139 e and 0.149 e, respectively), but in the unsubstituted aniline the H atoms are still less acidic (0.129 e). Conversely, the basicity of the nitrogen atom in the amino group is the largest in aniline ($q(N) = -0.185$ e), and the smallest one in the coplanar conformation of PNA ($q(N) = -0.168$ e).

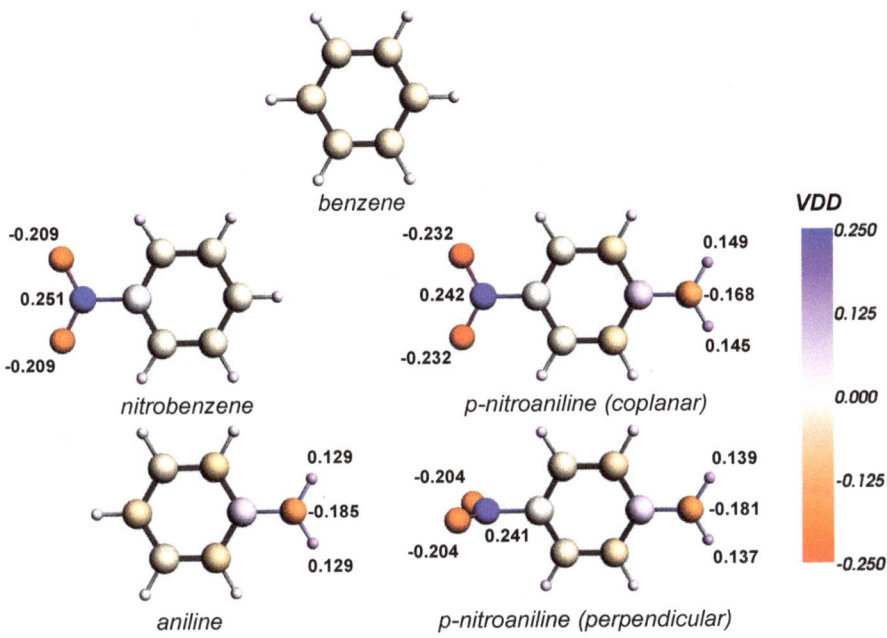

Figure 8. Voronoi deformation density (VDD) atomic charges (in electrons) for benzene and its derivatives (C_s symmetry for all systems has been imposed).

The use of the idea of cSAR(X) enables a deeper look at the substituent effects [44,45]. The application of this concept makes the comparison of group characteristics available and easier to understand. Table 1 presents the cSAR(X) data for all groups (nitro, amino and CH fragments) in the studied nitro and amino benzene derivatives.

The changes in cSAR(X) values as a result of NO_2 rotation in PNA are almost equivalent for the NO_2 and NH_2 groups. The nitro group loses 0.049 e whereas the amino group only 0.042 e. When absolute cSAR(X) values of these groups in coplanar PNA are compared with the data for nitrobenzene and aniline, then there results a difference of 0.082 e for the NO_2 group, whereas for the amino group this is 0.069 e. Thus, both groups have comparable flexibility in their accepting/donating properties upon activity of another group located in the *para* position in benzene.

Table 1. cSAR(X) values (in electrons) for substituents in *ortho-*, *meta-* and *para-*positions of benzene and its nitro and amino derivatives [1].

Molecule	cSAR(H) (*ortho-*, *meta-* and *para*-Positions)	cSAR(NO_2)	cSAR(NH_2)
benzene	0.0	-	-
nitrobenzene	0.033, 0.022, 0.036	−0.146	-
aniline	−0.044, −0.001, −0.039	-	0.129
p-nitroaniline 0	−0.017, 0.032, -	−0.219	0.198
p-nitroaniline 90	−0.019, 0.029, -	−0.170	0.156
p-nitroanilide anion	−0.063/−0.085,[2] −0.026, -	−0.509	-
p-nitroanilide anion 90	−0.080/−0.104,[2] −0.045, -	−0.348	-

Notes: [1] The cSAR(NO_2) and cSAR(NH_2) values taken from [44]; [2] Two numbers correspond to different *ortho-* positions in anions; second number characterizes the position closer to H from NH^-.

The dependence of cSAR(X) values on the Hammett sigma constants for different *p*-nitrobenzene-X derivatives has a negative slope with correlation coefficient $cc = -0.985$ [44] indicating that the more negative cSAR(X), the higher the electron-accepting power of X. It should be stressed that cSAR(X) describes the electronic state of the C-X fragment representing the state after the substituent effect has acted. As can be seen from Table 1, the nitro group has more electron-attracting power in coplanar PNA, then in its perpendicular conformation and finally in nitrobenzene, with cSAR(NO_2) equal to −0.219, −0.170 and −0.146, respectively.

As stated above, rotation of the NO_2 group around the C-N bond in PNA modifies the accepting/donating character of both substituents. Figure 9 shows how the rotation of the nitro group increases the relative energy, E_{rel}, of the system. Interestingly, in the range 30 < φ < 70 degrees the changes in energy are larger than for small and large values of the rotation angle.

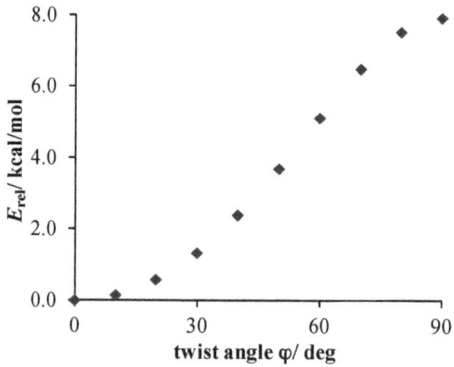

Figure 9. Dependence of the relative energy, E_{rel}, of *p*-nitroaniline on rotation angle of NO_2 group, φ.

The characteristics of both substituents obtained for the fully applied range of the rotation angle φ are collected in Table S1, whereas the relationships between cSAR(NO$_2$), cSAR(NH$_2$) and φ are shown in Figure 10. Rotation of the NO$_2$ group in PNA leads to a decrease in both the electron-accepting and electron-donating power of the nitro and amino groups, respectively (see also [44]).

Modification of the nitro substituent properties, due to its rotation, causes appropriate changes in cSAR(X) values of both substituents in the *para* position (Figure 10). Changes in cSAR(NO$_2$) and cSAR(NH$_2$) correlate well with $cc = -0.985$, see Figure S1. This means that the increase in the electron-accepting ability of the NO$_2$ group is directly associated with an increase in electron-donating ability of NH$_2$ group. Therefore, an augmentation of the aromaticity of the ring with an increase of φ may be expected. In this direction, the contribution of the quinoid form of PNA decreases, and hence an increase of the HOMA index is observed (Figure 11), in line with the results obtained for the data retrieved from CSD for *para*-substituted nitrobenzene derivatives.

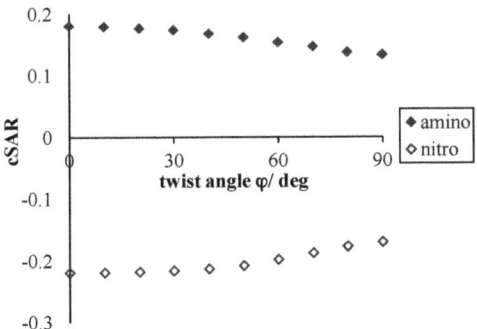

Figure 10. Dependence of cSAR(NO$_2$) and cSAR(NH$_2$) on rotation angle of NO$_2$ group, φ, for PNA.

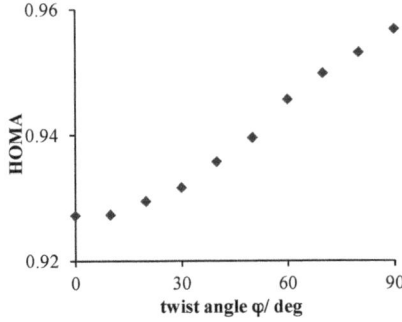

Figure 11. Dependences of HOMA values for *p*-nitroaniline (PNA) on rotation angle of NO$_2$ group, φ.

3.2.2. Consequences of the H-Bond Formation at the Amino Group in p-Nitroaniline

Four types of intermolecular interactions of the amino group with F^- and HF were taken into account, see Scheme 2. However, for the studied complexes, only three of them (**a**, **b** and **d**) exist in equilibrium systems since in the fourth one (**c**) the proton transfer reaction (HNH$\cdots$F$^-$ → HN$^-\cdots$HF) occurs and another kind of equilibrium system appears. Therefore, in the case of interaction **c** (Scheme 2) complexes can only be modeled by gradually varying the distance between the heavy atoms of H-bond from 2.65 Å up to 4.0 Å (in this way, the strength of H-bonding is gradually changed). Data for the strongest studied complexes of aniline and PNA are collected in Table 2, whereas dependences of H-bond energy on elongation of H-bond length are presented in Figure S2.

Table 2. Energy decomposition analysis (ΔE in kcal/mol) for H-bonded complexes of aniline and p-nitroaniline (X = F for N$\cdots$H interactions, X = N for H$\cdots$F ones). For HNH$\cdots$F$^-$ interactions $d_{N\ldots F}$ = 2.65 Å (no proton transfer).

Complex	Angle φ/deg	d_{X-H}/Å	ΔE_{def}	ΔE_{int}	ΔE_{Pauli}	ΔV_{elstat}	ΔE_{disp}	ΔE_{oi}	ΔE_{oi}^{σ}	ΔE_{oi}^{π}	ΔE_{HB}
H$_2$N$\cdots$HF											
aniline		0.971	1.37	−12.72	21.14	−17.91	−1.55	−14.40			−11.35
PNA	0	0.967	1.42	−9.67	17.49	−13.42	−1.58	−12.16			−8.25
	30	0.967	1.41	−9.87	17.78	−13.76	−1.58	−12.31			−8.46
	60	0.968	1.36	−10.43	18.53	−14.66	−1.58	−12.71			−9.07
PNA90	90	0.970	1.29	−10.97	19.20	−15.52	−1.58	−13.06			−9.68
HN$^-\cdots$HF											
aniline		1.069	8.70	−40.69	31.63	−41.29	−1.48	−29.54	−27.48	−2.07	−31.99
PNA	0	1.065	8.43	−35.61	50.65	−45.59	−1.54	−39.12	−36.02	−3.10	−27.18
	30	1.067	8.50	−36.07	51.26	−46.05	−1.54	−39.74			−27.57
	60	1.075	9.42	−37.65	53.46	−47.61	−1.53	−41.98			−28.23
PNA90	90	1.175	22.26 [1]	−53.60	78.08	−61.37	−1.47	−68.86	−63.72	−5.14	−31.34
HNH$\cdots$F$^-$											
aniline		1.174	12.24	−54.36	64.88	−58.37	−0.97	−59.90	−51.22	−8.67	−42.12
PNA	0	1.180	12.82	−67.83	34.89	−54.94	−0.98	−46.80	−37.55	−9.25	−55.01
	30	1.174	12.32	−66.68	34.36	−54.08	−0.98	−45.97			−54.36
	60	1.161	11.30	−63.51	33.19	−51.71	−0.98	−44.01			−52.21
PNA90	90	1.133	8.08	−56.37	30.90	−46.72	−0.99	−39.56	−32.61	−6.95	−48.29
HNH$\cdots$FH											
aniline		1.007	0.01	−1.94	1.78	−2.22	−0.64	−0.86	−0.79	−0.07	−1.93
PNA	0	1.009	0.01	−2.91	2.11	−3.20	−0.69	−1.12	−1.02	−0.10	−2.90
	30	1.012	0.00	−2.78	2.09	−3.09	−0.71	−1.08			−2.78
	60	1.013	0.00	−2.63	2.07	−2.96	−0.71	−1.04			−2.63
PNA90	90	1.008	0.01	−2.54	1.95	−2.82	−0.67	−1.01	−0.92	−0.09	−2.53

Note: [1] Proton transfer takes place: HN$^-\cdots$HF → HNH$\cdots$F$^-$.

An interesting effect is observed for dependence of E_{HB} on the rotation angle φ. For the HNH$\cdots$F$^-$ interactions, the H-bond becomes weaker with an increase of φ, since weakening of the electron-attracting power of the NO$_2$ group due to its rotation causes a decrease in N-H acidity (see Figure 12). Conversely, for HN$^-\cdots$HF and $H_2N\cdots HF$ interactions, an increase of φ is associated with an increase of the H-bonding strength. A similar situation is observed for the dependence of E_{HB} on σ_p^- constants of the NO_2 group estimated for various twist angles of the group [7].

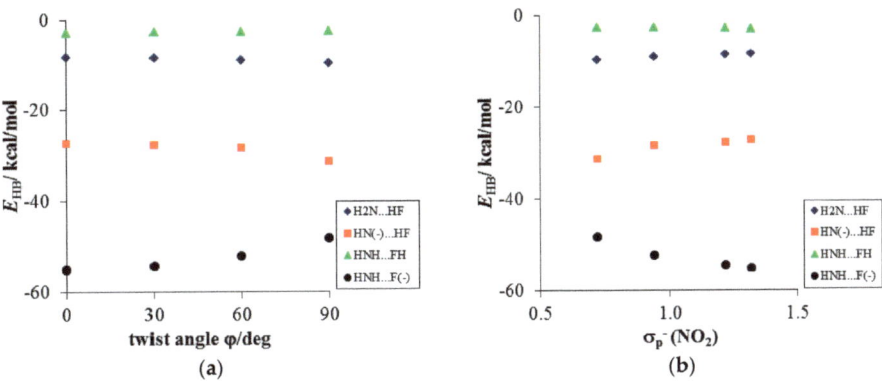

Figure 12. Dependences of H-bond energy, E_{HB}, on (a) rotation of NO_2 group and (b) its σ_p^- constants [7] (for equilibrium complexes, except for $HNH\cdots F^-$ interactions).

Figure S2 presents typical, close to the linear, dependences of H-bond strength on its distance for the complexes of PNA, its derivative PNA90 and aniline. The effect of the substituent on the strength of intermolecular interactions is consistent with the above-described substituent influence on acidity and basicity of the amino group. For complexes with $N\cdots HF$ interactions, both neutral and charge-assisted, the strongest H-bonds are observed for aniline and the weakest one for PNA. The opposite changes are found in the case of $NH\cdots F$ H-bonds.

The presence of electron-accepting substituents in the *para*-position to the amino group leads to a decrease in the strength of the $N\cdots HF$ hydrogen bond in comparison with aniline by 3.1 and 4.8 kcal/mol for the equilibrium structure of neutral and charged complexes, respectively (**a, d** in Scheme 2, Table 2). At the same time, if the amino group acts as a proton donor (**b, c** in Scheme 2), then $NH\cdots F$ bond becomes stronger by 1.0 and 13.0 kcal/mol for $HNH\cdots FH$ and $HNH\cdots F^-$ interactions, respectively. These differences decrease (almost half) when the NO_2 group is turned 90 degrees out of the ring plane because π-electron delocalization is partly switched off as we computed previously [4]. Since the rotation decreases the resonance effect between NO_2 and NH_2 groups, $HN^-\cdots HF$ interaction will become stronger, whereas $HNH\cdots F^-$ one will weaken. This is illustrated by Figure 12, σ_p^- taken from ref. [7]. Interestingly, the slopes of the regressions in Figure 12 are the steepest for stronger interactions and much less steep for the weaker ones.

To investigate the nature of the hydrogen bonds in the studied complexes the decomposition of the interaction energy into different components is very helpful. We performed the EDA of the interaction energy to determine information about the importance of specific energy terms (electrostatic, Pauli repulsion, orbital interaction, and dispersion). The results for H-bonded complexes of aniline and PNA are presented in Table 2. It is obvious that for all complexes the most important stabilizing term is the electrostatic one, showing more than 50% of total attractive interactions (i.e., $\Delta E_{oi} + \Delta V_{elstat} + \Delta E_{disp}$), followed by the orbital interaction and dispersion. For all complexes of PNA, the contribution of the orbital interaction term, ΔE_{oi}, is also very important (more than 40%), except for the weakest H-bonding, $HNH \cdots FH$, where the percentage of ΔE_{oi} is only 22% and is comparable to the magnitude of the dispersion term (15%). For the planar complexes, further decomposition of ΔE_{oi} into σ and π components shows that ΔE_{oi}^{σ} provides the main contribution, thereby indicating a partially covalent character of the corresponding H-bonds.

Variation of VDD charges at nitrogen and hydrogen atoms participating in H-bonding on rotation angle φ is presented in Figure S3. For comparison, appropriate values for "free" PNA and HF are added. It was shown that the rotation of the NO_2 group increases the negative charge at the nitrogen atom in the amino group. This effect depends on the strength of H-bonding: $H_2N \cdots HF$ vs. $HN^- \cdots HF$ (10 kcal/mol vs. 30 kcal/mol). In a stronger H-bonded system, the VDD(N) charge becomes more negative. A similar picture (decrease of the positive charge as the angle φ increases) is observed for VDD charges at the hydrogen atom of amino group involved in the interaction. Moreover, participation of this atom in intermolecular interaction increases its positive charge in comparison with free PNA and this effect depends also on the H-bond strength. The same is observed for the HF molecule participating in H-bonding and the free HF.

Due to the rotation of the NO_2 group, the atoms of the amino group involved in H-bonding become more negative (N) or less positive (H) (Figure S3). In the case of the $N \cdots H$ hydrogen bond, the increase of negative charge at the N atom causes a slight strengthening of the H-bond improving the electrostatic attraction between the PNA fragment and HF ($\Delta Q(N)$ = 0.010 for neutral complexes and 0.057 for charged ones). However, for the $H \cdots F$ hydrogen bond, a reverse situation is observed. The H-bond weakens because of the positive charge at H atom of the PNA fragment and the electrostatic contribution is reduced ($\Delta Q(H) \approx -0.01$ for both systems).

The electron-accepting ability of the substituent and the strength of H-bonding have both a significant impact on π-electron delocalization and hence on the aromaticity of the phenyl ring (as it is the transmitting moiety), expressed by the HOMA index (Figure 13 and Figure S4, respectively). The changes in the aromaticity are due to changes of the quinoid-like structure contribution and depend on the charge transfer from the electron-donating amino group involved in intermolecular interactions (H-bonding) and on the electron-attracting power of NO_2 which is governed by the rotation angle φ.

For equilibrium complexes, the HOMA values increase with the rotation of the NO_2 group for all types of H-bonded complexes of *p*-nitroaniline (Figure 13). For all PNA complexes with the NO_2 group in the plane of the ring, the values of the HOMA index are lower than those for aniline complexes (ΔHOMA ~ 0.02 for neutral system and 0.10 for charged ones). When this resonance effect is broken by rotation of the NO_2 group, an increase of aromaticity is observed. HOMA values for systems with a perpendicular NO_2 group are higher or similar to those for aniline complexes. Furthermore, it should also be noted that, as illustrated in Figure 13, the changes in aromaticity are consistent with those obtained for experimental data retrieved from CSD (Figure 7).

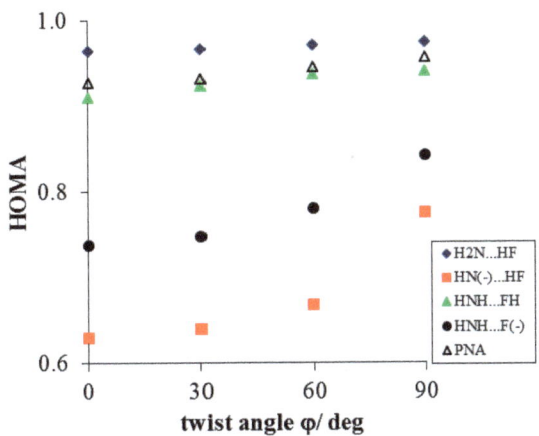

Figure 13. Dependences of HOMA values on rotation angle φ of NO_2 group (for equilibrium structures, except for HNH···F^- interactions).

When considering the effect of H-bonding on π-electron delocalization of the ring, it can be seen that the aromaticity behaves in two different ways: (i) the formation of H-bond promotes the increase of aromaticity only for complexes with H_2N···HF and HN^-···HF interactions whereas (ii) for HNH···F^- charged complexes the HOMA values are much lower. Finally, the very weak HNH···FH bonds almost do not affect the aromaticity (Figure S4).

Furthermore, in cases of PNA and anilines involved in H-bonding, one may expect changes in pyramidality of the NH_2 group and also changes in the C7-N14 bond length due to resonance effect, similarly as in the case of *para*-substituted aniline derivatives [46]. Indeed, Figure S5 shows the dependence of pyramidality of the NH_2 group on the C7-N14 bond length, which follows a linear relation with the correlation coefficient $cc = 0.998$ for H-bonded complexes of aniline and p-nitroaniline. Moreover, the pyramidality of the NH_2 group depends also on the rotation angle of the nitro group in PNA complexes, as presented in Figure S6. As a result of the $H_2N\cdots HF$ bond formation, the deformation (pyramidalization) of the NH_2 group becomes larger from 10.2 to 23.9 degrees when the NO_2 group is coplanar and from 16.3 to 26.4 degrees when the NO_2 group is perpendicular to the plane of the phenyl ring.

4. Conclusions

The nitro group is one of the most common substituents. Results of the analyses of geometrical and electronic parameters of the group extracted from CSD for *para*-substituted nitrobenzene derivatives as well as results obtained quantum chemically reveal a sensitivity of the properties of this group to intra- and intermolecular interactions. It has been shown that:

(1) Considering CSD data, deformations of the ONO angle are small, with the mean value (123.4 deg) close to the value for nitrobenzene in the crystal (123.2(1) deg) with a small value of esd (*ca.* 0.7). This indicates a rather small effect of the chaotic action of the neighboring molecules in the crystal lattice on this angle of the NO_2 group. Similar results are found for the NO bond lengths.

(2) Much stronger deformations are observed for the twist angle, with a maximum value of ~40 deg, but a mean value of 7.3 deg with a rather large esd (5.6 deg). This is in line with QM results for PNA showing that rotation of the NO_2 group with respect to the benzene ring up to 30 degrees is associated with very small changes in energy of the system, namely less than 2.0 kcal/mol, whereas to achieve a perpendicular orientation *ca.* 8.0 kcal/mol is required.

(3) Substantial range of the CN bond length is observed in the crystals of nitrobenzene derivatives (from 1.410 to 1.535 Å). Shortening of the CN bond length indicates an increased contribution of the quinoid structure and as a consequence a lower aromaticity of the ring. Such changes can be caused either by distortion from coplanar orientation between the NO_2 group and the phenyl ring or by the effect of a substituent located in the *para*-position, as was confirmed by QM results for PNA and its H-bonded complexes.

(4) Energy decomposition analysis of the H-bonds in the modeled complexes of aniline and PNA revealed that the amino group can form hydrogen bonds in a wide range of strengths (from -2.0 to -55.0 kcal/mol). The results support the view that these interactions are mostly provided by electrostatic attraction, however a significant covalent contribution has to be considered.

(5) Application of the VDD atomic charges and the cSAR concept allows an estimation of the magnitude of the intramolecular charge transfer. For all of the studied systems, a mutually dependent change in characteristics of both

groups (NO_2 and NH_2) was observed. It was shown that the electron-attracting ability of the nitro group dramatically depends on the moiety to which it is attached (see Table 1) as well as on intermolecular interactions occurring even at distant parts of the studied systems.

Supplementary Materials: The following are available online at http://www.mdpi.com/2073-4352/6/3/29/s001. Table S1: Characteristics of substituents in *p*-nitroaniline. Figure S1: Dependence of cSAR(NO_2) on cSAR (NH_2) for free PNA. Figure S2: Relation between hydrogen bond energy, E_{HB}, and its length, d_{HB}, for the H-bonded complexes of aniline and *p*-nitroaniline (NO_2 group is coplanar and perpendicular to the ring plane) with HF or F^-. Figure S3: Dependences of VDD atomic charge at N and H atoms of amino group on NO_2 rotation, φ, for fragment of *p*-nitroaniline. Figure S4: Dependences of HOMA values on H-bond length for different types of aniline and *p*-nitroaniline complexes. Figure S5: Correlation between angle of pyramidality of NH_2 group, Φ, and C7-N14 bond length, d_{CN}. Figure S6: Dependence of pyramidality angle, Φ, of NH_2 group on rotation of NO_2 group, φ, for $H_2N \cdots HF$ interactions.

Acknowledgments: Célia Fonseca Guerra acknowledges the financial support from the Netherlands Organization for Scientific Research NWO. H.S. and T.M.K. thank the National Science Centre and Ministry of Science and Higher Education of Poland for supporting this work under the Grant No. UMO-2013/11/B/ST4/00531.

Author Contributions: The paper has been conceived and designated by all authors. Olga A. Stasyuk and Célia Fonseca Guerra performed calculations and analysis results obtained, Halina Szatylowicz retrieved structural data from CSD and carried out an analysis of the results, Tadeusz M. Krygowski and Halina Szatylowicz wrote the paper.

Conflicts of Interest: The authors declare no conflict of interest.

References

1. Smith, M.B. *March's Advanced Organic Chemistry: Reactions, Mechanisms, and Structure*, 7th ed.; Wiley: Hoboken, NJ, USA, 2013.
2. Allen, F.H. The Cambridge Structural Database: A quarter of a million crystal structures and rising. *Acta Crystallogr. Sect. B Struct. Sci.* **2002**, *58*, 380–388.
3. Hansch, C.; Leo, A.; Taft, R.W. A survey of Hammett substituent constants and resonance and field parameters. *Chem. Rev.* **1991**, *91*, 165–195.
4. Stasyuk, O.A.; Szatylowicz, H.; Krygowski, T.M.; Fonseca Guerra, C. How amino and nitro substituents direct electrophilic aromatic substitution in benzene: An explanation with Kohn–Sham molecular orbital theory and Voronoi deformation density analysis. *Phys. Chem. Chem. Phys.* **2016**.
5. Campanelli, A.R.; Domenicano, A.; Ramondo, F.; Hargittai, I. Group electronegativities from benzene ring deformations: A quantum chemical study. *J. Phys. Chem. A* **2004**, *108*, 4940–4948.
6. Exner, O. A critical compilation of substituent constants. Chapter 10. In *Correlation Analysis in Chemistry: Recent Advances*; Chapman, N.B., Shorter, J., Eds.; Plenum Press: New York, NY, USA, 1978; pp. 439–540.

7. Dobrowolski, M.A.; Krygowski, T.M.; Cyrański, M.K. Substituent constants (σ_p^-) of the rotated nitro group. The interplay between the substituent effect of a rotated $-NO_2$ group and H-bonds affecting π-electron delocalization in 4-nitrophenol and 4-nitrophenolate complexes: A B3LYP/6-311+G** study. *Croat. Chim. Acta* **2009**, *82*, 139–147.

8. Xu, J. Synthesis of β-lactams with π electron-withdrawing substituents. *Tetrahedron* **2012**, *68*, 10696–10747.

9. Zhang, C. Review of the establishment of nitro group charge method and its applications. *J. Hazard. Mater.* **2009**, *161*, 21–28.

10. Badgujar, D.M.; Talawar, M.B.; Asthana, S.N.; Mahulikar, P.P. Advances in science and technology of modern energetic materials: An overview. *J. Hazard. Mater.* **2008**, *151*, 289–305.

11. Rouchaud, J.; Neus, O.; Cools, K.; Bulcke, R. Dissipation of the triketone mesotrione herbicide in the soil of corn crops grown on different soil types. *Toxicol. Environ. Chem.* **2000**, *77*, 31–40.

12. Irle, S.; Krygowski, T.M.; Niu, J.E.; Schwarz, W.H.E. Substituent effects of $-NO$ and $-NO_2$ groups in aromatic systems. *J. Org. Chem.* **1995**, *60*, 6744–6755.

13. Cambridge Crystallographic Data Centre (CCDC). *Vista—A Program for the Analysis and Display of Data Retrieved from the CSD*; Cambridge Crystallographic Data Centre: Cambridge, UK, 1994.

14. Te Velde, G.; Bickelhaupt, F.M.; van Gisbergen, S.J.A.; Fonseca Guerra, C.; Baerends, E.J.; Snijders, J.G.; Ziegler, T. Chemistry with ADF. *J. Comput. Chem.* **2001**, *22*, 931–967.

15. *ADF 2012*; SCM, Theoretical Chemistry, Vrije Universiteit: Amsterdam, The Netherlands, 2012.

16. Becke, A.D. Density-functional exchange-energy approximation with correct asymptotic-behavior. *Phys. Rev. A* **1988**, *38*, 3098–3100.

17. Lee, C.; Yang, W.; Parr, R.G. Development of the Colle-Salvetti correlation-energy formula into a functional of the electron-density. *Phys. Rev. B Condens. Matter* **1988**, *37*, 785–789.

18. Grimme, S.; Antony, J.; Ehrlich, S.; Krieg, H. A consistent and accurate *ab initio* parametrization of density functional dispersion correction (DFT-D) for the 94 elements H-Pu. *J. Chem. Phys.* **2010**, *132*.

19. Goerigk, L.; Grimme, S. A thorough benchmark of density functional methods for general main group thermochemistry, kinetics, and noncovalent interactions. *Phys. Chem. Chem. Phys.* **2011**, *13*, 6670–6688.

20. Snijders, J.G.; Baerends, E.J.; Vernooijs, P. Roothaan-Hartree-Fock-Slater atomic wave functions. Single-zeta, double-zeta, and extended Slater-type basis sets for $_{87}Fr-_{103}Lr$. *Atom. Data Nucl. Data Tables* **1981**, *26*, 483–509.

21. Ziegler, T.; Rauk, A. On the calculation of bonding energies by the Hartree Fock Slater method: I. The transition state method. *Theor. Chim. Acta* **1977**, *46*, 1–10.

22. Ziegler, T.; Rauk, A. A theoretical study of the ethylene-metal bond in complexes between Cu^+, Ag^+, Au^+, Pt^0, or Pt^{2+} and ethylene, based on the Hartree-Fock-Slater transition-state method. *Inorg. Chem.* **1979**, *18*, 1558–1565.

23. Ziegler, T.; Rauk, A. CO, CS, N$_2$, PF$_3$, and CNCH$_3$ as σ donors and π acceptors. A theoretical study by the Hartree-Fock-Slater transition-state method. *Inorg. Chem.* **1979**, *18*, 1755–1759.
24. Grimme, S. Accurate description of van der Waals complexes by density functional theory including empirical corrections. *J. Comput. Chem.* **2004**, *25*, 1463–1473.
25. Grimme, S. Semiempirical GGA-type density functional constructed with a long-range dispersion correction. *J. Comput. Chem.* **2006**, *27*, 1787–1799.
26. Steiner, T. The hydrogen bond in the solid state. *Angew. Chem. Int. Ed.* **2002**, *41*, 48–76.
27. Grabowski, S.J. What is the covalency of hydrogen bonding? *Chem. Rev.* **2011**, *111*, 2597–2625.
28. Bickelhaupt, F.M.; van Eikema Hommes, N.J.R.; Fonseca Guerra, C.; Baerends, E.J. The carbon-lithium electron pair bond in (CH$_3$Li)$_n$ (n = 1, 2, 4). *Organometallics* **1996**, *15*, 2923–2931.
29. Fonseca Guerra, C.; Handgraaf, J.-W.; Baerends, E.J.; Bickelhaupt, F.M. Voronoi deformation density (VDD) charges: Assessment of the Mulliken, Bader, Hirshfeld, Weinhold, and VDD Methods for charge analysis. *J. Comput. Chem.* **2004**, *25*, 189–210.
30. Sadlej-Sosnowska, N. On the way to physical interpretation of Hammett constants: How substituent active space impacts on acidity and electron distribution in p-substituted benzoic acid molecules. *Pol. J. Chem.* **2007**, *81*, 1123–1134.
31. Sadlej-Sosnowska, N. Substituent active region—A gate for communication of substituent charge with the rest of a molecule: Monosubstituted benzenes. *Chem. Phys. Lett.* **2007**, *447*, 192–196.
32. Krygowski, T.M.; Sadlej-Sosnowska, N. Towards physical interpretation of Hammett constants: Charge transferred between active regions of substituents and a functional group. *Struct. Chem.* **2011**, *22*, 17–22.
33. Krygowski, T.M.; Szatylowicz, H.; Stasyuk, O.A.; Dominikowska, J.; Palusiak, M. Aromaticity from the Viewpoint of Molecular Geometry: Application to Planar Systems. *Chem. Rev.* **2014**, *114*, 6383–6422.
34. Kruszewski, J.; Krygowski, T.M. Definition of aromaticity basing on the harmonic oscillator model. *Tetrahedron Lett.* **1972**, 3839–3842.
35. Krygowski, T.M. Crystallographic studies of inter- and intramolecular interactions reflected in aromatic character of π-electron systems. *J. Chem. Inf. Comput. Sci.* **1993**, *33*, 70–78.
36. Boese, R.; Blaser, D.; Nussbaummer, M.; Krygowski, T.M. Low temperature crystal and molecular structure of nitrobenzene. *Struct. Chem.* **1992**, *3*, 363–368.
37. Domenicano, A.; Schultz, G.; Hargittai, I.; Colapietro, M.; Portalone, G.; George, P.; Bock, C.W. Molecular structure of nitrobenzene in the planar and orthogonal conformations—A concerted study by electron diffraction, X-ray crystallography, and molecular orbital calculations. *Struct. Chem.* **1989**, *1*, 107–122.
38. Di Rienzo, F.; Domenicano, A.; Riva di Sanseverino, L. Structural studies of benzene-derivatives. 8. Refinement of the crystal-structure of para-dinitrobenzene. *Acta Cryst. Sect. B* **1980**, *B36*, 586–591.

39. Tonogaki, M.; Kawata, T.; Ohba, S.; Iwata, Y.; Shibuya, I. Electron-density distribution in crystals of p-nitrobenzene derivatives. *Acta Crystallogr. Sect. B-Struct. Sci.* **1993**, *49*, 1031–1039.
40. Cantillo, D.; Avalos, M.A.; Babiano, R.; Cintas, P.; Jimenez, J.L.; Light, M.E.; Palacios, J.C.; Rodriguez, V. Push-pull 1,3-thiazolium-5-thiolates. Formation via concerted and stepwise pathways, and theoretical evaluation of NLO properties. *Org. Biomol. Chem.* **2010**, *8*, 5367–5374.
41. Krygowski, T.M.; Wisiorowski, M.; Nakata, K.; Fujio, M.; Tsuno, Y. Changes of the aromatic character of the ring in exocyclically substituted derivatives of benzylic cation as a result of varying charge at the exo-carbon atom. *Bull. Chem. Soc. Jpn.* **1996**, *69*, 2275–2279.
42. Krygowski, T.M.; Ejsmont, K.; Stepien, M.K.; Poater, J.; Sola, M. Relation between the substituent effect and aromaticity. *J. Org. Chem.* **2004**, *69*, 6634–6640.
43. Krygowski, T.M.; Stepien, B.T.; Cyranski, M.K.; Ejsmont, K. Relation between resonance energy and substituent resonance effect in P-phenols. *J. Phys. Org. Chem.* **2005**, *18*, 886–891.
44. Stasyuk, O.A.; Szatylowicz, H.; Fonseca Guerra, C.; Krygowski, T.M. Theoretical study of electron-attracting ability of the nitro group: Classical and reverse substituent effects. *Struct. Chem.* **2015**, *26*, 905–913.
45. Szatylowicz, H.; Siodla, T.; Stasyuk, O.A.; Krygowski, T.M. Towards physical interpretation of substituent effects: The case of meta- and para-substituted anilines. *Phys. Chem. Chem. Phys.* **2016**.
46. Szatylowicz, H.; Krygowski, T.M.; Hobza, P. How the shape of the NH_2 group depends on the substituent effect and H-bond formation in derivatives of aniline. *J. Phys. Chem. A* **2007**, *111*, 170–175.

Non-Covalent Interactions in Hydrogen Storage Materials LiN(CH$_3$)$_2$BH$_3$ and KN(CH$_3$)$_2$BH$_3$

Filip Sagan, Radosław Filas and Mariusz P. Mitoraj

Abstract: In the present work, an in-depth, qualitative and quantitative description of non-covalent interactions in the hydrogen storage materials LiN(CH$_3$)$_2$BH$_3$ and KN(CH$_3$)$_2$BH$_3$ was performed by means of the charge and energy decomposition method (ETS-NOCV) as well as the Interacting Quantum Atoms (IQA) approach. It was determined that both crystals are stabilized by electrostatically dominated intra- and intermolecular M···H–B interactions (M = Li, K). For LiN(CH$_3$)$_2$BH$_3$ the intramolecular charge transfer appeared (B–H→Li) to be more pronounced compared with the corresponding intermolecular contribution. We clarified for the first time, based on the ETS-NOCV and IQA methods, that homopolar BH···HB interactions in LiN(CH$_3$)$_2$BH$_3$ can be considered as destabilizing (due to the dominance of repulsion caused by negatively charged borane units), despite the fact that some charge delocalization within BH···HB contacts is enforced (which explains H···H bond critical points found from the QTAIM method). Interestingly, quite similar (to BH···HB) intermolecular homopolar dihydrogen bonds CH···HC appeared to significantly stabilize both crystals—the ETS-NOCV scheme allowed us to conclude that CH···HC interactions are dispersion dominated, however, the electrostatic and σ/σ*(C–H) charge transfer contributions are also important. These interactions appeared to be more pronounced in KN(CH$_3$)$_2$BH$_3$ compared with LiN(CH$_3$)$_2$BH$_3$.

Reprinted from *Crystals*. Cite as: Sagan, F.; Filas, R.; Mitoraj, M.P. Non-Covalent Interactions in Hydrogen Storage Materials LiN(CH$_3$)$_2$BH$_3$ and KN(CH$_3$)$_2$BH$_3$. *Crystals* **2016**, *6*, 28.

1. Introduction

An increase in energy consumption as well as the environmental harmfulness of current coal or hydrocarbon based fuels has led to intensive search for alternative energy sources [1–3]. Therefore, various hydrogen storage materials, that contain significant amounts of hydrogen, have been recently proposed. Boranes are probably one of the best known group among numerous hydrogen storage materials [4–10]. For example one can present ammonia borane (NH$_3$BH$_3$) [4–10]—the attractiveness of this material stems from its high stability, even at higher temperature (the melting point is 104 °C), as well as its large hydrogen storage capacity (19.6 wt% H$_2$). It has been demonstrated that the former feature of ammonia borane crystal originates predominantly from the existence of polar dihydrogen bonds N–H$^{\delta+}$⋯$^{-\delta}$H–B between monomers [11–16]. Furthermore, it has been proven that the presence of N–H$^{\delta+}$⋯$^{-\delta}$H–B as well as other non-covalent interactions not only determines the stability, but it can also facilitate various steps of dehydrogenation [5–8,11–23].

It has been shown that incorporation of alkali metals into boranes might accelerate thermolitic dehydrogenation as well as reduce formation of volatile byproducts [24–26]. Therefore various hybrid type materials have also been proposed and extensively studied—as examples one can present LiBH$_4$/NH$_3$BH$_3$ [27], M[Zn(BH$_4$)$_3$], M = Li, Na, K [28] or Al(BH$_4$)$_3$·NH$_3$BH$_3$ [25].

Quite recently McGrady and coworkers published the cutting-edge article in which they had synthesized and characterized two further hydrogen rich crystals LiN(CH$_3$)$_2$BH$_3$ and KN(CH$_3$)$_2$BH$_3$ [29]. They are depicted in Figure 1. In addition, the authors performed topological electron density based study by means of the QTAIM method of Bader [30]—it was reported that mainly M⋯H–B (M = Li, K) interactions stabilize the crystals. Remarkably, the authors also emphasized that, apart from the mentioned non-covalent interactions, one observes untypical homopolar dihydrogen interactions of the types BH⋯HB and CH⋯HC which are found to determine the chain-like 1D architecture of LiN(CH$_3$)$_2$BH$_3$ and 2D layers in KN(CH$_3$)$_2$BH$_3$ crystal [29], Figure 1 and Figure S1. It is noteworthy that these types of connections are intuitively considered as destabilizing due to the lack of electrostatic attraction between hydrogen atoms—homopolar dihydrogen interactions (especially the intramolecular ones) are still a matter of debate in the literature [31–46]. Very recently more and more evidence has been reported in the literature that highlight the stabilizing nature of homopolar dihydrogen interactions [17,21,29,36–46].

Accordingly, in this work we provide complementary results which shed light on energetic, quantitative and qualitative characteristics of non-covalent interactions that contribute to the stability of LiN(CH$_3$)$_2$BH$_3$ and KN(CH$_3$)$_2$BH$_3$ crystals. It is an important goal as it is known that purely topological QTAIM analysis might provide bond paths between atoms (or fragments) even in situations where the overall interaction energies are positive (destabilizing) [36]. Therefore, we applied the charge and energy decomposition scheme (ETS-NOCV) [47–49] which has been proven to provide compact, qualitative and quantitative, descriptions of various types of chemical bonds starting from strong covalent bonds, going through dative connections [47–49] and ending up at various non-covalent interactions [7,8,37,50,51]. We applied the ADF program [52–54] in which the ETS-NOCV scheme is implemented. In order to achieve our goal we chose two types of cluster models—the first type (containing four monomers), marked by a blue dashed line in Figure 1A, is suitable for extraction of M⋯H–B (M = Li, K), and BH⋯HB interactions, whereas the second one (containing eight monomers), depicted by a red dashed line (Figure 1B), contains the CH⋯HC contacts. For selected models we also plotted the reduced density gradient of the NCI (Non-Covalent Index) method [55] in order to qualitatively characterize non-covalent interactions. In addition, the Interacting Quantum Atoms (IQA) energy decomposition scheme [56] was applied for a quantitative description of selected non-covalent interactions in LiNMe$_2$BH$_3$.

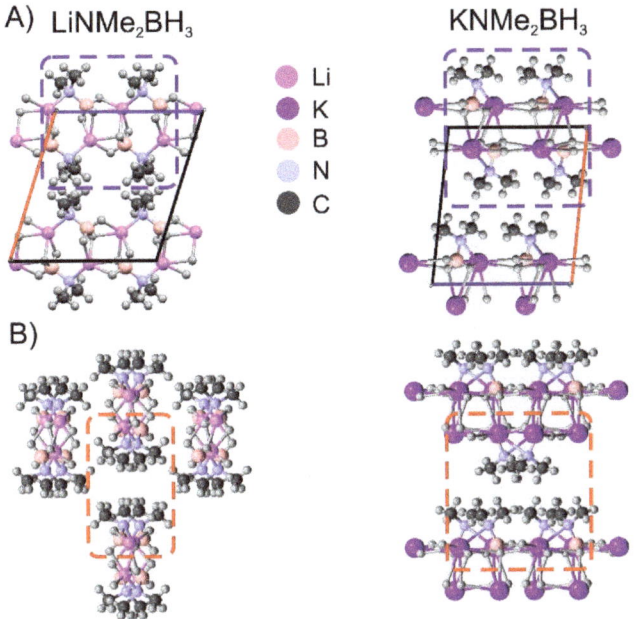

Figure 1. The crystal structures of LiN(CH$_3$)$_2$BH$_3$ and KN(CH$_3$)$_2$BH$_3$. In addition the cluster models used in the charge and energy decomposition method (ETS-NOCV) analysis are marked by blue (part A) and red dotted lines (part B). The unit cell is also highlighted in the part A.

2. Materials and Methods

Our calculations were performed by means of the Amsterdam Density Functional (ADF) program [52–54]. We used DFT/BLYP-D3/TZP because it has been proven many times in the past that they provide satisfactory results for non-covalent interactions [57–59]. The empirical Grimme correction (D3) [60] was used as implemented in the ADF program. We did not calculate basis set superposition errors (BSSE) because these effects are captured in the empirical correction D3 [60]. It is important to emphasize that we also performed additional test calculations for the tetrameric cluster of LiN(CH$_3$)$_2$BH$_3$ using the following methods: PBE-D3/TZP, BP86-D3/TZP (ADF program) as well as MP2/6-311 + G**, PBE-D3/6-311 + G**, BP86-D3/6-311 + G**, MO6-2X/6-311 + G**, wB97XD/6-311 + G** (based on the

Gaussian package) [61]—the results of total bonding energies appeared to be very similar among all these methods, Tables S1,S2 in ESI file. Accordingly, in the main text we have only discussed the results from ADF/DFT/BLYP-D3/TZP.

In the next paragraph the main formulas of charge and energy decomposition scheme ETS-NOCV are outlined.

2.1. ETS-NOCV

ETS-NOCV method [49] is a merger of the energy decomposition scheme by Ziegler-Rauk [53,62] with the Natural Orbitals for Chemical Valence (NOCV) [47,48].

The natural orbitals for chemical valence (NOCV) are eigenvectors that diagonalize the deformation density matrix:

$$\Delta P C_i = v_i C_i, \Psi_i = \sum_j^N C_{i,j} \lambda_j \quad (1)$$

where C_i is a vector of coefficients, expanding Ψ_i in the basis of fragment orbitals λ_j; N is a total number of fragment λ_j orbitals. It was shown that the natural orbitals for chemical valence pairs (ψ_{-k}, ψ_k) decompose the differential density $\Delta \rho$ into NOCV-contributions ($\Delta \rho_k$):

$$\Delta \rho(r) = \sum_{k=1}^{M/2} v_k \left[-\psi_{-k}^2(r) + \psi_k^2(r) \right] = \sum_{k=1}^{M/2} \Delta \rho_k(r) \quad (2)$$

where v_k and M stand for the NOCV eigenvalues and the number of basis functions, respectively. Visual inspection of deformation density plots ($\Delta \rho_k$) helps to attribute symmetry and the direction of the charge flow. In addition, these pictures are enriched by providing the energetic estimations, $\Delta E_{orb}(k)$, for each $\Delta \rho_k$ within the ETS-NOCV scheme.

The exact formula, which links the ETS and NOCV methods, is given in the next paragraph, after we briefly present the basic concept of the ETS scheme. In this method the *total bonding energy*, ΔE_{total}, between interacting fragments is divided into four components:

$$\Delta E_{total} = \Delta E_{dist} + \Delta E_{elstat} + \Delta E_{Pauli} + \Delta E_{orb} = \Delta E_{dist} + \Delta E_{int} \quad (3)$$

One could add that the negative value $-\Delta E_{total}$ is a bond dissociation energy. The first contribution of Equation (3), ΔE_{dist}, describes the amount of energy required to promote fragments from their equilibrium geometry to the conformations they adopt in the final optimized molecule. The second term, ΔE_{elstat}, corresponds to the classical electrostatic interaction between the promoted fragments as they are brought to their positions in the final complex. The third term, ΔE_{Pauli}, accounts for the repulsive

Pauli interaction between occupied orbitals on the two fragments in the combined molecule. Finally, the last stabilizing term, ΔE_{orb}, represents interactions between the occupied molecular orbitals of one fragment with the unoccupied molecular orbitals of the other fragment as well as mixing of occupied and virtual orbitals within the same fragment (inner-fragment polarization). This energy term is linked to the electronic bonding effect coming from the formation of a chemical bond. In the combined ETS-NOCV scheme [49] the orbital interaction term (ΔE_{orb}) is expressed in terms of NOCV's eigenvalues (v_k) as:

$$\Delta E_{orb} = \sum_k \Delta E_{orb}(k) = \sum_{k=1}^{M/2} v_k \left[-F_{-k,-k}^{TS} + F_{k,k}^{TS} \right] \tag{4}$$

where $F_{i,i}^{TS}$ are diagonal Kohn-Sham matrix elements defined over NOCV with respect to the transition state (TS) density at the midpoint between the density of the molecule and the sum of fragment densities. The above components $\Delta E_{orb}(k)$ provide the energetic estimation of $\Delta \rho_k$ that may be related to the importance of a particular electron flow channel for the bonding between the considered molecular fragments. Finally, in this work we applied a dispersion corrected functional, so this term (ΔE_{disp}) enters additionally into Equation (3). The ETS-NOCV analysis was done based on the Amsterdam Density Functional (ADF) package in which this scheme was implemented.

For ETS-NOCV analyses the crystal coordinates (not reoptimized) have been predominantly used in order to reflect the real structures of $LiN(CH_3)_2BH_3$ and $KN(CH_3)_2BH_3$. Accordingly, the distortion energy term, ΔE_{dist} (Equation (3)), was not calculated. Hence, *an interaction energy ΔE_{int} was mostly analyzed in this work*. We found for the tetrameric lithium model that the reoptimized geometry as well as bonding properties corresponds well to the crystal structure—accordingly, for this system we also considered the ΔE_{dist} term (Figure 2A) and discussed the corresponding total bonding ΔE_{total} and interaction ΔE_{int} energies. It needs to be added that our efforts to reoptimize the remaining models provided geometries which do not correspond to the crystal structures of $LiN(CH_3)_2BH_3$ and $KN(CH_3)_2BH_3$. For monomers the optimized structures are considered for ETS-NOCV analysis as they are similar to crystal geometries—in addition, it allows for discussion of the overall stability of monomers (ΔE_{total} values).

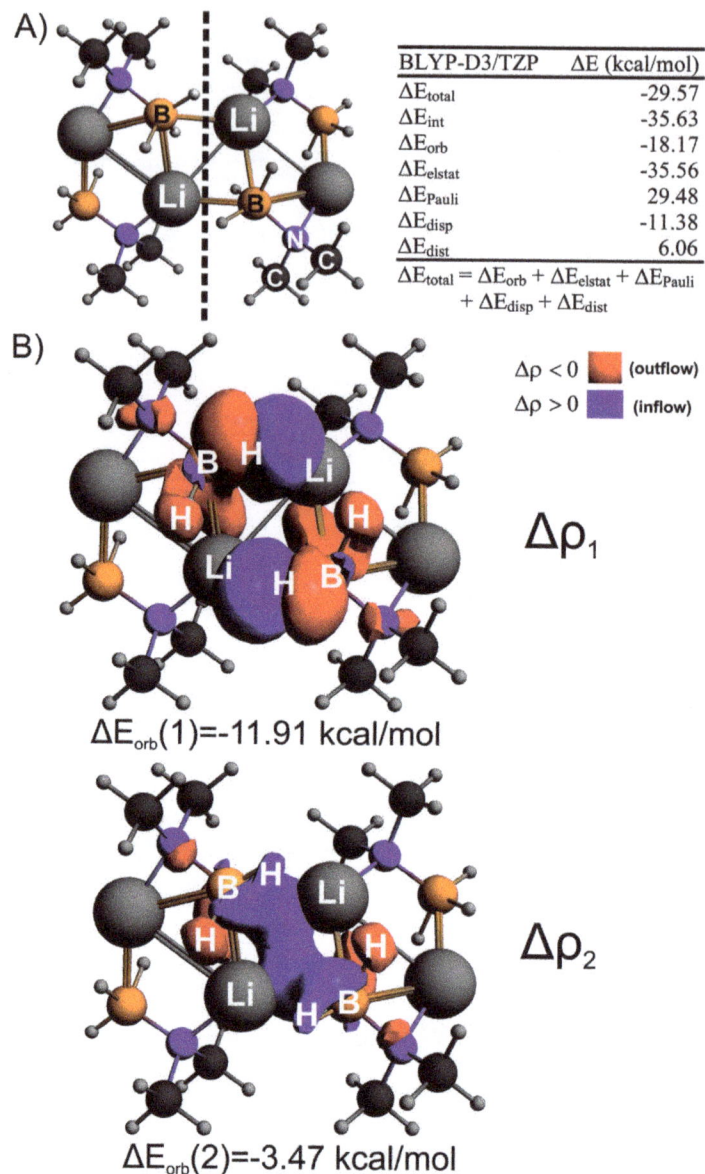

Figure 2. The optimized tetrameric cluster model of LiN(CH$_3$)$_2$BH$_3$ along with energy decomposition results describing the interaction between two dimeric fragments in LiN(CH$_3$)$_2$BH$_3$ (**A**). The fragmentation pattern used in ETS-NOCV analysis is indicated by black dashed line. (**B**) displays the most relevant deformation density contributions describing Li···H–B interactions. The red color of deformation densities shows charge depletion, whereas the blue an electron accumulation due to Li···H–B interaction.

2.2. NCI Technique

It has been shown that the reduced density gradient:

$$s = \frac{1}{2(3\pi^2)^{1/3}} \frac{|\nabla \rho|}{\rho^{4/3}} \tag{5}$$

appeared to be a useful quantity for a description of non-covalent interactions. In order to obtain information about the type of bonding, plot of reduced density gradient s against molecular density ρ is very often examined. When a weak inter- or intramolecular interaction is present, there exists a characteristic spike lying at low values of both density ρ and reduced density gradient s. To distinguish between attractive and repulsive interactions the eigenvalues (λ_i) of the second derivative of density (Hessian, $\nabla^2 \rho$) are used, $\nabla^2 \rho = \lambda_1 + \lambda_2 + \lambda_3$. Namely, bonding interactions are characterized by $\lambda_2 < 0$, whereas $\lambda_2 > 0$ indicates that the atoms are in non-bonded contact. Therefore, within the NCI technique, one can draw information about non-covalent interactions from the plots of sign(λ_2)ρ vs. s. In such plots the low gradient spike, an indicator of stabilizing interaction, is located within the region of negative values of the density. On the contrary, the repulsive interactions are characterized by positive values of sign(λ_2)ρ. One can also plot the contour of s colored by the value of sign(λ_2)ρ providing a pictorial representation of non-covalent interactions.

2.3. IQA (Interacting Quantum Atoms) Energy Decomposition Scheme

The Interacting Quantum Atoms (IQA) approach of Blanco and coworkers [56] allows to partition an electronic energy E into atomic (E_{self}^A) and diatomic contributions (E_{int}^{AB}):

$$E = \sum_A E_{self}^A + \frac{1}{2} \sum_A \sum_{B \neq A} E_{int}^{AB} \tag{6}$$

The interatomic interaction energy E_{int}^{AB} covers all inter-particle interactions: nucleus-nucleus, V_{nn}^{AB}, nucleus-electron, V_{ne}^{AB}, electron-nucleus, V_{en}^{AB}, and electron-electron, V_{ee}^{AB}, coming from interatomic interaction energies of particles ascribed to atom A with particles ascribed to atom B:

$$E_{int}^{AB} = V_{nn}^{AB} + V_{en}^{AB} + V_{ne}^{AB} + V_{ee}^{AB} = V_{nn}^{AB} + V_{en}^{AB} + V_{ne}^{AB} + V_{eeC}^{AB} + V_{eeX}^{AB} \tag{7}$$

The V_{ee}^{AB} term can be further divided into exchange (V_{eeX}^{AB}) and Coulomb (V_{eeC}^{AB}) contributions. The AIMALL program was used for the IQA calculations [63]. Due to the fact that we are interested in interaction energies in crystals we focused our attention on E_{int}^{AB}. More details and numerous applications of the IQA technique can be found in Reference [56].

2.4. Molecular Electrostatic Potential (MEP)

The electrostatic potential $V(r)$ of a molecule at point "r", due to nuclei and electrons, is defined as:

$$V(r) = \sum_A \frac{Z_A}{|R_A - r|} - \int \frac{\rho(r\prime)dr\prime}{|r - r\prime|} \tag{8}$$

where Z_A is the charge of nucleus at position R_A and $\rho(r)$ is the total electronic density. The sign of $V(r)$ depends upon whether the positive contribution of the nuclei or negative from the electrons is dominant. Negative values of $V(r)$ correspond to nucleophilic areas of the molecule, whereas the positive to electrophilic regions. It has been demonstrated in numerous works that MEP is a very useful quantity for in depth description of electron density distribution [64–67].

3. Results and Discussion

Let us start by discussing factors determining the stability of $LiN(CH_3)_2BH_3$. It can be seen from Figure 2A that the overall interaction energy between fragments, each consisting of the two $LiN(CH_3)_2BH_3$ monomers, is $\Delta E_{int} = -35.63$ kcal/mol (-17.81 kcal/mol permonomer-monomer interaction). This is in very good agreement with ΔE_{int} obtained for the crystal (non-optimized) geometry, Figure S2. An inclusion of the geometry distortion term leads to an overall bonding energy, $\Delta E_{total} = -29.57$ kcal/mol. It is noteworthy that other computational protocols provide very similar ΔE_{total} values, Tables S1 and S2.

The dominating contribution (55%) to ΔE_{total} stems from the electrostatic stabilization $\Delta E_{elstat} = -35.56$ kcal/mol, followed by the orbital interaction $\Delta E_{orb} = -18.17$ kcal/mol (28%) and the dispersion components $\Delta E_{disp} = -11.38$ kcal/mol (17%), Figure 2A. The molecular electrostatic potential (MEP) of the monomer demonstrates (Figure 3A) that the borane group is negatively charged, whereas the electrophilic region (positive MEP values) is seen around Li which explains the dominance of the electrostatic term in the intermolecular stabilization of $LiN(CH_3)_2BH_3$.

Further decomposition of ΔE_{orb} into deformation density contributions according to the ETS-NOCV scheme is shown in Figure 2B. It can be seen that the leading deformation densities, $\Delta\rho_1$ and Δ_2, depict the formation of Li⋯H–B bonds. They originate from an outflow of electron density from σ (B–H) bonds to Li^+ ions, Figure 2B, and correspond to the following stabilization, $\Delta E_{orb}(1) = -11.91$ kcal/mol, $\Delta E_{orb}(2) = -3.47$ kcal/mol, respectively. Such outflow leads to the elongation of B-H bonds by ~0.1 Å (as compared with non-bonding monomers). It is important to emphasize that Li⋯H–B interactions also lead to some charge delocalization within the "bay" containing two Li ions and two borane units, see $\Delta\rho_2$ in Figure 2B. This is fully consistent with the presence of QTAIM bond critical points between hydrogen atoms in homopolar bridges BH···HB as found by McGrady and coworkers [29]. However, an interesting question emerges at this point: is the LiN(CH$_3$)$_2$BH$_3$ dimer consisting of two monomers exposed to each other via pure BH···HB interaction stable? Such a situation, *i.e.*, the dimer in the geometry of the optimized tetrameric cluster model, is depicted in Figure 4B. Our results clearly indicate that in such a case the overall monomer-monomer interaction energy is positive due to significant Pauli and electrostatic repulsions (with total of +9.06 kcal/mol) that overcome subtle stabilization from charge transfer ΔE_{orb} and dispersion ΔE_{disp} (summing up to −4.16 kcal/mol)—accordingly, pure BH···HB interactions would destabilize the LiN(CH$_3$)$_2$BH$_3$ crystal. The same conclusions are valid when considering the dimer in crystal geometry, Figure S3.

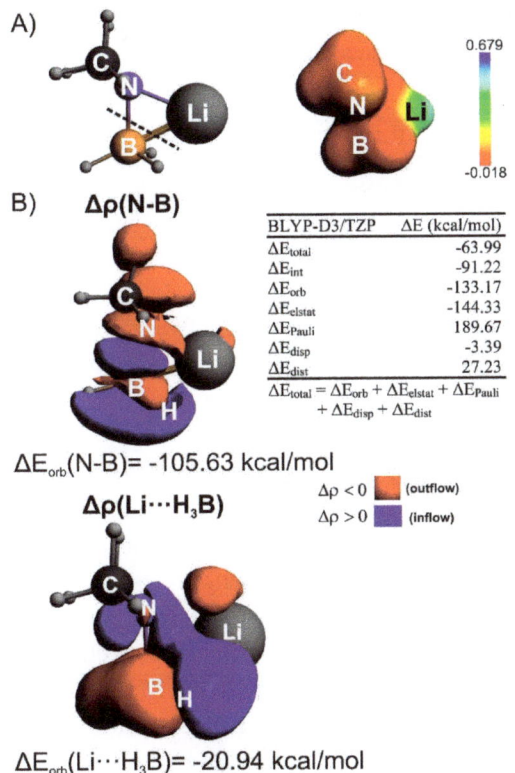

Figure 3. Monomer of LiN(CH$_3$)$_2$BH$_3$ along with the corresponding molecular electrostatic potential (**A**). In (**B**) the results of ETS-NOCV analysis are presented that describe bonding between the BH$_3$ unit and the LiN(CH$_3$)$_2$ fragment.

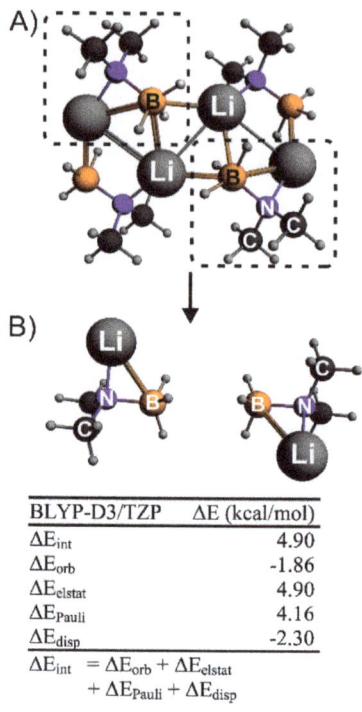

BLYP-D3/TZP	ΔE (kcal/mol)
ΔE_{int}	4.90
ΔE_{orb}	-1.86
ΔE_{elstat}	4.90
ΔE_{Pauli}	4.16
ΔE_{disp}	-2.30

$$\Delta E_{int} = \Delta E_{orb} + \Delta E_{elstat} + \Delta E_{Pauli} + \Delta E_{disp}$$

Figure 4. Dimer of LiN(CH$_3$)$_2$BH$_3$ consisting of BH···HB interactions together with results of ETS-NOCV analysis (**B**). The dimer was cut from the optimized tetramer model and it is marked with black dotted lines (**A**).

Furthermore, we have not found any local minimum energy structure on the potential energy surface for a dimer that would contain solely BH···HB interaction (a dimer where BH$_3$ units are exposed to each other). Summarizing, the major stabilization in the LiN(CH$_3$)$_2$BH$_3$ crystal arises from electrostatically dominated intermolecular Li···H–B interactions, as previously reported by McGrady et al. [29]. The formation of such bonds additionally enforce some charge delocalization within BH···HB contacts ($\Delta\rho_2$ in Figure 2B), however, the overall BH···HB interaction is destabilizing. Our conclusions are fully in line with the other, mostly experimental studies, in which the destabilizing role of BH···HB interactions were also suggested [68–72]. McGrady and coworkers reported the opposite in their series of recent articles [17,20,21,29]. Due to the fact that such interactions are clearly a matter of debate, in addition, we performed the Interacting Quatum Atoms (IQA) based study for the tetrameric models of LiNMe$_2$BH$_3$. The results are gathered in Table 1.

Table 1. The interacting quatum atoms (IQA) energy decomposition results (in kcal/mol) describing the two atomic interactions X···Y (X=Li, H; Y=H) in LiNMe$_2$BH$_3$.

IQA(X···Y)	V^{AB}_{ne}	V^{AB}_{en}	V^{AB}_{nn}	V^{AB}_{ee}	V^{AB}_{eeC}	V^{AB}_{eeX}	V^{AB*}_{int}
Li···H(B)	−805.9	−338.4	480.1	565.4	568.2	−2.8	−98.8
(C)H···H(C)	−140.6	−138.2	130.8	147.2	148.6	−1.4	−0.80
(B)H···H(B)	−198.9	−198.6	119.7	327.2	330.0	−2.7	+49.4

* $E^{AB}_{int} = V^{AB}_{nn} + V^{AB}_{en} + V^{AB}_{ne} + V^{AB}_{ee} = V^{AB}_{nn} + V^{AB}_{en} + V^{AB}_{ne} + V^{AB}_{eeC} + V^{AB}_{eeX}$.

It is visible from Table 1 that the overall diatomic interaction energies, E^{AB}_{int}, for Li···H(B) and (C)H···H(C) are negative, $E^{Li···H(B)}_{int} = -98.8$ kcal/mol, $E^{(C)H···H(C)}_{int} = -0.80$ kcal/mol, which indicates the stabilizing interactions as opposed to the homopolar (B)H···H(B) contacts which appeared to be significantly destabilizing, $E^{(B)H···H(B)}_{int} = +49.4$ kcal/mol. This is due to the significantly positive electron-electron repulsion term, $V^{AB}_{ee} = 327.2$ kcal/mol, caused in turn by the Coulomb contribution $V^{AB}_{eeC} = 330.0$ kcal/mol, Table 1. We also looked at the partial charges in the monomer and tetramer of LiN(CH$_3$)$_2$BH$_3$ and found that in both cases the borane units are negatively charged, which conforms to the significant value of the Coulomb repulsion V^{AB}_{eeC} found from the IQA analysis, Figure S4 and Table 1. We further calculated the electrostatic interaction between the monomers bonded via BH···HB contacts in the presence of the two other monomers (Figure S5). Interestingly, we found the electrostatic repulsion, *ca.* 15.6 kcal/mol, which is even more pronounced than the repulsion noted for the dimer without the neighboring monomers, 7.08 kcal/mol, Figures S3,S5—we determined that it is due to the closely located Li ions which act as electron density attractors making BH$_3$ units more nucleophilic compared with the monomer, Figure S6. All these results based on the ETS-NOCV, IQA methods, atomic charges and molecular electrostatic potentials allow to conclude on the destabilizing nature of BH···HB contacts in LiN(CH$_3$)$_2$BH$_3$—it is important to admit that our results are based on cluster models which might lead to omission of some bonding features in real crystals especially as far as weak non-covalent interactions are taken into account—accordingly, further studies based on other methods within the periodic calculations are necessary in order to fully delineate the nature of homopolar BH···HB interactions. Unfortunately, the ETS-NOCV and IQA schemes are not yet available for public use in periodic calculations codes. On the other hand, it seems rational to comment, based on the very huge positive value of the (B)H···H(B) interaction energy $E^{(B)H···H(B)}_{int} = +49.4$ kcal/mol in the tetrameric Li-model, that it is to be expected that in the real Li-crystal the BH···HB interactions are likely to be destabilizing.

It is probably important to add that the electron density accumulation in the inter-hydrogen region of BH···HB is indeed sufficient to observe a bond path (as it has been noted by McGrady *et al.* [29])—however, it does not necessarily imply the overall stabilizing interactions: for example, as nicely demonstrated by Cukrowski et al. [36], two water molecules enforced to approach each other via oxygen atoms leads to formation of the oxygen-oxygen bond critical point, however, the overall binding energy is as expected positive (destabilizing) due to the fact that some subtle stabilization arising from the electron-exchange channel (resulting also in the negative ΔE_{orb} values) is diminished by the Pauli and electrostatic repulsion. A similar situation is observed in dimers of M_2X_2 (for M = Li, K, X = H, Cl) where M-X stabilization outweighs the repulsion stemming from M-M and X-X interactions [73].

In Figure 3A the structure of the $LiN(CH_3)_2BH_3$ monomer is presented. It can be seen that lithium ion forms a chemical bond not only with the nitrogen atom, but also with the BH_3 unit through intramolecular Li$\cdots$H–B interactions. The binding energy of BH_3 to NMe_2Li appears to be significant, $\Delta E_{total} = -63.99$ kcal/mol, Figure 3B. Just for comparison, the binding energy of BH_3 to ammonia in NH_3BH_3 is only -30.3 kcal/mol [13]. Such a difference is related to the existence of strong intramolecular Li$\cdots$H–B interactions. ETS-NOCV allowed us to conclude that, apart from strong dative bonds, described by $\Delta\rho(N-B)$ and the corresponding $\Delta E_{orb}(B-N) = -105.63$ kcal/mol, additional intramolecular Li$\cdots$H–B interactions are formed, $\Delta\rho(Li\cdots H-B)$. The latter component corresponds to significant charge transfer stabilization, $\Delta E_{orb}(Li\cdots H-B) = -20.94$ kcal/mol, Figure 3B. It is noteworthy that the intramolecular Li$\cdots$H–B charge transfer component (Figure 3B) appears to be stronger than the intermolecular one, $\Delta E_{orb}(Li\cdots H-B) = -15.38$ kcal/mol (Figure 2B)—this is related to the fact that in the latter case B–H bonds interact with the single Li ion, whereas in the former one with multiple Li ions.

We further performed similar calculations for the tetrameric model of $KN(CH_3)_2BH_3$ as well as for the monomer—the results of ETS-NOCV calculations are shown in Figures 5 and 6. Similarly to $LiN(CH_3)_2BH_3$, intermolecular interactions in the tetramer of $KN(CH_3)_2BH_3$ are dominated by electrostatic forces (60% of total stabilizing contributions), followed by dispersion (22%) and orbital interaction components (18%), Figure 5. The overall intermolecular K$\cdots$H–B interaction energy is slightly less pronounced compared to the corresponding Li$\cdots$H–B due to the larger size of potassium compared to lithium. One should point out the lack of BH$\cdots$HB electron delocalization within the "bay" between the two potassium atoms and BH_3 groups, which is present in $LiN(CH_3)_2BH_3$ ($\Delta\rho_2$ in Figure 2b). It has been already noted by McGrady and coworkers [29]. The B–N dative bond in $KN(CH_3)_2BH_3$ monomer is of similar strength $\Delta E_{orb}(B-N) = -104.30$ kcal/mol with respect to $LiN(CH_3)_2BH_3$. The intramolecular K$\cdots$H–B charge delocalization is of similar magnitude to the intermolecular-one, Figures 5B and 6B. Finally, the overall bonding energy of BH_3 to $KN(CH_3)_2$ in the monomer, $\Delta E_{total} = -67.34$ kcal/mol, appeared to be more negative compared to the corresponding $\Delta E_{total} = -63.99$ kcal/mol for

LiN(CH$_3$)$_2$BH$_3$ predominantly due to a lower Pauli repulsion contribution, Figures 3 and 6.

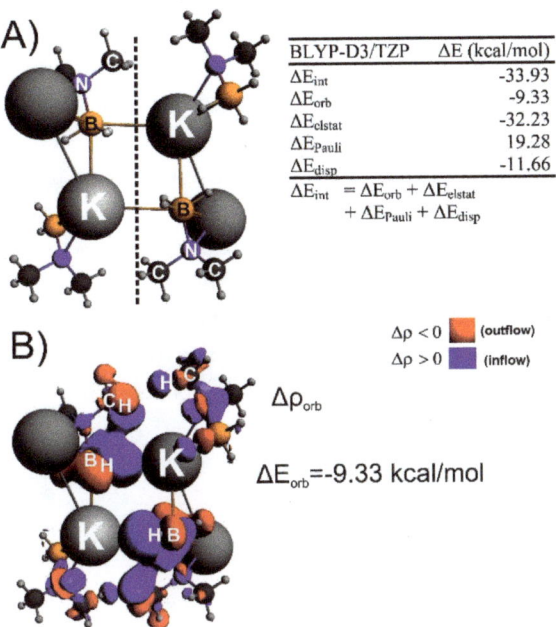

Figure 5. The tetrameric cluster model of KN(CH$_3$)$_2$BH$_3$ along with energy decomposition results describing the interaction between two dimeric fragments in KN(CH$_3$)$_2$BH$_3$ (**A**). The fragmentation pattern used in ETS-NOCV analysis is indicated by a black line. Part (**B**) displays the most relevant deformation density contributions describing K···H–B interactions. The red color of the deformation densities shows charge depletion, whereas the blue an electron accumulation due to K···H–B interaction.

Finally, we consider the two remaining cluster models of LiN(CH$_3$)$_2$BH$_3$ and KN(CH$_3$)$_2$BH$_3$ where homopolar CH···HC interactions are involved, Figures 1B, 7A and 8A.

The results of energy decomposition analyses demonstrate the stabilizing nature of CH···HC interactions in both crystals—the overall interaction energy is $\Delta E_{int} = -17.45$ kcal/mol for KN(CH$_3$)$_2$BH$_3$, whereas it is only $\Delta E_{int} = -4.34$ kcal/mol for LiN(CH$_3$)$_2$BH$_3$, Figures 7A and 8A. Significantly more pronounced stabilization in potassium crystal originates from larger number of CH···HC contacts as compared to the lithium analog. It constitutes the 2D layers architecture of the potassium crystal as compared to rather 1D chain-like structure in LiN(CH$_3$)$_2$BH$_3$. In both cases the main contribution (64%–68%) to overall stabilization is dispersion,

ΔE_{disp} = −4.63 kcal/mol for LiN(CH$_3$)$_2$BH$_3$ and ΔE_{disp} = −18.78 kcal/mol for KN(CH$_3$)$_2$BH$_3$, Figures 7A and 8A. The same has been already suggested by McGrady and coworkers [29]. We found in addition that the electrostatic ΔE_{elstat} and orbital interaction ΔE_{orb} terms are also important, Figures 7A and 8A. Figures 7B and 8B shows that the formation of the CH···HC interactions is accompanied by a charge outflow from the occupied σ(C–H) bonds and the electron density accumulation is visible in the inter-hydrogen region of CH···HC units. Considering the classical language of a molecular orbital theory one can summarize that the "electronic" part of the CH···HC bonding is based on both donation from the occupied σ (C–H) bonds into the empty σ*(C–H) of methyl groups as well as polarization of the C–H bonds (mixing of σ/σ*(C–H)). These stabilizing contributions (the polarization and charge transfer) are clearly mixed—at this point one can reference other important and interesting works in which the meaning of both contributions is debated in terms of non-covalent interactions [67,68,74–76].

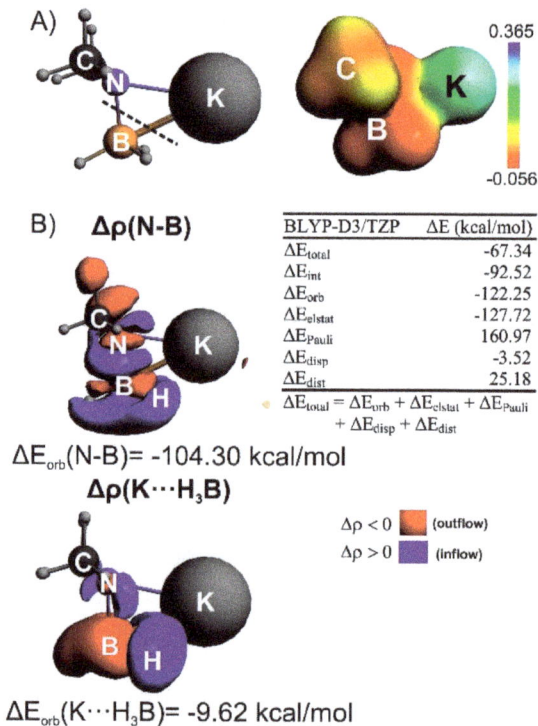

Figure 6. Monomer of KN(CH$_3$)$_2$BH$_3$ along with the corresponding molecular electrostatic potential (part (**A**)). In part (**B**) the results of ETS-NOCV analysis are presented that describe bonding between the BH$_3$ unit and the NKMe$_2$ fragment.

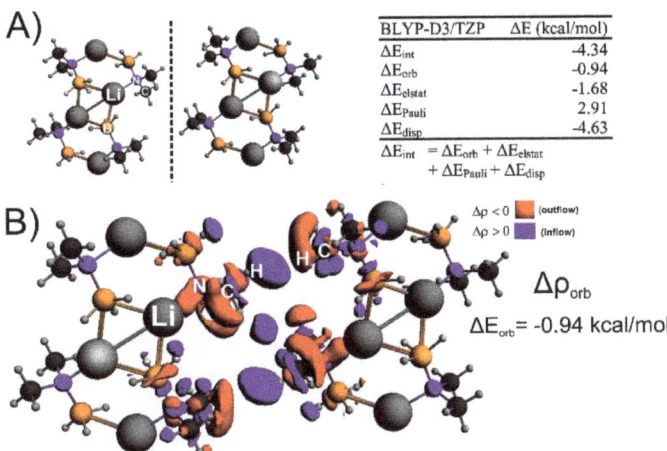

Figure 7. Cluster model containing eight monomers of LiN(CH$_3$)$_2$BH$_3$ along with energy decomposition results describing CH···HC interactions between the two selected fragments (marked by the black line), part (**A**). In part (**B**) the overall deformation density $\Delta\rho_{orb}$ is depicted together with the corresponding stabilization ΔE_{orb}.

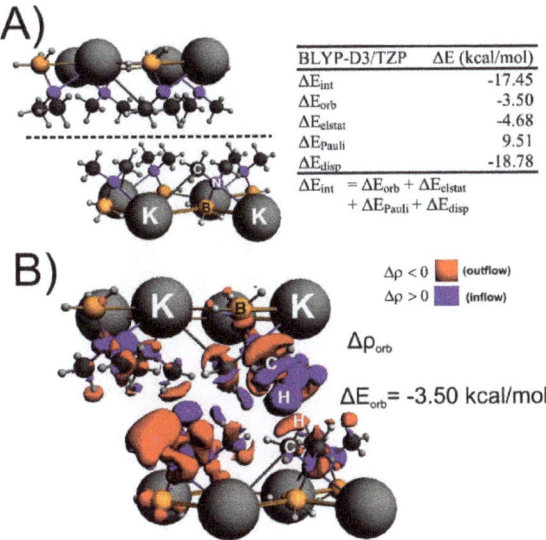

Figure 8. Cluster model containing eight monomers of KN(CH$_3$)$_2$BH$_3$ along with energy decomposition results describing CH···HC interactions between the two selected fragments (marked by the black line), part (**A**). In part (**B**) the overall deformation density $\Delta\rho_{orb}$ is depicted together with the corresponding stabilization ΔE_{orb}.

It is important to reference other works where dispersion dominated CH···HC interactions have been found—Echeverría and coworkers found, based on MP2 studies, the "subtle but not faint" stabilizing CH···HC interactions between alkanes [77]. Further Valence Bond studies have revealed [78], in line with our conclusions, that, apart from the crucial dispersion term, also the σ/σ*(C–H) polarization/charge transfer and electrostatic contributions are important. Recently, numerous reports are present in the literature on the stabilizing nature of CH···HC interactions in various types of hydrocarbons [37,38,79–85], as well as their importance in catalysis [42,43,86].

In order to complement and confirm the conclusions obtained by the ETS-NOCV method we additionally plotted the contour of the reduced density gradient of the NCI (Non-Covalent Index) method [55] for $KN(CH_3)_2BH_3$, Figure 9. It was demonstrated that this method is well suited for visualization of non-covalent interactions [55]. It is clearly seen, in line with ETS-NOCV based study, that the crystal of $KN(CH_3)_2BH_3$ is stabilized by numerous inter and intramolecular K···H–B as well as additionally by CH···HC interactions. The same is valid for $LiN(CH_3)_2BH_3$. It is to be anticipated that the existence of strong M···H–B interactions in both crystals might affect the mechanism of dehydrogenation similarly as has been already shown for the parent compound $LiNH_2BH_3$ for which dehydrogenation is initiated by Li···H–B interactions and proceeds further through formation of LiH hydride and NH_2BH_2 as intermediates [87–89].

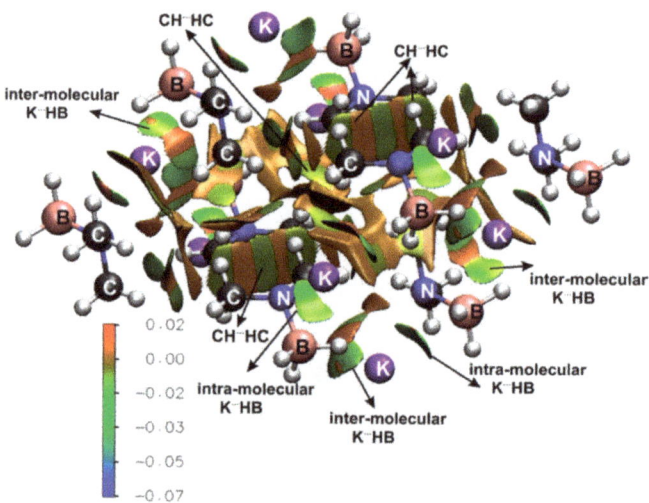

Figure 9. The surfaces describing the reduced density gradient at an isovalue of 0.5 a.u. for the octamer of $KN(CH_3)_2BH_3$. The surfaces are coloured on a blue-green-red scale according to the values of $sign(\lambda_2)\rho$, ranging from –0.05 to 0.02 au.

4. Conclusions

In the present study non-covalent interactions in the hydrogen storage materials $LiN(CH_3)_2BH_3$ and $KN(CH_3)_2BH_3$ are for the first time quantitatively (and qualitatively) described by means of the charge and energy decomposition method ETS-NOCV as well as the Interacting Quantum Atoms (IQA) approach.

It was found, in line with the pioneering work of McGrady et al. [29], that both crystals are stabilized by numerous intra- and intermolecular M⋯H–B interactions (M = Li, K). The ETS-NOCV calculations indicated that these bonds are electrostatically dominated, followed by charge transfer and dispersion contributions. Interestingly, the intramolecular charge transfer contributing to Li⋯H–B interaction appeared to be more pronounced than the corresponding intermolecular delocalization. We further noticed in $LiN(CH_3)_2BH_3$ that formation of intermolecular Li⋯H–B interactions enforces charge delocalization within the homopolar BH⋯HB contacts which explains the presence of the QTAIM bond critical points between clashing hydrogen atoms as was found by McGrady et al. [29]. In contrast, our investigations allowed us to conclude that monomers of $LiN(CH_3)_2BH_3$ are not likely to spontaneously form any stable aggregates via "pure" stabilizing BH⋯HB interactions (BH_3 to BH_3 orientation) due to the presence of negatively charged borane units—from a fundamental point of view it demonstrates that BH⋯HB interactions can be rather considered as destabilizing in this type of compounds. We further confirmed this conclusion by the Interacting Quantum Atoms (IQA) energy decomposition calculations—namely, the dihydrogen interaction energy in BH⋯HB appeared to be significantly positive (destabilizing), $E_{int}^{B(H)\cdots H(B)} = +49.4$ kcal/mol as opposed to Li⋯H–B interactions, $E_{int}^{Li\cdots H(B)} = -98.8$ kcal/mol.

Contrary to homopolar BH⋯HB interactions, the ETS-NOCV and IQA methods allowed us to identify stabilizing homopolar dihydrogen interactions CH⋯HC in both $LiN(CH_3)_2BH_3$ and $KN(CH_3)_2BH_3$—the presence of such stabilization has already been suggested by McGrady and coworkers based on a topological QTAIM study [29]. We found herein quantitatively that these interactions are dispersion dominated (64% for $LiN(CH_3)_2BH_3$ and 69% for $KN(CH_3)_2BH_3$), followed by charge transfer (13% for both $LiN(CH_3)_2BH_3$ and $KN(CH_3)_2BH_3$) and electrostatic (23% for $LiN(CH_3)_2BH_3$ and 17% for $KN(CH_3)_2BH_3$) terms. It was confirmed that these interactions are far stronger in the potassium crystal due to the larger number of CH⋯HC contacts compared to the lithium analogue. Moreover, the NOCV-based deformation density contributions allowed to state that the "electronic" part of the CH⋯HC interaction is based on both donation from the occupied σ(C–H) bonds into the empty σ*(C–H) of methyl groups as well as polarization of the C–H bonds (mixing of σ/σ*(C–H)).

Briefly summarizing, our in-depth theoretical investigations performed by means of the ETS-NOCV and IQA energy decomposition methods, electrostatic potentials and charges, allowed us to confirm most of the findings that have been already reported in the pioneering work of McGrady *et al.* [29]. Furthermore, we provided for the first time the energetic description of non-covalent interactions contributing to the stability of LiNMe$_2$BH$_3$ and KNMe$_2$BH$_3$ as well demonstrating, contrary to McGrady *et al.* [29], the repulsive nature of the homopolar interactions BH···HB. The latter is in line with numerous experimental papers [68–72]. Due to the fact that our calculations are based on the cluster approach as well as the fact that all theoretical methods are not free from approximations and very often not from arbitrariness, we believe that further works are needed from both theoretical and experimental laboratories in order to fully uncover the nature of homopolar BH···HB interactions.

Supplementary Materials: The supplementary materials are available at http://www.mdpi.com/2073-4352/6/3/28/s1.

Acknowledgments: Results presented in this work were partially obtained using PL-Grid Infrastructure and resources provided by ACC Cyfronet AGH.

Author Contributions: Filip Sagan performed the majority of the calculations presented in this work. In addition, Filip Sagan contributed to the interpretation of the results and the final manuscript form. Radosław Filas initiated the work on bonding in LiN(CH$_3$)$_2$BH$_3$ and KN(CH$_3$)$_2$BH$_3$. Mariusz P. Mitoraj interpreted the results as well as writing the manuscript text.

Conflicts of Interest: The authors declare no conflict of interest.

References

1. Christopher, K.; Dimitrios, R. A review on energy comparison of hydrogen production methods from renewable energy sources. *Energy Environ. Sci.* **2012**, *5*, 6640–6651.
2. Dincer, I.; Acar, C. Review and evaluation of hydrogen production methods for better sustainability. *Int. J. Hydrog. Energy* **2015**, *40*, 11094–11111.
3. Dutta, S. A review on production, storage of hydrogen and its utilization as an energy resource. *J. Ind. Eng. Chem.* **2014**, *20*, 1148–1156.
4. Staubitz, A.; Robertson, A.P.M.; Manners, I. Ammonia-Borane and Related Compounds as Dihydrogen Sources. *Chem. Rev.* **2010**, *110*, 4079–4124.
5. Hamilton, C.W.; Baker, R.T.; Staubitz, A.; Manners, I. B–N compounds for chemical hydrogen storage. *Chem. Soc. Rev.* **2009**, *38*, 279–293.
6. Kim, S.-K.; Han, W.-S.; Kim, T.-J.; Kim, T.-Y.; Nam, S.W.; Mitoraj, M.; Piękoś, Ł.; Michalak, A.; Hwang, S.-J.; Kang, S.O. Palladium Catalysts for Dehydrogenation of Ammonia Borane with Preferential B–H Activation. *J. Am. Chem. Soc.* **2010**, *132*, 9954–9955.

7. Parafiniuk, M.; Mitoraj, M.P. Origin of Binding of Ammonia-Borane to Transition-Metal-Based Catalysts: An Insight from the Charge and Energy Decomposition Method ETS-NOCV. *Organometallics* **2013**, *32*, 4103–4113.
8. Parafiniuk, M.; Mitoraj, M.P. On the origin of internal rotation in ammonia borane. *J. Mol. Model.* **2014**, *20*, 2272–2281.
9. Huang, Z.; Autrey, T. Boron–nitrogen–hydrogen (BNH) compounds: Recent developments in hydrogen storage, applications in hydrogenation and catalysis, and new syntheses. *Energy Environ. Sci.* **2012**, *5*, 9257–9268.
10. Umegaki, T.; Yan, J.M.; Zhang, X.B.; Shioyama, H.; Kuriyama, N.; Xu, Q. Boron- and nitrogen-based chemical hydrogen storage materials. *Int. J. Hydrog. Energy* **2009**, *34*, 2303–2311.
11. Custelcean, R.; Jackson, J.E. Dihydrogen Bonding: Structures, Energetics, and Dynamics. *Chem. Rev.* **2001**, *101*, 1963–1980.
12. Bakhmutov, V.I. *Dihydrogen Bonds: Principles, Experiments and Applications*; Wiley-Interscience: Hoboken, NJ, USA, 2008.
13. Mitoraj, M. Bonding in Ammonia Borane: An Analysis Based on the Natural Orbitals for Chemical Valence and the Extended Transition State Method (ETS-NOCV). *J. Phys. Chem. A* **2011**, *115*, 14708–14716.
14. Jonas, V.; Frenking, G.; Reetz, M.T. Comparative Theoretical Study of Lewis Acid-Base Complexes of BH3, BF3, BCl3, AlCl3, and SO2. *J. Am. Chem. Soc.* **1994**, *116*, 8741–8753.
15. Merino, G.; Bakhmutov, V.I.; Vela, A. Do Cooperative Proton-Hydride explain the Gas-Solid Structural Difference of BH3NH3? *J. Phys. Chem. A* **2002**, *106*, 8491–8994.
16. Popelier, P.L.A. Characterization of a Dihydrogen Bond on the Basis of the Electron Density. *J. Phys. Chem. A* **1998**, *102*, 1873–1878.
17. Wolstenholme, D.J.; Traboulsee, K.T.; Hua, Y.; Calhoun, L.A.; McGrady, G.S. Thermal desorption of hydrogen from ammonia borane: Unexpected role of homopolar B–H···H–B interactions. *Chem. Commun.* **2012**, *48*, 2597–2599.
18. Guerra, D.; David, J.; Restrepo, A. (H$_3$N–BH$_3$)$_4$: The ammonia borane tetramer. *Phys. Chem. Chem. Phys.* **2012**, *14*, 14892–14897.
19. Chen, X.; Yuan, F.; Tan, Y.; Tang, Z.; Yu, X. Improved Dehydrogenation Properties of Ca(BH$_4$)$_2$·nNH$_3$ (n = 1, 2, and 4) Combined with Mg(BH$_4$)$_2$. *J. Phys. Chem. C* **2012**, *116*, 21162–21168.
20. Wolstenholme, D.J.; Dobson, J.L.; McGrady, G.S. Homopolar dihydrogen bonding in main group hydrides: Discovery, consequences, and applications. *Dalton Trans.* **2015**, *44*, 9718–9731.
21. Wolstenholme, D.J.; Fradsham, E.J.; McGrady, G.S. Supramolecular interactions in boron hydrides: How non-classical bonding directs their crystal architecture. *CrystEngComm* **2015**, *17*, 7623–7627.
22. Chen, X.; Zhao, J.C.; Shore, S.G. The Roles of Dihydrogen Bonds in Amine Borane Chemistry. *Acc. Chem. Res.* **2013**, *46*, 2666–2675.

23. Sagan, F.; Piękoś, Ł.; Andrzejak, M.; Mitoraj, M.P. From Saturated BN–compounds to Isoelectronic BN/CC Counterparts—An Insight from Computational Perspective. *Chemistry* **2015**, *21*, 15299–15307.
24. Grochala, W.; Edwards, P.P. Thermal Decomposition of the Non-Interstitial Hydrides for the Storage and Production of Hydrogen. *Chem. Rev.* **2004**, *104*, 1283–1316.
25. Dovgaliuk, I.; Le Duff, C.S.; Robeyns, K.; Devillers, M.; Filinchuk, Y. Mild Dehydrogenation of Ammonia Borane Complexed with Aluminum Borohydride. *Chem. Mater.* **2015**, *27*, 768–777.
26. Orimo, S.; Nakamori, Y.; Eliseo, J.R.; Züttel, A.; Jensen, C.M. Complex Hydrides for Hydrogen Storage. *Chem. Rev.* **2007**, *107*, 4111–4132.
27. Luoa, J.; Wub, H.; Zhoub, W.; Kang, X.; Fang, Z.; Wang, P. $LiBH_4 \cdot NH_3BH_3$: A new lithium borohydride ammonia borane compound with a novel structure and favorable hydrogen storage properties. *Int. J. Hydrog. Energy* **2012**, *37*, 10750–10757.
28. Jaroń, T.; Orłowski, P.A.; Wegner, W.; Fijałkowski, K.J.; Leszczyński, P.J.; Grochala, W. Hydrogen Storage Materials: Room-Temperature Wet-Chemistry Approach toward Mixed-Metal Borohydrides. *Angew. Chem. Int. Ed.* **2015**, *54*, 1236–1239.
29. Wolstenholme, D.J.; Flogeras, J.; Che, F.N.; Decken, A.; McGrady, G.S. Homopolar Dihydrogen Bonding in Alkali Metal Amidoboranes: Crystal Engineering of Low-Dimensional Molecular Materials. *J. Am. Chem. Soc.* **2013**, *135*, 2439–2442.
30. Bader, R.F.W. *Atoms in Molecules: A Quantum Theory*; Oxford University Press: Oxford, UK, 1990.
31. Weinhold, F.; Schleyer, P.R.; McKee, W.C. Bay-Type H–H "Bonding" in cis-2-Butene and Related Species: QTAIM Versus NBO Description. *J. Computat. Chem.* **2014**, *35*, 1499–1508.
32. Poater, J.; Solà, M.; Bickelhaupt, F.M. Hydrogen–Hydrogen Bonding in Planar Biphenyl, Predicted by Atoms-In-Molecules Theory, Does Not Exist. *Chemistry* **2006**, *12*, 2889–2895.
33. Grimme, S.; Mück-Lichtenfeld, C.; Erker, G.; Kehr, G.; Wang, H.; Beckers, H.; Willner, H. When Do Interacting Atoms Form a Chemical Bond? Spectroscopic Measurements and Theoretical Analyses of Dideuteriophenanthrene. *Angew. Chem. Int. Ed.* **2009**, *48*, 2592–2595.
34. Jacobsen, H. Kinetic energy density and covalent bonding—A complementary analysis at the border of bond and no bond. *Dalton Trans.* **2010**, *39*, 5426–5428.
35. Hancock, R.D.; Nikolayenko, I.V. Do Nonbonded H–H Interactions in Phenanthrene Stabilize It Relative to Anthracene? A Possible Resolution to this Question and Its Implications for Ligands such as 2,2′-Bipyridyl. *J. Phys. Chem. A* **2012**, *116*, 8572–8583.
36. Cukrowski, I.; de Lange, J.H.; Adeyinka, A.S.; Mangondo, P. Evaluating common QTAIM and NCI interpretations of the electron density concentration through IQA interaction energies and 1D cross-sections of the electron and deformation density distributions. *Comput. Theor. Chem.* **2015**, *1053*, 60–76.

37. Safin, D.A.; Babashkina, M.G.; Robeyns, K.; Mitoraj, M.P.; Kubisiak, P.; Garcia, Y. Influence of the Homopolar Dihydrogen Bonding CH···HC on Coordination Geometry: Experimental and Theoretical Studies. *Chemistry* **2015**, *21*, 16679–16687.
38. Cukrowski, I. IQA-embedded fragment attributed molecular system energy change in exploring intramolecular interactions. *Comput. Theor. Chem.* **2015**, *1066*, 62–75.
39. Cukrowski, I.; Govender, K.K.; Mitoraj, M.P.; Srebro, M. QTAIM and ETS-NOCV Analyses of Intramolecular CH···HC Interactions in Metal Complexes. *J. Phys. Chem. A* **2011**, *115*, 12746–12757.
40. Cukrowski, I.; de Lange, J.H.; Mitoraj, M.P. Physical Nature of Interactions in ZnII Complexes with 2,2′-Bipyridyl: Quantum Theory of Atoms in Molecules (QTAIM), Interacting Quantum Atoms (IQA), Noncovalent Interactions (NCI), and Extended Transition State Coupled with Natural Orbitals for Chemical Valence (ETS-NOCV) Comparative Studies. *J. Phys. Chem. A* **2014**, *118*, 623–637.
41. Safin, D.A.; Babashkina, M.G.; Kubisiak, P.; Mitoraj, M.P.; Le Duff, C.S.; Robeyns, K.; Garcia, Y. Crucial influence of the intramolecular hydrogen bond on the coordination mode of RC(S)NHP(S)(OiPr)$_2$ in homoleptic complexes with Ni^{II}. *Eur. J. Inorg. Chem.* **2013**, *2013*, 545–555.
42. Lyngvi, E.; Sanhueza, I.A.; Schoenebeck, F. Dispersion Makes the Difference: Bisligated Transition States Found for the Oxidative Addition of Pd(PtBu3)2 to Ar-OSO2R and Dispersion-Controlled Chemoselectivity in Reactions with Pd[P(iPr)(tBu2)]2. *Organometallics* **2015**, *34*, 805–812.
43. Wolters, L.P.; Koekkoek, R.; Bickelhaupt, F.M. Role of Steric Attraction and Bite-Angle Flexibility in Metal-Mediated C–H Bond Activation. *ACS Catal.* **2015**, *5*, 5766–5775.
44. Bader, R.F.W. Pauli Repulsions Exist Only in the Eye of the Beholder. *Chemistry* **2006**, *12*, 2896–2901.
45. Hernández-Trujillo, J.; Matta, C.F. Hydrogen–hydrogen bonding in biphenyl revisited. *Struct. Chem.* **2007**, *18*, 849–857.
46. Matta, C.F.; Sadjadi, S.A.; Braden, D.A.; Frenking, G. The Barrier to the Methyl Rotation in Cis-2-Butene and its Isomerization Energy to Trans-2-Butene, Revisited. *J. Comput. Chem.* **2016**, *37*, 143–154.
47. Mitoraj, M.P.; Michalak, A. Natural orbitals for chemical valence as descriptors of chemical bonding in transition metal complexes. *J. Mol. Model.* **2007**, *13*, 347–355.
48. Michalak, A.; Mitoraj, M.P.; Ziegler, T. Bond Orbitals from Chemical Valence Theory. *J. Phys. Chem. A* **2008**, *112*, 1933–1939.
49. Mitoraj, M.P.; Michalak, A.; Ziegler, T. A Combined Charge and Energy Decomposition Scheme for Bond Analysis. *J. Chem. Theory Comput.* **2009**, *5*, 962–975.
50. Mitoraj, M.P.; Michalak, A. Theoretical description of halogen bonding—An insight based on the natural orbitals for chemical valence combined with the extended-transition-state method (ETS-NOCV). *J. Mol. Model.* **2013**, *19*, 4681–4688.
51. Mitoraj, M.P.; Janjic, G.V.; Medakovic, V.B.; Veljkovic, D.Z.; Michalak, A.; Zaric, S.D.; Milcic, M.K. Nature of the Water/Aromatic Parallel Alignment Interactions. *J. Comput. Chem.* **2015**, *36*, 171–180.

52. Te Velde, G.; Bickelhaupt, F.M.; Baerends, E.J.; Fonseca Guerra, C.; van Gisbergen, S.J.A.; Snijders, J.G.; Ziegler, T. Chemistry with ADF. *J. Comput. Chem.* **2001**, *22*, 931–967.
53. Bickelhaupt, F.M.; Baerends, E.J. Kohn-Sham Density Functional Theory: Predicting and Understanding Chemistry. *Rev. Comput. Chem.* **2007**, *15*, 1–86.
54. Baerends, E.J.; Autschbach, J.; Bashford, D.; Bérces, A.; Bickelhaupt, F.M.; Bo, C.; Boerrigter, P.M.; Cavallo, L.; Chong, D.P.; Deng, L.; *et al. SCM Theoretical Chemistry*; ADF2012.01; Vrije Universiteit: Amsterdam, The Netherlands, 2012.
55. Johnson, E.R.; Keinan, S.; Mori-Sánchez, P.; Contreras-García, J.; Cohen, A.J.; Yang, W. Revealing Noncovalent Interactions. *J. Am. Chem. Soc.* **2010**, *132*, 6498–6506.
56. Blanco, M.A.; Pendás, A.M.; Francisco, E. Interacting Quantum Atoms: A Correlated Energy Decomposition Scheme Based on the Quantum Theory of Atoms in Molecules. *J. Chem. Theory Comput.* **2005**, *1*, 1096–1109.
57. Gao, W.; Feng, H.; Xuan, X.; Chen, L. The assessment and application of an approach to noncovalent interactions: The energy decomposition analysis (EDA) in combination with DFT of revised dispersion correction (DFT-D3) with Slater-type orbital (STO) basis set. *J. Mol. Model.* **2012**, *18*, 4577–4589.
58. Van der Wijst, T.; Fonseca Guerra, C.; Swart, M.; Bickelhaupt, F.M.; Lippert, B. A Ditopic Ion-Pair Receptor Based on Stacked Nucleobase Quartets. *Angew. Chem. Int. Ed.* **2009**, *48*, 3285–3287.
59. Fonseca Guerra, C.; van der Wijst, T.; Poater, J.; Swart, M.; Bickelhaupt, F.M. Adenine *versus* guanine quartets in aqueous solution: Dispersion-corrected DFT study on the differences in *p*-stacking and hydrogen-bonding behavior. *Theor. Chem. Acc.* **2010**, *125*, 245–252.
60. Grimme, S. Semiempirical GGA-Type Density Functional Constructed with a Long-Range Dispersion Correction. *J. Comput. Chem.* **2006**, *27*, 1787–1799.
61. Frisch, M.J.; Trucks, G.W.; Schlegel, H.B.; Scuseria, G.E.; Robb, M.A.; Cheeseman, J.R.; Scalmani, G.; Barone, V.; Mennucci, B.; Petersson, G.A.; *et al. Gaussian 09*; Gaussian Inc.: Wallingford, CT, USA, 2009.
62. Ziegler, T.; Rauk, A. On the calculation of bonding energies by the Hartree Fock Slater method. *Theor. Chim. Acta* **1977**, *46*, 1–10.
63. AIMAll, Version 13.05.06 Professional. Todd A. Keith, TK Gristmill Software, Overland Park, KS, USA, 2016.
64. Politzer, P.; Murray, J.S.; Clark, T. Halogen bonding: An electrostatically-driven highly directional noncovalent interaction. *Phys. Chem. Chem. Phys.* **2010**, *12*, 7748–7757.
65. Politzer, P.; Murray, J.S.; Janjić, G.V.; Zarić, S.D. σ-Hole Interactions of Covalently-Bonded Nitrogen, Phosphorus and Arsenic: A Survey of Crystal Structures. *Crystals* **2014**, *4*, 12–31.
66. Clark, T.; Murray, J.S.; Politzer, P. The Role of Polarization in a Halogen Bond. *Aust. J. Chem.* **2014**, *67*, 451–456.
67. Politzer, P.; Murray, J.S.; Clark, T. Mathematical modeling and physical reality in noncovalent interactions. *J. Mol. Model.* **2015**, *52*, 21–31.

68. Schouwink, P.; Hagemann, H.; Embs, J.P.; Anna, V.D.; Černý, R. Di-hydrogen contact induced lattice instabilities and structural dynamics in complex hydride perovskites. *J. Phys. Condens. Matter* **2015**, *27*, 265403–265415.
69. Černý, R.; Kim, K.C.; Penin, N.; D'Anna, V.; Hagemann, H.; Sholl, D.S. AZn$_2$(BH$_4$)$_5$ (A = Li, Na) and NaZn(BH4)3: Structural Studies. *J. Phys. Chem. C* **2010**, *114*, 19127–19133.
70. Černý, R.; Ravnsbæk, D.B.; Schouwink, P.; Filinchuk, Y.; Penin, N.; Teyssier, J.; Smrčok, L.; Jensen, T.R. Potassium Zinc Borohydrides Containing Triangular [Zn(BH$_4$)$_3$] and Tetrahedral [Zn(BH$_4$)Cl$_4$]$_2$ Anions. *J. Phys. Chem. C* **2012**, *116*, 1563–1571.
71. Ravindran, P.; Vajeeston, P.; Vidya, R.; Kjekshus, A.; Fjellvåg, H. Violation of the Minimum H–H Separation "Rule" for Metal Hydrides. *Phys. Rev. Lett.* **2002**, *89*, 106403–106407.
72. Ravnsbæk, D.; Filinchuk, Y.; Cerenius, Y.; Jakobsen, H.J.; Besenbacher, F.; Skibsted, J.; Jensen, T.R. A Series of Mixed-Metal Borohydrides. *Angew. Chem. Int. Ed.* **2009**, *48*, 6659–6663.
73. Demyanov, P.; Polestshuk, P. A Bond Path and an Attractive Ehrenfest Force Do Not Necessarily Indicate Bonding Interactions: Case Study on M2X2 (M = Li, Na, K; X = H, OH, F, Cl). *Chemistry* **2012**, *18*, 4982–4993.
74. Wang, C.; Danovich, D.; Mo, Y.; Shaik, S. On The Nature of the Halogen Bond. *J. Chem. Theory Comput.* **2014**, *10*, 3726–3737.
75. Grabowski, S.J. Hydrogen and Halogen Bonds Are Ruled by the Same Mechanisms. *Phys. Chem. Chem. Phys.* **2013**, *15*, 7249–7259.
76. Mulliken, R.S. Structures of Complexes Formed by Halogen Molecules with Aromatic and with Oxygenated Solvents. *J. Am. Chem. Soc.* **1950**, *72*, 600–608.
77. Echeverría, J.; Aullón, G.; Danovich, D.; Shaik, S.; Alvarez, S. Dihydrogen contacts in alkanes are subtle but not faint. *Nat. Chem.* **2011**, *3*, 323–330.
78. Danovich, D.; Shaik, S.; Neese, F.; Echeverría, J.; Aullón, G.; Alvarez, S. Understanding the Nature of the CH···HC Interactions in Alkanes. *J. Chem. Theory Comput.* **2013**, *9*, 1977–1991.
79. Grimme, S.; Schreiner, P.R. Steric Crowding Can Stabilize a Labile Molecule: Solving the Hexaphenylethane Riddle. *Angew. Chem. Int. Ed.* **2011**, *50*, 12639–12642.
80. Fokin, A.; Chernish, L.V.; Gunchenko, P.A.; Tikhonchuk, E.Y.; Hausmann, H.; Serafin, M.; Dahl, J.E.P.; Carlson, R.M.K.; Schreiner, P.R. Stable Alkanes Containing Very Long Carbon–Carbon Bonds. *J. Am. Chem. Soc.* **2012**, *134*, 13641–13650.
81. Zhang, J.; Dolg, M. Dispersion Interaction Stabilizes Sterically Hindered Double Fullerenes. *Chemistry* **2014**, *20*, 13909–13912.
82. Grabowski, S.J.; Pfitzner, A.; Zabel, M.; Dubis, A.T.; Palusiak, M. Intramolecular H···H Interactions for the Crystal Structures of [4-((E)-But-1-enyl)-2,6-dimethoxyphenyl] pyridine-3-carboxylate and [4-((E)-Pent-1-enyl)-2,6-dimethoxyphenyl] pyridine-3-carboxylate; DFT Calculations on Modeled Styrene Derivatives. *J. Phys. Chem. B* **2004**, *108*, 1831–1837.
83. Grabowski, S.J. Dihydrogen bond and X–H···σ interaction as sub-classes of hydrogen bond. *J. Phys. Org. Chem.* **2013**, *26*, 452–459.

84. Grabowski, S.J.; Sokalski, W.A.; Leszczynski, J. Nature of X–H$^{+\delta}\cdots{}^{-\delta}$H–Y dihydrogen bonds and X–H$\cdots\sigma$ Interactions. *J. Phys. Chem. A* **2004**, *108*, 5823–5830.
85. Grabowski, S.J.; Sokalski, W.A.; Leszczynski, J. How short can the H$\cdots$H intermolecular contact can be? New findings that reveal the covalent nature of extremely strong interactions. *J. Phys. Chem. A* **2005**, *109*, 4331–4341.
86. Krapp, A.; Frenking, G.; Uggerud, E. Nonpolar Dihydrogen Bonds—On a Gliding Scale from Weak Dihydrogen Interaction to Covalent H–H in Symmetric Radical Cations [H$_n$E-H-H-EH$_n$]$^+$. *Chemistry* **2008**, *14*, 4028–4038.
87. Kim, D.Y.; Singh, N.J.; Lee, H.M.; Kim, K.S. Hydrogen-Release Mechanisms in Lithium Amidoboranes. *Chemistry* **2009**, *15*, 5598–5604.
88. Lee, T.B.; McKee, M.L. Mechanistic Study of LiNH$_2$BH$_3$ Formation from (LiH)$_4$ + NH$_3$BH$_3$ and Subsequent Dehydrogenation. *Inorg. Chem.* **2009**, *48*, 7564–7575.
89. Luedtke, A.T.; Autrey, T. Hydrogen Release Studies of Alkali Metal Amidoboranes. *Inorg.Chem.* **2010**, *49*, 3905–3910.

[FHF]⁻—The Strongest Hydrogen Bond under the Influence of External Interactions

Sławomir J. Grabowski

Abstract: A search through the Cambridge Structural Database (CSD) for crystal structures containing the [FHF]⁻ anion was carried out. Forty five hydrogen bifluoride structures were found mainly with the H-atom moved from the mid-point of the F ... F distance. However several [FHF]⁻ systems characterized by $D_{\infty h}$ symmetry were found, the same as this anion possesses in the gas phase. The analysis of CSD results as well as the analysis of results of *ab initio* calculations on the complexes of [FHF]⁻ with Lewis acid moieties show that the movement of the H-atom from the central position depends on the strength of interaction of this anion with external species. The analysis of the electron charge density distribution in complexes of [FHF]⁻ was performed with the use of the Quantum Theory of Atoms in Molecules (QTAIM) approach and the Natural Bond Orbitals (NBO) method.

Reprinted from *Crystals*. Cite as: Grabowski, S.J. [FHF]⁻—The Strongest Hydrogen Bond under the Influence of External Interactions. *Crystals* **2016**, *6*, 3.

1. Introduction

The [FHF]⁻ system has been the subject of numerous studies [1–3] since it is an example of one of the strongest hydrogen bonds (if not the strongest known) [4–7]; it is often classified as a short strong hydrogen bond (SSHB) [5,8]. The electronic structure of the [FHF]⁻ anion and of other hydrogen bihalides, [XHX]⁻, is often interpreted in terms of a three-center four-electron bond [9]; these bihalides are linear with a symmetrically shared proton [10]. In numerous studies the bond dissociation energy, D_0, of the reaction [FHF]⁻ → F⁻ + HF was evaluated and is located over a broad range between 35 and 60 kcal/mol [3,10]. Recent benchmark fc-CCSD(T) calculations with aug-cc-pVnZ basis sets up to cardinal number $n = 7$ and followed by extrapolation to the complete basis set limit, with the consideration of zero point vibration energy and contribution of the mass-dependent diagonal Born-Oppenheimer correction, results in D_0 value of 43.39 kcal/mol [3].

It seems that the hydrogen bihalides, [XHX]⁻, described briefly here, are the only well documented systems with a symmetrical position of the H-atom, exactly in the middle of X ... X distance. The $H_5O_2^+$ and $(N_2)H^+(N_2)$ species, or others often known as proton-bound homodimers [11], were analyzed and the high level calculations show that the H-atom is slightly moved from the mid-point of the distance between the neighboring heavier centers [12]. It is situated closer to one of the neighboring electronegative centers but the potential barrier height for the proton transfer process is very low, even negligible [12]. This explains why studies on the symmetrical and unsymmetrical [FHF]⁻ systems are very interesting. The latter anion is characterized by the $D_{\infty h}$ symmetry in a gas phase [1–3] but in crystal structures shifts of the proton from the mid-point are often observed [4,13].

A brief survey of the properties of the [FHF]⁻ anion and the corresponding deuterated species, [FDF]⁻, was performed by Jeffrey [4]—these anions possess $D_{\infty h}$ symmetry in crystal structures containing Na⁺ and K⁺ cations. However for the crystal structure with the pCH₃C₆H₅NH₂⁺ cation the H-atom is shifted from the centric position which results in the F-H distances for the [FHF]⁻ ion amounting 1.025 and 1.235 Å. Analysis on the NMR and neutron diffraction data for the [FHF]⁻ anion was performed by Panich [13]. The shorter fluorine-hydrogen distance in this species is designated as F-H, the longer one as H ... F while the distance between fluorine atoms as F-F. These designations are retained for the results discussed by Panich as well as for the results of this study discussed herein. The relationship between the F-H and F-F distances was found and discussed [13]; one can observe shortening of the F-H distance with elongation of the F-F distance; *i.e.*, a greater H-atom shift from the central position for longer F-F distances is observed. It is worth mentioning that systems containing [FHF]⁻ ions of the $D_{\infty h}$ symmetry have been observed in LiHF₂, NaHF₂, KHF₂, CsHF₂, and Ca(HF₂)₂ structures [13].

The similar relationships between the A-H bond length and the A ... B distance were found in numerous samples of the A-H ... B hydrogen bonds analyzed both experimentally and theoretically [4]. Such relationships were described early for the O-H ... O hydrogen bonded systems in crystal structures [14–17]. It is interesting that the correlations between A-H and H ... B (or A ... B) distances are often in agreement with the simple model known as the bond number conservation rule [18,19] which is based on adaptation for triatomic systems of the concept of bond number introduced early by Pauling [20]. Numerous studies on the interrelation between this rule for triatomic A-H ... B hydrogen bonded systems and experimental or theoretical dependencies have been reported [21–25]. The rule may be briefly described for hydrogen bonds in the following way; for the A-H single bond not involved in any interaction the bond number of A-H is equal to unity; for the A-H ... B systems, the A-H bond is elongated and its bond number decreases. However the latter is compensated by H ... B contact because the sum of the A-H and H ... B bond numbers is equal to unity (the bond number conservation rule). It is worth mentioning that the bond number conservation rule was also applied to systems which are the subject of this study, *i.e.*, for the FHF hydrogen bonds [26]; the dependence derived from this rule is in excellent agreement with the geometrical experimental parameters [26].

Numerous other important studies on the [FHF]$^-$ systems may be mentioned. For example, liquid state ^{1}H and ^{19}F NMR studies were performed at different temperatures for hydrogen bonded complexes of the fluoride anion, F$^-$, with HF molecules (from one to four HF molecules) [27]. At low concentration of the HF species the bifluoride ion, [FHF]$^-$, was observed, while at higher concentration of HF, [F(HF)$_n$]$^-$ complexes (n = 2, 3, 4), are formed [27]. The low-temperature (the temperature range 94–170 K) ^{1}H, ^{19}F, and ^{15}N NMR spectra were also measured for collidine-HF mixtures in CDF$_3$/CDF$_2$Cl solvent [26]. The chemical shifts and scalar coupling constants were measured for the analyzed species which led to the determination of their composition [26]; one can mention that collidinium hydrogen difluoride is an ionic salt linked through a strong hydrogen bond [26].

The FHF hydrogen bonds were also analyzed in clusters of [F(HF)$_n$]$^-$, [RF(HF)$_n$], and [XF(HF)$_n$] where R is alkyl while X = H, Br, Cl, and F [28]. The MP2/6-31++G(d,p) calculations were performed on twenty five such complexes and it was found that the bonding character in the complexes changes from covalent to van der Waals interaction [28]. Of note is the study on the H/D isotope effects on the hydrogen bond geometries of [F(HF)$_n$]$^-$ complexes [29], or the studies on the significance of the hydrogenic zero-point vibrations which give rise to structural changes and to geometric H/D isotope effects in the [FHF]$^-$ ion [30,31], the latter effects were analyzed also for the low barrier N-H . . . N hydrogen bonds [25]. In the study mentioned earlier [27] one-bond couplings between a hydrogen bond proton and two heavy atoms (fluorine atoms) of a hydrogen bridge were observed in the [F(HF)$_n$]$^-$ clusters. Further analyses of spin-spin coupling constants were performed for the [F(HF)$_n$]$^-$ species [32] as well as for the [FHF]$^-$ and [FDF]$^-$ ions [33].

It can be seen that there are numerous experimental and theoretical studies on the [FHF]$^-$ ion and the related complexes and clusters. For the purposes of this study the most interesting are those where changes of the geometry of the hydrogen bifluoride are considered; related studies were mentioned earlier here [26–31]. The other example concerns the CID level calculations with a [3s2p1d/2s1p] basis set where the transition from a centrosymmetric single minimum to a pair of equivalent minima as the inter-fluorine distance increases was observed; this transition occurs at the F-F distance approximately equal to 2.4 Å [34]. Similar conclusions were stated in the other study where the proton transfer process was also analyzed [35]. More recently the (ROH)$_n$ . . . [FHF]$^-$ clusters (n = 1, 2 and R = H, CH$_3$, C$_2$H$_5$) were analyzed theoretically and the influence of the external O-H . . . F hydrogen bonds on the position of H-atom in [FHF]$^-$ was observed [36]. The other recent and interesting theoretical study is concerned with the interrelations between the [FHF]$^-$ system and the pnicogen bonded complex [37].

The aim of this study is to analyze the geometry of the [FHF]$^-$ anion, particularly the factors influencing the asymmetry of the H-atom position are considered. There have been earlier studies where the symmetric and asymmetric hydrogen bifluoride anions were considered [26–37]. However in this study the emphasis is put on electron charge redistribution being the result of complexation. That is why the Quantum Theory of Atoms in Molecules (QTAIM) [38] and the Natural Bond Orbitals (NBO) method [39,40] have been applied here. The parameters which may be treated as indicators of the hydrogen bifluoride anion asymmetry are also discussed and the examples of the crystal structures containing [FHF]$^-$ anion are presented for comment. However the main topic not analyzed previously which concerns this study, is the decomposition of the [FHF]$^-$ anion due to its strong, covalent in nature interactions with Lewis acid centers.

2. Experimental and Computational Section

The Cambridge Structural Database (CSD) [41] search was performed to find structures containing the [FHF]$^-$ anion. The following search criteria were used: no disordered structures, no structures with unresolved errors, no powder structures, no polymeric structures, e.s.d.'s $\leqslant 0.005$ Å, R $\leqslant 7.5\%$ and geometrical restrictions of F-F distance less than 2.6 Å and the F-H-F angle in the range (140°–180°). Thirty four crystal structures containing 45 unique [FHF]$^-$ geometries were found. Additionally the search was performed for the sample mentioned above of crystal structures to find the Lewis acid—fluorine (F-atom belongs to the [FHF]$^-$ anion) contacts, shorter than the corresponding sum of the van der Waals radii. Two hundred and ninety H . . . F contacts were found corresponding to the hydrogen bond interactions and 34 other contacts (for details see the next section).

MP2/aug-cc-pVTZ calculations were performed with the Gaussian09 set of codes [42] on the [FHF]$^-$ anion and its complexes with the HF, HCl, H$_2$O, C$_2$H$_2$, HCN, and H$_2$S species. The complexes with the NH$_4^+$, PH$_4^+$, H$_3$O$^+$, and H$_3$S$^+$ cations were also calculated since strong hydrogen bonds between them and [FHF]$^-$ are expected. Additionally interactions with Li$^+$ and Na$^+$ cations are considered for comparison with the sample of the H-bonded systems. Figure 1 presents the molecular graphs (based on the electron charge distributions derived from the Bader theory [38]) of selected complexes considered in this study; the molecular graph of the [FHF]$^-$ ion not involved in any interaction is included for comparison.

The geometry optimizations performed on these moieties as well as on the monomeric Lewis acids led to energetic minima since no imaginary frequencies were observed for them. The interaction energies and binding energies were calculated for the complexes analyzed. The interaction energy, E_{int}, of the A . . . B complex is usually defined according to the supermolecular approach [43] by the following expression—Equation (1).

$$E_{int} = E_{A\ldots B}(A\ldots B)^{A\cup B} - E_{A\ldots B}(A)^{A} - E_{A\ldots B}(B)^{B} \qquad (1)$$

The designations in parentheses correspond to systems for which energies are calculated, the superscripts correspond to the basis sets used and the subscripts relate to the optimized geometry. Thus the interaction energy is the difference between the energy of the A ... B complex and the energies of the A and B monomers. The geometry for the complex was optimized with the use of the complex A∪B basis set. The monomers are characterized by geometries taken from the complex—energies for them were calculated with the A and B monomers' basis sets.

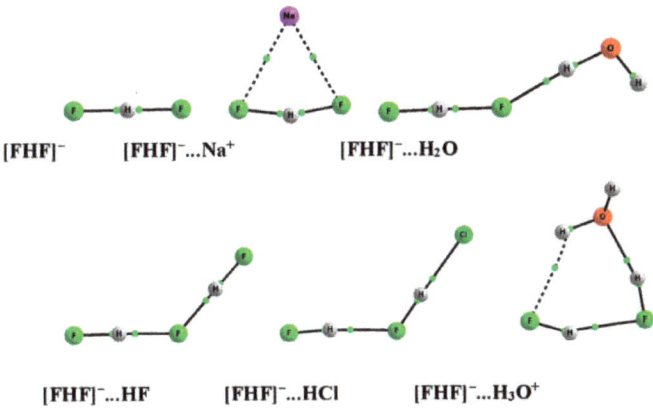

[FHF]⁻   [FHF]⁻...Na⁺   [FHF]⁻...H₂O

[FHF]⁻...HF   [FHF]⁻...HCl   [FHF]⁻...H₃O⁺

Figure 1. Molecular graphs of selected systems analyzed here, big circles correspond to atoms, continuous and broken lines to bond paths while small, green circles to bond critical points.

The binding energy, E_{bin}, takes into account the deformation energy (E_{def}) being the result of complexation.

$$E_{bin} = E_{int} + E_{def} = E_{A\ldots B}(A\ldots B)^{A\cup B} - E_A(A)^A - E_B(B)^B \qquad (2)$$

For the binding energy (Equation (2)) the energies of separately optimized monomers are considered (see the appropriate subscripts). The deformation energy defined by Equation (3) is positive since the separate molecules having geometries taken from the complex are not in energetic minima.

$$E_{def} = E_{A\ldots B}(A)^A + E_{A\ldots B}(B)^B - E_A(A)^A - E_B(B)^B \qquad (3)$$

The Counterpoise (CP) correction is applied here to calculate the basis set superposition errors (BSSE) [44]. Hence the BSSE corrected interaction and binding energies, E_{int}BSSE and E_{bin}BSSE, for complexes analyzed here are also presented.

The Natural Bond Orbitals (NBO) method [39,40] implemented in the NBO 5.0 program [45] and incorporated into GAMESS set of codes [46] was applied. The

NBO method was used to calculate the atomic charges of the systems considered. The Quantum Theory of "Atoms in Molecules" (QTAIM) [38] was applied for the localization of bond paths and corresponding critical points in the species analyzed. The following characteristics of BCPs corresponding to the intermolecular interactions are considered here; the electron density at BCP (ρ_{BCP}), its Laplacian ($\nabla^2\rho_{BCP}$) and the total electron energy density at BCP (H_{BCP}). The QTAIM calculations were carried out with the use of the AIMAll program [47].

3. Results and Discussion

3.1. [FHF]$^-$ Anion in Crystal Structures

Figure 2 presents two crystal structures of bis(tetramethylammonium) di-fluoro-dioxoiodide hydrogen difluoride, and tetramethylammonium hydrogen difluoride (refcodes: FAJHAA and KELRIC01, respectively) where the $D_{\infty h}$ symmetry of the [FHF]$^-$ anion results from the crystal symmetry. It means that the symmetrical environment of the fluorine atoms of the hydrogen difluoride does not enforce the change of the $D_{\infty h}$ anion symmetry. Another situation is observed in the crystal structure of diphenylguanidinium hydrogen difluoride (Figure 3, refcode: IBOWOL) where the non-symmetrical environment of the [FHF]$^-$ anion results in the break of the $D_{\infty h}$ symmetry since the F-H and H...F distances are equal to 1.14 and 1.17 Å, respectively, and the F-H-F angle amounts to 165.9°.

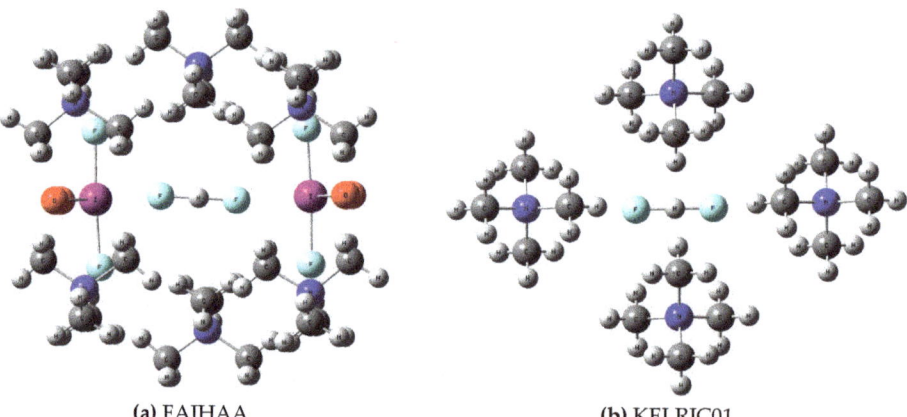

(a) FAJHAA (b) KELRIC01

Figure 2. The fragments of the crystal structures of (**a**) bis(tetramethylammonium) di-fluoro-dioxoiodide (FAJHAA); (**b**) tetramethylammonium hydrogen difluoride (KELRIC01).

Deviations from the above-mentioned $D_{\infty h}$ symmetry are observed for other crystal structures; Forty five different hydrogen bifluoride structures in 34 crystal structures were found through the Cambridge Structural Database (CSD) [41] search (see details in the former section). For example, for tetramethylammonium dihydrogen trifluoride (Figure 4, refcode: GIBGOB01) the F-H and H ... F distances are equal to 0.89 and 1.43 Å, respectively, and the F-H-F angle amounts to 177.9°. X-ray crystal structures are considered here since only one structure of the search used results from neutron diffraction measurement. For the X-ray crystal structures the positions of the H-atoms are not as accurate as the positions of heavier atoms. It is worth mentioning that several neutron diffraction [FHF]⁻ structures [4,13] discussed briefly in the introduction are usually not classified as organic or organometallic structures and thus not inserted in the CSD [41].

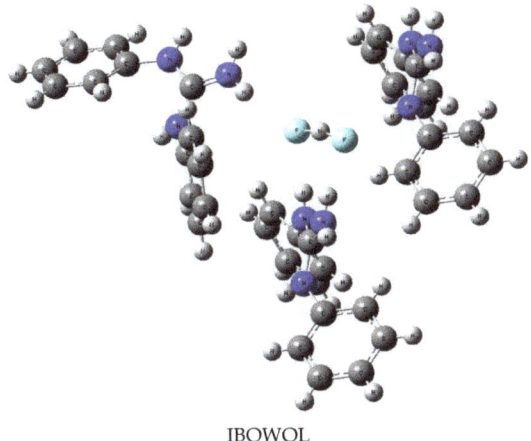

IBOWOL

Figure 3. The fragment of the crystal structure of diphenylguanidinium hydrogen difluoride (IBOWOL).

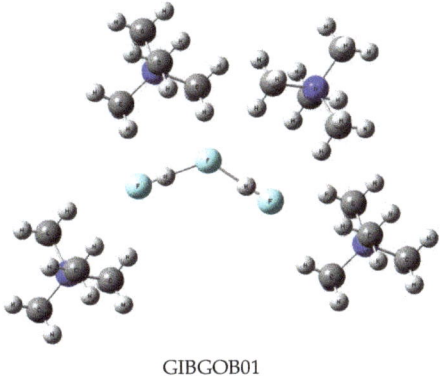

GIBGOB01

Figure 4. The fragment of the crystal structure of tetramethylammonium dihydrogen trifluoride (GIBGOB01).

One can see that for the GIBGOB01 structure (Figure 4) one of the F-centers of the [FHF]$^-$ anion is strongly affected by the proximity of the HF molecule. In principle the [FHF]$^-$ ... HF system may be treated as an $F_3H_2^-$ anion since it is characterized by C_{2v} symmetry with symmetry equivalent positions of the H-atoms.

For the 45 hydrogen bifluoride anions found in the crystal structures the F-centers' intermolecular contacts were analyzed here; the rough criterion for contacts to be shorter than the sum of the corresponding van der Waals radii was applied. It is interesting that 290 F ... H contacts were found and only 34 contacts between fluorine centers and the other non-hydrogen atoms (Z-atoms). However for the latter F ... Z contacts, Z is mainly nitrogen or oxygen belonging to O-H or N-H proton donating bond or a cation of a transition metal. Only in one case, an F ... S interaction with a positively charged sulfur atom is observed. The contacts with the N or O centers result from the N-H ... F and O-H ... F hydrogen bonds being far from linearity thus they correspond to the sample of 290 F ... H interactions mentioned earlier. One can see that all short intermolecular contacts of the [FHF]$^-$ ions concern interactions with the Lewis acid centers, mainly with H-atoms. It means that in crystal structures the [FHF]$^-$ anion forms mainly hydrogen bonds with proton donating moieties and interactions with other Lewis acid centers are rather rare. The strong Lewis base properties of the [FHF]$^-$ anion may be explained by the negative electrostatic potential of the whole isosurface of this anion (0.001 au electron density isosurface is considered). A minimum electrostatic potential for this isosurface is equal to -0.234 hartrees, *i.e.*, -614.4 kJ/mol (calculated at MP2/aug-cc-pVTZ level).

3.2. The Geometry of the [FHF]$^-$ Anion

The results of the CSD search presented in the former section show that in the crystal structures the [FHF]$^-$ anion interacts with the Lewis acid centers, in particular through the hydrogen bonds since F ... H contacts are most often observed. This is why the MP2/aug-cc-pVTZ calculations were performed here on the [FHF]$^-$ complexes with the proton donating species; the interactions with the Li$^+$ and Na$^+$ cations are also considered for comparison (Figure 1). If in the complexes only the geometry of the hydrogen bifluoride is considered, one can consequently divide them into three groups. Equal distances between the H-atom and the fluorine atoms are observed for the first group containing the isolated [FHF]$^-$ anion and its two complexes with Li$^+$ and Na$^+$ (Figure 1). It was mentioned earlier that free [FHF]$^-$ is characterized by the $D_{\infty h}$ symmetry; however in the complexes with lithium and sodium cations this anion is not linear and it possesses C_{2v} symmetry, the same as for the whole corresponding complexes.

For the second group of complexes the hydrogen bifluoride anion preserves its identity since it may be treated as an integral unit. These are the complexes with H_2O, C_2H_2, HCN, and HF which act as Lewis acids but particularly they may be considered here as Brønsted acids since they are the proton donating species in the hydrogen bonds formed. The [FHF]$^-$ structure for the latter complexes is not destroyed since interactions between the hydrogen bifluoride and the Brønsted acid moieties occurring here are not so strong—discussed in the next sections. One can see that the [FHF]$^-$ ... HF complex of the C$_{2v}$ symmetry (Figures 1 and 4) may be also treated as interaction of two HF molecules with the central F$^-$ anion.

The strong [FHF]$^-$–Brønsted acid interactions enforce great structural changes of the [FHF]$^-$ moiety for the third group of complexes. These are complexes with HCl, H$_2$S, NH$_4^+$, PH$_4^+$ H$_3$O$^+$ and H$_3$S$^+$ which may be also considered as clusters where the HF ... HF dimer interacts with the Cl$^-$, SH$^-$, NH$_3$, PH$_3$, H$_2$O, and H$_2$S species, respectively (Figure 1 presents the complexes with H$_3$O$^+$ and HCl). In this group of complexes the [FHF]$^-$ ion does not preserve its identity.

Table 1 presents the geometrical parameters of the [FHF]$^-$ anion in the analyzed complexes. The linearity of the hydrogen bifluoride is only slightly disturbed by external interactions with neutral species; in the case of interactions with cations there are greater deviations from linearity because the F-H-F angle amounts here to ~151–167°. Figure 5 presents the relation between the F-H and H ... F distances; there is a second order polynomial regression for the theoretical results (R^2 = 0.983). The experimental results taken from CSD are included for comparison (the geometrical parameters of [FHF]$^-$ anion in the crystal structures are collated in the Supplementary Material, Table S1). In general, the experimental results are in agreement with the results of the calculations. One can see that a greater disagreement between experimental and theoretical results occurs for greater F ... H distances. It may be an effect of packing forces; intermolecular distances in crystals are usually shorter than in a gas phase [48,49] and this effect may result in the compression of the [FHF]$^-$ anion. Additionally the X-ray results are analyzed here and the X-ray bonds containing H-atoms are shorter than the neutron diffraction counterparts [50]. In the former case the bond length is determined as the distance between electron density maxima while in the latter case as the distance between nuclei [50], thus these experimental techniques present different physical properties of the structures analyzed. One can see that the neutron diffraction bond length is in accordance with the accepted understanding of the chemical bond length. However the X-ray results are presented in Figure 5 only to show that the experimental tendency is roughly in agreement with the theoretical relationship. Figure 5 presents also the continuous line passing through the points corresponding to the hydrogen bifluoride anions possessing D$_{\infty h}$ and C$_{2v}$ symmetries; theoretical results are presented (free [FHF]$^-$ anion and its complexes with Na$^+$, Li$^+$ cations) as well as species found in the crystal structures.

Table 1. The geometrical parameters (Å, degrees), F-H and H...F are distances within [FHF]$^-$ anion, F-H-F is the angle of this anion, F...H(Li$^+$,Na$^+$) is the distance between the F-center of the anion and the external Lewis acid center.

Complex	F-H	H...F	F-F	F-H-F	F...H(Li$^+$, Na$^+$)
[FHF]$^-$	1.143	1.143	2.286	180.0	-
[FHF]$^-$ ... H$_2$O	1.036	1.284	2.320	179.1	1.565
[FHF]$^-$ ... C$_2$H$_2$	1.049	1.263	2.311	179.7	1.672
[FHF]$^-$ ... HCN	1.013	1.333	2.345	179.5	1.469
[FHF]$^-$ ... HF	1.008	1.343	2.351	178.4	1.343
[FHF]$^-$ ... HCl	0.956	1.535	2.488	174.6	1.026
[FHF]$^-$ ... H$_2$S	0.963	1.499	2.460	176.2	1.077
[FHF]$^-$ ... NH$_4^+$	0.943	1.648	2.555	160.2	0.995
[FHF]$^-$ ... H$_3$O$^+$	0.939	1.721	2.593	152.9	0.959
[FHF]$^-$ ... PH$_4^+$	0.933	1.734	2.651	167.0	0.947
[FHF]$^-$ ... H$_3$S$^+$	0.935	1.744	2.641	159.7	0.947
[FHF]$^-$ ... Li$^+$	1.148	1.148	2.224	151.2	1.801
[FHF]$^-$ ... Na$^+$	1.151	1.151	2.274	162.4	2.289

Table 1 shows the F...H contacts corresponding to the interactions of the [FHF]$^-$ anion with Brønsted acids. The cation....F distances for complexes with Li$^+$ and Na$^+$ are also presented, they are equal to 1.80 and 2.29 Å, respectively and are greater than the F...H distances. This may be connected with the greater radii of these cations than of the H-atom. However it may also result from the nature of interactions; for complexes with Li$^+$ and Na$^+$ cations electrostatic interactions are the most important attractive ones while for the remaining complexes charge transfer and polarization interactions play the major role.

The external F...H distance is situated between 0.95 Å and 1.67 Å. For the third sub-group of complexes mentioned earlier the F...H distances are close to 1 Å (or even less). It means that for this group two HF molecules are formed—the hydrogen bifluoride does not preserve its identity. It also means that for this group of complexes the F...H external distance is just the H-F bond length for the newly formed hydrogen fluoride molecule. Such short distances were not detected for 290 F...H contacts found here in the crystal structures. Only in four cases were the F...H distances situated in the (1.20 Å; 1.43 Å) range. However these distances are observed in the [FHF]$^-$...HF complex. This complex was observed in earlier studies [27–29] and is also analyzed here; it possesses C$_{2v}$ symmetry and may be considered as the F$_3$H$_2^-$ anion. The (MP2/aug-cc-pVTZ level) F-H and H...F distances calculated here for this complex are equal to 1.008 and 1.343 Å, respectively (Table 1) which is in agreement with the previous MP2/6-31+G** calculations where 1.012 and 1.349 Å distances were found [27].

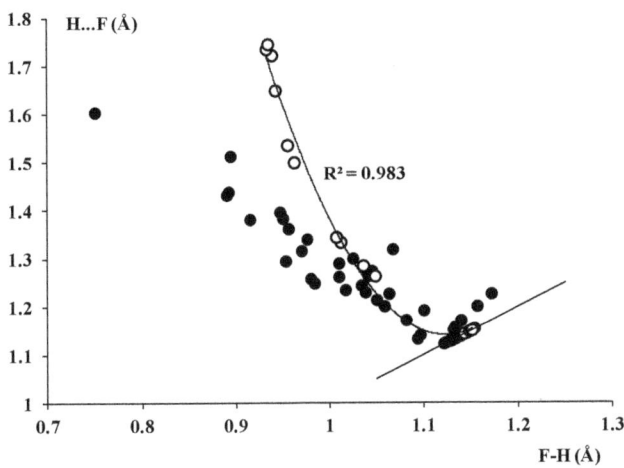

Figure 5. The dependence between the F-H and H...F distances in the [FHF]⁻ species interacting with Lewis acid centers; open circles—theoretical results, full circles—experimental data taken from the Cambridge Structural Database (CSD); R^2 is the squared correlation coefficient (for this figure and other ones presented in this study). The solid line passing through the structures of the $D_{\infty h}$ and C_{2v} symmetries is presented.

3.3. Interactions of the [FHF]⁻ Anion

Table 2 presents the binding energies and interaction energies, defined earlier, for the analyzed complexes; these energies corrected for BSSE are also included as well as the deformation energies. The deformation energy is the loss of energy from the result of the change of geometries of the interacting species after complexation [51,52]. This loss of energy is compensated by the stabilizing, attractive interaction in the complex which as a whole is in the energetic minimum.

One can see (Table 2) that the strongest interactions (characterized by "the most negative values") exist for the mentioned earlier third group of complexes as well as for complexes with Li⁺ and Na⁺ cations. However for the former complexes the interactions correspond to the F...H contacts which possess characteristics of covalent bonds (the F...H distances of ~1Å); the [FHF]⁻ anions are destroyed here followed by the formation of two HF molecules. This formation is connected with the large deformation energies between 137 and 355 kJ/mol (Table 2). The structural changes observed are connected with large electron density redistributions which usually correspond to large attractive charge transfer and polarization interactions. The latter interactions are often attributed in the literature to covalency [53]. For complexes with sodium and lithium cations much lower deformation energies are observed, 9.6 and 26.8 kJ/mol, respectively, and the structure of hydrogen bifluoride is preserved; it may be connected with much lower contributions to the interaction energies coming from the polarization and charge transfer effects. Hence for the two latter complexes, with Li^+ and Na^+ cations, electrostatic interactions should be dominant. For the remaining complexes the interaction energy (corrected for BSSE) is situated approximately in the range between -57 kJ/mol to -144 kJ/mol; thus these interactions are not so weak. One can see (Table 2) that the binding energies are "less negative" than the corresponding interaction energies. The latter results from the positive deformation energies' contributions included in the binding energies.

Table 2. The interaction and binding energies, E_{int}'s and E_{bin}'s, for complexes analyzed here, the basis set superposition errors (BSSE) corrected values are also given, E_{int}BSSE's and E_{bin}BSSE's, respectively; the deformation energies, E_{def}'s, are included (all energies in kJ/mol).

Complex	E_{int}	E_{int}BSSE	E_{bin}	E_{bin}BSSE	E_{def}
$[FHF]^- \ldots H_2O$	−80.8	−77.4	−72.4	−69.0	8.8
$[FHF]^- \ldots C_2H_2$	−60.7	−57.3	−54.4	−50.6	6.7
$[FHF]^- \ldots HCN$	−124.7	−120.1	−106.3	−101.7	18.4
$[FHF]^- \ldots HF$	−148.1	−143.5	−121.8	−117.2	26.4
$[FHF]^- \ldots HCl$	−332.2	−324.3	−146.9	−139.3	184.9
$[FHF]^- \ldots H_2S$	−212.1	−204.6	−74.9	−67.4	137.2
$[FHF]^- \ldots NH_4^+$	−860.6	−852.3	−592.9	−584.5	267.8
$[FHF]^- \ldots H_3O^+$	−1077.4	−1068.6	−741.8	−733.0	335.6
$[FHF]^- \ldots PH_4^+$	−975.8	−967.4	−620.7	−612.4	355.1
$[FHF]^- \ldots H_3S^+$	−1021.7	−1012.9	−708.6	−699.7	313.2
$[FHF]^- \ldots Li^+$	−701.7	−698.7	−674.9	−672.0	26.8
$[FHF]^- \ldots Na^+$	−566.5	−564.8	−556.9	−554.8	9.6

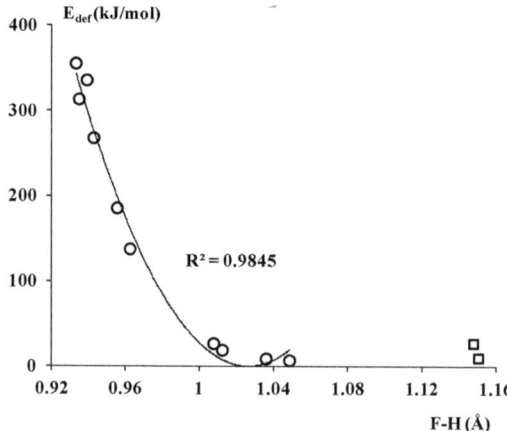

Figure 6. The second order polynomial relationship between the F-H distance and the deformation energy. Two cases of $[FHF]^- \ldots Li^+$ and $[FHF]^- \ldots Na^+$ are not included in this relation, however they are presented in the figure (squares).

Figure 6 presents the second order polynomial relationship between the F-H length in the [FHF]$^-$ anion and the deformation energy. The F-H geometrical parameter shows the movement of the H-atom from its central position, this distance is shorter for stronger interactions of the hydrogen bifluoride ion with the Brønsted acids (Table 1). The greater H-atom movement, *i.e.*, the shorter F-H distance, is connected with the greater deformation of this anion. In extreme cases of very strong hydrogen bonds, such as complexes with HCl, NH_4^+ or other complexes of this group, this movement leads to the destruction of the anion and to the formation of two HF molecules. Extremely short F-H distances in the latter complexes are typical for covalent bonds. Hence the relationship presented in Figure 6 illustrates the nature of the above mentioned deformation in that it is connected mainly with the position of the H-atom within the [FHF]$^-$ anion. Two complexes with Na$^+$ and Li$^+$ are presented in Figure 6 for comparison with the H-bonded systems. They are characterized by the central position of the H-atom (only the F-H-F angle is slightly out of linearity) thus the deformation energy is low as was described earlier.

3.4. The Electron Charge Redistribution

Table 3 presents the Natural Bond Orbitals (NBO) [39,40] atomic charges as well as the NBO electron charge, El$_{trans}$, transferred in complexes from the [FHF]$^-$ anion to the Lewis acid. One can see that the lowest El$_{trans}$ values occur for complexes with lithium and sodium cations, values of 22 and 9 milielectrons, respectively. It means that the charge transfer and polarizations effects for these complexes are not as important as for the remaining complexes; the latter confirms the conclusions from the previous sections. The greatest El$_{trans}$ values, between 230 and 380 milielectrons, are observed for those complexes where the hydrogen bifluoride structure is not preserved.

Table 3 shows that for the [FHF]$^-$ anion the fluorine atom being in contact with the Lewis acid is "more negative" than the further placed F-atom. The greater differences between F-atom charges occur for the group of complexes with weaker interactions where the structure of [FHF]$^-$ is preserved; the greatest difference of 100 milielectrons occurs for the complex with water. For the group of complexes with stronger interactions, there the fluorine being in contact with the H-atom Brønsted acid center practically belongs to the new HF molecule, formed as a result of complexation. The symmetrical distribution of the electron density is observed for complexes with Li$^+$ and Na$^+$ thus the F-atoms' charges are equal to each other, and there is only a slight electron charge shift, El$_{trans}$, from [FHF]$^-$ thus the Li$^+$ and Na$^+$ ions possess charges very close to unity, +0.978 au and +0.991 au, respectively. Hence for the latter complexes the interactions may be classified as those between ions: between the [FHF]$^-$ anion and the sodium or lithium cation. It is worth mentioning that for all [FHF]$^-$ species (Table 3) the H-atom charge is situated in the range between +0.572 and +0.602 au, this means that the hydrogen bifluoride does not have the composition of two fluorine anions and a proton since within this species the large electron charge density shift is observed; from the terminal F-atoms to the H-atom center.

Table 3. The Natural Bond Orbitals (NBO) atomic charges (in au) in the [FHF]$^-$ anion, Q_{F1}, Q_H and Q_{F2} (free anion and its complexes are presented); Q_{F2} designates the charge of fluorine being in contact with external Lewis acid, Q_{Hext} is the charge of the external H-atom (or the charge of the Li$^+$, Na$^+$ ion), El$_{trans}$ is the electron charge transfer from [FHF]$^-$ to the Lewis acid (in au).

Complex	Q_{F1}	Q_H	Q_{F2}	Q_{Hext}	El$_{trans}$
[FHF]$^-$	−0.790	0.581	−0.790	-	0.000
[FHF]$^-$ ... H$_2$O	−0.725	0.595	−0.825	0.544	0.045
[FHF]$^-$ ... C$_2$H$_2$	−0.733	0.593	−0.820	0.323	0.040
[FHF]$^-$ ... HCN	−0.706	0.598	−0.805	0.341	0.088
[FHF]$^-$ ... HF	−0.702	0.599	−0.794	0.599	0.103
[FHF]$^-$ ... HCl	−0.644	0.601	−0.688	0.533	0.269
[FHF]$^-$ ... H$_2$S	−0.654	0.602	−0.711	0.492	0.237
[FHF]$^-$ ... NH$_4^+$	−0.617	0.590	−0.657	0.574	0.316
[FHF]$^-$ ... H$_3$O$^+$	−0.608	0.591	−0.626	0.596	0.357
[FHF]$^-$ ... PH$_4^+$	−0.598	0.581	−0.616	0.569	0.367
[FHF]$^-$ ... H$_3$S$^+$	−0.599	0.583	−0.608	0.575	0.377
[FHF]$^-$ ... Li$^+$	−0.775	0.572	−0.775	0.978	0.022
[FHF]$^-$ ... Na$^+$	−0.782	0.573	−0.782	0.991	0.009

3.5. The Analysis of QTAIM Parameters

The Quantum Theory of Atoms in Molecules (QTAIM) [38] calculations were performed here. Table 4 presents characteristics of the bond critical points of the analyzed complexes. The bond critical points (BCPs) within the [FHF]$^-$ anion are analyzed; for the F-H and H ... F interactions, as well as the critical point corresponding to the interaction with the Lewis acid, *i.e.*, corresponding to the F ... H contact (or to the contacts with Na$^+$ and Li$^+$). All F-H interactions within anions, that is those which concern the shorter fluorine-hydrogen distances, may be treated as covalent bonds. The latter is supported by the characteristics of the F-H BCP; large values of the electron density at BCP, ρ_{BCP}, as well as negative values of the laplacian of the electron density at BCP, $\nabla^2\rho_{BCP}$. The negative value of $\nabla^2\rho_{BCP}$ is connected with the concentration of the electron density in the interatomic region [38] thus it is often treated as a signature of the covalent character of interaction. In the case of free [FHF]$^-$ and complexes with monatomic cations the same characteristics of the second interaction are observed due to the symmetry of the anion. For the remaining systems the further H ... F interaction in the anion is characterized by the positive value of $\nabla^2\rho_{BCP}$; however the total electron energy density at BCP, H$_{BCP}$, is negative in this case. It means that these interactions may be classified at least as partly covalent in nature [54,55].

Table 4. The Quantum Theory of Atoms in Molecules (QTAIM) characteristics of bond critical points (BCPs) (in au). The ρ, $\nabla^2\rho$ and H values designate the electron density at the BCP, its laplacian and the total electron energy density at the BCP, respectively. F-H and H...F subscripts correspond to distances within the [FHF]$^-$ anion, F...H subscript stands for the contact between the F-atom and the external Lewis acid center (H, Li$^+$ or Na$^+$).

Complex	ρ_{F-H}	$\nabla^2\rho_{F-H}$	H_{F-H}	$\rho_{H...F}$	$\nabla^2\rho_{H...F}$	$H_{H...F}$	$\rho_{F...H}$	$\nabla^2\rho_{F...H}$	$H_{F...H}$
[FHF]$^-$	0.176	−0.572	−0.257	0.176	−0.572	−0.257	-	-	-
[FHF]$^-$ ... H$_2$O	0.241	−1.537	−0.489	0.116	0.047	−0.086	0.055	0.145	−0.015
[FHF]$^-$ ... C$_2$H$_2$	0.232	−1.412	−0.460	0.124	0.000	−0.102	0.043	0.137	−0.006
[FHF]$^-$ ... HCN	0.261	−1.807	−0.552	0.101	0.112	−0.061	0.074	0.157	−0.026
[FHF]$^-$ ... HF	0.264	−1.870	−0.567	0.098	0.121	−0.057	0.098	0.121	−0.057
[FHF]$^-$ ... HCl	0.318	−2.706	−0.767	0.056	0.146	−0.016	0.250	−1.680	−0.516
[FHF]$^-$ ... H$_2$S	0.309	−2.581	−0.737	0.063	0.150	−0.020	0.217	−1.191	−0.400
[FHF]$^-$ ... NH$_4^+$	0.335	−2.889	−0.815	0.043	0.129	−0.008	0.280	−2.008	−0.605
[FHF]$^-$ ... H$_3$O$^+$	0.340	−2.952	−0.829	0.036	0.120	−0.004	0.316	−2.622	−0.749
[FHF]$^-$ ... PH$_4^+$	0.348	−3.022	−0.850	0.033	0.113	−0.003	0.331	−2.754	−0.783
[FHF]$^-$ ... H$_3$S$^+$	0.346	−3.000	−0.843	0.033	0.113	−0.003	0.332	−2.769	−0.786
[FHF]$^-$ ... Li$^+$	0.178	−0.540	−0.253	0.178	−0.540	−0.253	0.038	0.340	0.013
[FHF]$^-$ ... Na$^+$	0.174	−0.539	−0.248	0.174	−0.539	−0.248	0.020	0.153	0.007

It is interesting that for complexes with NH$_4^+$, PH$_4^+$, H$_3$S$^+$ and H$_3$O$^+$ cations, the H$_{BCP}$ values for H...F are still negative but very close to zero, between −0.008 and −0.003 au. The latter shows the interaction possesses low covalent character. In fact for these complexes the strongest interactions of hydrogen bifluoride with Lewis acid moieties are observed (Table 2) and the H...F BCPs discussed here are concerned rather with the intermolecular H...F contacts between two HF molecules since the complexation leads to the decomposition of the [FHF]$^-$ species.

For the BCPs corresponding to the F...Li$^+$ and F...Na$^+$ interactions in the [FHF]$^-$...Li$^+$ and [FHF]$^-$...Na$^+$ complexes, the H$_{BCP}$ values are positive which indicate interactions between closed-shell systems. The lack of covalency for these interactions (or only its low contribution) and the strong total interactions (see Table 2 where interaction energies corrected for BSSE are equal to −699 and −565 kJ/mol) suggest a great contribution of electrostatic interactions to the stabilization of these systems. For the remaining complexes the H$_{BCP}$ values of F...H contacts are negative thus these interactions are at least partly covalent in nature. In the case of the [FHF]$^-$...HCCH complex a rather low 0.043 au value of ρ_{BCP} is observed for the external F...H interaction. The corresponding H$_{BCP}$ value is very close to zero, −0.006 au which may suggest the weakest interaction occurs here. It is supported by a still strong interaction but the weakest interaction in the sample analyzed; the energy of interaction amounts to −57 kJ/mol (Table 2).

Figure 7 shows how the electron charge transfer, El$_{trans}$, from the [FHF]$^-$ anion to the Brønsted acid influences the electron density at the BCPs discussed above. The complexes with Na$^+$ and Li$^+$ are excluded from the relationships but the free [FHF]$^-$ anion is included in Figure 7 for comparison.

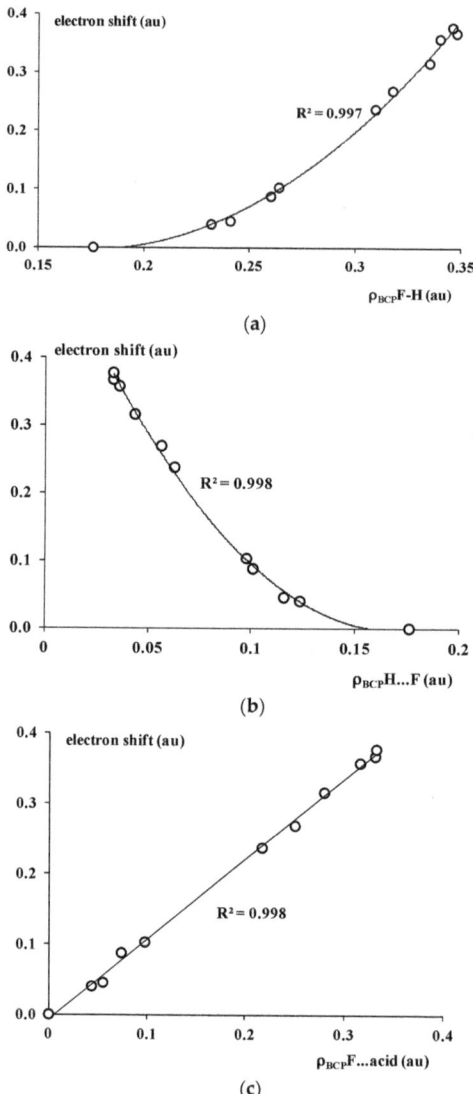

Figure 7. The relationships between the electron shift, El$_{trans}$ (in au), and (a) ρ_{FH---} electron density at the BCP of the F-H of [FHF]$^-$; (b) $\rho_{H\ldots F}$ electron density at BCP of the H $\ldots$ F of [FHF]$^-$; (c) $\rho_{F\ldots H}$ electron density at BCP of the external [FHF]$^-$ — Brønsted acid contact.

It was discussed earlier that a greater El_{trans} is connected with a greater deformation of the hydrogen bifluoride structure, thus the greatest El_{trans} values are observed for those systems where the $[FHF]^-$ anion structure is destroyed. Figure 7a shows that for the F-H interaction the greater ρ_{BCP} value is observed for greater El_{trans} (second order polynomial correlation); this is connected with the strengthening of the F-H interaction and the shortening of the corresponding F-H distance. The further interaction in the anion, *i.e.*, the H ... F one, is weakened for greater El_{trans} which is reflected by the lower values of ρ_{BCP} (Figure 7b, second order polynomial relationship). Finally greater El_{trans} is connected with stronger F ... H interactions between the $[FHF]^-$ anion and the external Brønsted acid, the ρ_{BCP} value increases for such interactions with the increase of El_{trans} (Figure 7c, linear correlation) up to such cases where F-H covalent bonds are formed (complexes with H_2S, HCl, NH_4^+, PH_4^+, H_3O^+ or H_3S^+).

Figure 8 shows the molecular graphs of the selected species analyzed in this study; the isolines of laplacian of electron density are also presented, positive values of laplacian are depicted in solid lines and negative values in broken lines. The negative laplacian values show the regions of concentration of electron density while the positive values of its depletion [38]. It can be seen that for the $[FHF]^-$ anion the electron density concentration occurs around the fluorine and hydrogen atoms' nuclei. The concentration of the electron charge close to the H-atom nucleus confirms the earlier observations performed here of the electron charge shifts from F-atoms to the center of the hydrogen bifluoride anion. Besides, both BCPs are situated in the region of negative laplacian which confirms the covalent character of both fluorine-hydrogen interactions in the $[FHF]^-$ ion. A similar situation is observed for the $[FHF]^-$...Li^+ complex (and $[FHF]^-$...Na^+ not presented in Figure 8). However, as it was mentioned earlier, the $[FHF]^-$ anion is not linear in the latter complexes due to electrostatic interactions with the Li^+ (or Na^+) cation, and it is characterized by C_{2v} symmetry.

The movement of the H-atom to one of the fluorine centers in the $[FHF]^-$ ion is observed for its complex with acetylene (Figure 8). This movement is connected with the strengthening of one of the interactions (F-H) and the weakening of the second one (H...F). The linearity of the $[FHF]^-$ ion is preserved in this complex. A slightly different situation is observed for the $[FHF]^-$...HF complex. Figure 8 shows that the species may be considered as the symmetrical (C_{2v} symmetry) $F_3H_2^-$ anion; two H-atoms are moved to the terminal fluorines while the central F-atom may be treated as an F^- ion; in other words this system may be considered as a F^-...$(HF)_2$ cluster. A quite different situation is observed for the $[FHF]^-$ ... H_3O^+ complex where the hydrogen bifluoride structure is completely destroyed and the whole system may be considered as a H_2O ... HF ... HF cluster. Even continuous areas of negative laplacian values for H-F bonds of the hydrogen fluoride molecules are observed (Figure 8). A similar situation occurs for complexes with the H_2S, HCl, PH_4^+, and H_3S^+ Brønsted acids as well as for the $[FHF]^- \ldots NH_4^+$ complex which may be considered as complexes with a separated HF ... HF system. It is worth mentioning that the $H_3N \ldots HF \ldots HF$ cluster was analyzed both experimentally by rotational spectroscopy and theoretically by *ab initio* calculations [56]. Two well defined HF molecules were found for this system [56]; additionally this cluster was compared with the $H_3N \ldots HF$ complex since the cooperativity effects were analyzed. It was found that the $H_3N \ldots HF$ interaction is stronger in the triad than in the corresponding diad [56].

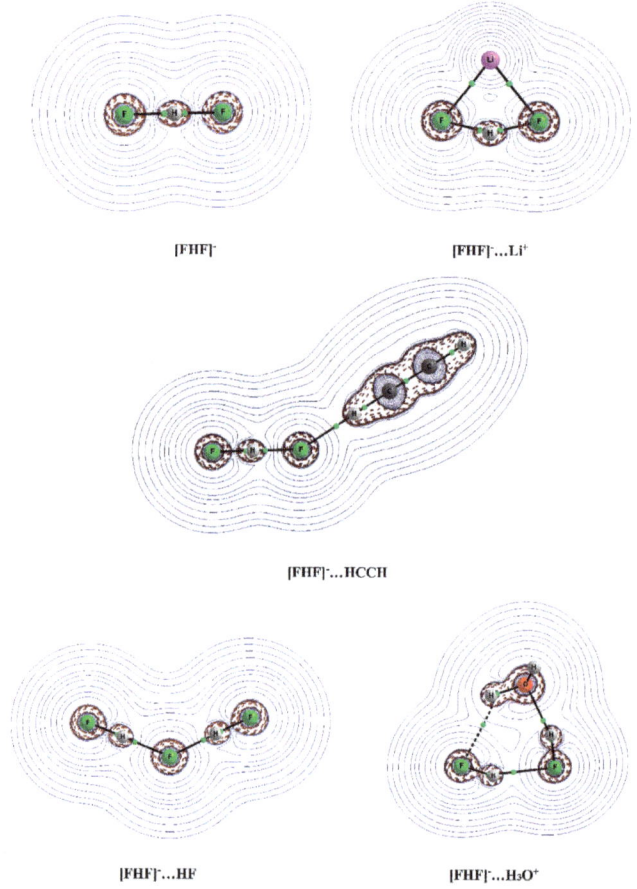

Figure 8. The molecular graphs of $[FHF]^-$, $[FHF]^-\ldots Li^+$, $[FHF]^-\ldots HCCH$, $[FHF]^-\ldots HF$ and $[FHF]^-\ldots H_3O^+$. Solid and broken lines correspond to bond paths, big circles to attractors, and small green circles to BCPs, isolines of laplacian of electron density are presented; positive values are depicted in solid lines and negative values in broken lines.

4. Conclusions

It was confirmed in this work that the [FHF]$^-$ anion, which is linear and centrosymmetric in the gas phase, is often deformed in crystal structures due to interactions with Lewis acids, mainly from external hydrogen bonds. The asymmetry of the hydrogen bifluoride anion has been observed and discussed in numerous earlier studies. Such deformation is characterized mainly by the movement of the H-atom in the anion towards one of the fluorine atoms situated further from the Lewis acid than the other F-center of the anion. This movement is accompanied by a slight deformation of the F-H-F angle from the linear.

It was found here that in the case of very strong interactions with an external moiety the decomposition of the [FHF]$^-$ anion is observed with the creation of two HF molecules interacting with the deprotonated species; the following systems are observed in such a case; FH ... FH ... B, where B designates the Lewis base center (N, P, O or S in this study). The interaction of hydrogen bifluoride with HF leads to system which may be considered as [FHF]$^-$ ... HF on the one hand and as a F$^-$ ... (HF)$_2$ cluster on the other hand. Such a system has been analyzed previously both experimentally and theoretically [27–29].

Numerous interrelations between energetic, geometrical, and topological parameters were found for the species considered here. The electron charge distribution, described by the QTAIM and NBO parameters, follows geometrical dependencies.

It is interesting that the decomposition described here of hydrogen bifluoride is typical for some moieties commonly known as super-acids [57,58]. For example HBF$_4$, HSbF$_6$, HPF$_6$ or HB(C$_6$F$_5$)$_4$ acids do not exist because of such decomposition, for the HB(C$_6$F$_5$)$_4$ moiety the cleavage of a B-C bond is observed, similarly HBF$_4$ should be treated as a HF ... BF$_3$ complex linked through a strong F ... B interaction. For the complexes analyzed here the decomposition of the hydrogen bifluoride anion followed by the formation of two HF molecules is connected with large electron charge density shifts from this anion to the Lewis acid moiety.

Supplementary Materials: The following are available online at www.mdpi.com/2073-4352/6/1/3/s001, Table S1: Geometrical parameters (Å, degrees) for the [FHF]$^-$ anions found in crystal structures through the CSD search; refcodes and R-factors for the crystal structures are included in the table.

Acknowledgments: Financial support comes from Eusko Jaurlaritza (GIC IT-588-13) and the Spanish Office for Scientific Research (CTQ2012-38496-C05-04). Technical and human support provided by Informatikako Zerbitzu Orokora—Servicio General de Informática de la Universidad del País Vasco (SGI/IZO-SGIker UPV/EHU), Ministerio de Ciencia e Innovación (MICINN), Gobierno Vasco Eusko Jaurlanitza (GV/EJ), European Social Fund (ESF) is gratefully acknowledged.

Conflicts of Interest: The author declares no conflict of interest.

References

1. Wenthold, P.G.; Squires, R.R. Bond Dissociation Energies of F_2^- and HF_2^-. A Gas-Phase Experimental and G2 Theoretical Study. *J. Phys. Chem.* **1995**, *99*, 2002–2005.
2. Sode, O.; Hirata, S. Second-order many-body perturbation study of solid hydrogen fluoride under pressure. *Phys. Chem. Chem. Phys.* **2012**, *14*, 7765–7779.
3. Stein, C.; Oswald, R.; Sebald, P.; Botschwina, P.; Stoll, H.; Peterson, K.A. Accurate bond dissociation energies (D_0) for FHF^- isotopologues. *Mol. Phys.* **2013**, *111*, 2647–2652.
4. Jeffrey, G.A. *An Introduction to Hydrogen Bonding*; Oxford University Press: New York, NY, USA, 1997.
5. Humbel, S. Short Strong Hydrogen Bonds: A Valence Bond Analysis. *J. Phys. Chem. A* **2002**, *106*, 5517–5520.
6. Sobczyk, L.; Grabowski, S.J.; Krygowski, T.M. Interrelation between H-Bond and Pi-Electron Delocalization. *Chem. Rev.* **2005**, *105*, 3513–3560.
7. Denisov, G.D.; Mavri, J.; Sobczyk, L. Potential Energy Shape for the Proton Motion in Hydrogen Bonds Reflected in Infrared and NMR Spectra, chapter in the book. In *Hydrogen Bonding—New Insights*; Grabowski, S.J., Ed.; Springer: Dordrecht, The Netherlands, 2006.
8. Guthrie, J.P. Short strong hydrogen bonds: Can they explain enzymic catalysis? *Chem. Biol.* **1996**, *3*, 163–170.
9. Landrum, G.A.; Goldberg, N.; Hoffmann, R. Bonding in the trihalides (X_3^-, mixed trihalides (X_2Y^-) and hydrogen bihalides (X_2H^-). The connection between hypervalent, electron-rich three-center, donor-acceptor and strong hydrogen bonding. *J. Chem. Soc. Dalton Trans.* **1997**, *19*, 3605–3613.
10. Klepeis, N.E.; East, A.L.L.; Császár, A.G.; Allen, W.D. The [FHCl]$^-$ molecular anion: Structural aspects, global surface, and vibrational eigenspectrum. *J. Chem. Phys.* **1993**, *99*, 3865–3897.
11. Chan, B.; del Bene, J.E.; Radom, L. Proton-Bound Homodimers: How Are the Binding Energies Related to Proton Affinities. *J. Am. Chem. Soc.* **2007**, *129*, 12179–12199.
12. Grabowski, S.J.; Ugalde, J.M. High-level ab initio calculations on low barrier hydrogen bonds and proton bound homodimers. *Chem. Phys. Lett.* **2010**, *493*, 37–44.
13. Panich, A.M. NMR study of the F-H ... F hydrogen bond. Relation between hydrogen atom position and F-H ... F bond length. *Chem. Phys.* **1995**, *196*, 511–519.
14. Nakamoto, K.; Margoshes, M.; Rundle, R.E. Stretching Frequencies as a Function of Distances in Hydrogen Bonds. *J. Am. Chem. Soc.* **1955**, *77*, 6480–6486.
15. Olovsson, I.; Jönsson, P.-G. *The Hydrogen Bond Recent Developments in Theory and Experiments*; Schuster, P., Zundel, G., Sandorfy, C., Eds.; North-Holland: Amsterdam, The Netherlands, 1976; pp. 393–455.
16. Chiari, G.; Ferraris, G. The water molecule in crystalline hydrates studied by neutron diffraction. *Acta Crystallogr. Sect. B* **1982**, *38*, 2331–2341.
17. Steiner, T.; Saenger, W. Lengthening of the covalent O-H bond in O-H ... O hydrogen bonds re-examined from low-temperature neutron diffraction data of organic compounds. *Acta Crystallogr. Sect. B* **1994**, *50*, 348–357.

18. Johnston, H.S. Large Tunnelling Corrections in Chemical reaction Rates. *Adv. Chem. Phys.* **1960**, *3*, 131–170.
19. Bürgi, H.-B. Stereochemistry of reaction paths as determined from cristal structure data—A relationship between structure and energy. *Angew. Chem. Int. Ed. Engl.* **1975**, *14*, 460–473.
20. Pauling, L. Atomic Radii and Interatomic Distances in Metals. *J. Am. Chem. Soc.* **1947**, *69*, 542–553.
21. Grabowski, S.J. The Bond Number Relationship for the O-H ... O Systems. *Croat. Chem. Acta* **1988**, *61*, 815–819.
22. Gilli, P.; Bertolasi, V.; Ferretti, V.; Gilli, G. Evidence for resonance-assisted hydrogen bonding. 4. Covalent nature of the strong homonuclear hydrogen bond. Study of the O-H ... O system by cristal structure correlation methods. *J. Am. Chem. Soc.* **1994**, *116*, 909–915.
23. Steiner, T. Lengthening of the Covalent X.H Bond in Heteronuclear Hydrogen Bonds Quantified from Organic and Organometallic Neutron Crystal Structures. *J. Phys. Chem. A* **1998**, *102*, 7041–7052.
24. Grabowski, S.J.; Krygowski, T.M. Estimation of the Proton Position and the Energy of O-H ... O Bridges in Crystals from X-Ray Diffraction Data. *Tetrahedron* **1998**, *54*, 5683–5694.
25. Benedict, H.; Limbach, H.-H.; Wehlan, M.; Fehlhammer, W.-P.; Golubev, N.S.; Janoschek, R. Solid State 15N NMR and Theoretical Studies of Primary and Secondary Geometric H/D Isotope Effects on Low-Barrier NHN – Hydrogen Bonds. *J. Am. Chem. Soc.* **1998**, *120*, 2939–2950.
26. Shenderovich, I.G.; Tolstoy, P.M.; Golubev, N.S.; Smirnov, S.N.; Denisov, G.S.; Limbach, H.-H. Low-Temperature NMR Studies of the Structure and Dynamics of a Novel Series of Acid-Base Complexes of HF with Collidine Exhibiting Scalar Couplings Across Hydrogen Bonds. *J. Am. Chem. Soc.* **2003**, *125*, 11710–11720.
27. Shenderovich, I.G.; Smirnov, S.N.; Denisov, G.S.; Gindin, V.A.; Golubev, N.S.; Dunger, A.; Reibke, R.; Kirpekar, S.; Malkina, O.L.; Limbach, H.-H. Nuclear magnetic Resonance of Hydrogen Bonded Clusters Between F^- and $(HF)_n$: Experiment and Theory. *Ber. Bunsenges. Phys. Chem.* **1998**, *102*, 422–428.
28. Kucherov, S.Y.; Bureiko, S.F.; Denisov, G.S. Anticooperativity of FHF hydrogen bonds in clusters of the type F^- x $(HF)_n$, RF x $(HF)_n$ and XF x $(HF)_n$, R = alkyl and X = H, Br, Cl, F. *J. Mol. Struct.* **2016**, *1105*, 246–255.
29. Shenderovich, I.G.; Limbach, H.-H.; Smirnov, S.N.; Tolstoy, P.M.; Denisov, G.S.; Golubev, N.S. H/D isotope effects on the low-temperature NMR parameters and hydrogen bond geometries of $(FH)_2F^-$ and $(FH)_3F^-$ disolved in CDF_3/CDF_2Cl. *Phys. Chem. Chem. Phys.* **2002**, *4*, 5488–5497.
30. Almlöf, J. Hydrogen bond studies. 71. Ab initio calculation of the vibrational structure and equilibrium geometry in HF_2^- and HD_2^-. *Chem. Phys. Lett.* **1972**, *17*, 49–52.

31. Golubev, N.S.; Melikova, S.M.; Shchepkin, D.N.; Shenderovich, I.G.; Tolstoy, P.M.; Denisov, G.S. Interpretation of Hydrogen/Deuterium Isotope Effects on NMR Chemical Shifts of [FHF]$^-$ Ion Base don Calculations of Nuclear Magnetic Shilding Tensor Surface. *Z. Phys. Chem.* **2003**, *217*, 1549–1563.
32. Perera, S.A.; Bartlett, R.J. NMR Spin-Spin Coupling for Hydrogen Bonds of [F(HF)$_n$]$^-$, n = 1–4, Clusters. *J. Am. Chem. Soc.* **2000**, *122*, 1231–1232.
33. Del Bene, J.E.; Jordan, M.J.T.; Perera, A.A.; Bartlett, R.J. Vibrational Effects on the F-F Spin-Spin Coupling Constant ($^{2h}J_{F-F}$) in FHF$^-$ and FDF$^-$. *J. Phys. Chem. A* **2001**, *105*, 8399–8402.
34. Epa, V.C.; Choi, J.H.; Klobukowski, M.; Thorson, W.R. Vibrational dynamics of the bifluoride ion. I. Construction of a model potential surface. *J. Chem. Phys.* **1990**, *92*, 466–472.
35. Latajka, Z.; Scheiner, S. Critical assessment of density functional methods for study of proton transfer processes. (FHF)$^-$. *Chem. Phys. Lett.* **1995**, *234*, 159–164.
36. Nieckarz, R.J.; Oldridge, N.; Fridgen, T.D.; Li, G.P.; Hamilton, I.P.; McMahon, T.B. Investigations of Strong Hydrogen Bonding in (ROH)$_n$ ... FHF$^-$ (n = 1,2 and R = H, CH$_3$, C$_2$H$_5$) Clusters via High-Pressure Mass Spectroscopy and Quantum Calculations. *J. Phys. Chem. A* **2009**, *113*, 644–652.
37. Alkorta, I.; Sánchez-Sanz, G.; Elguero, J. Interplay of F-H ... F Hydrogen Bonds and P ... N Pnicogen Bonds. *J. Phys. Chem. A* **2012**, *116*, 9205–9213.
38. Bader, R.F.W. *Atoms in Molecules, a Quantum Theory*; Oxford University Press: Oxford, UK, 1990.
39. Weinhold, F.; Landis, C. *Valency and Bonding, a Natural Bond Orbital Donor—Acceptor Perspective*; Cambridge University Press: Cambridge, UK, 2005.
40. Reed, E.; Curtiss, L.A.; Weinhold, F. Intermolecular interactions from a natural bond orbital, donor-acceptor viewpoint. *Chem. Rev.* **1988**, *88*, 899–926.
41. Wong, R.; Allen, F.H.; Willett, P. The scientific impact of the Cambridge Structural Database: A citation-based study. *J. Appl. Cryst.* **2010**, *43*, 811–824.
42. Frisch, M.J.; Trucks, G.W.; Schlegel, H.B.; Scuseria, G.E.; Robb, M.A.; Cheeseman, J.R.; Scalmani, G.; Barone, V.; Mennucci, B.; Petersson, G.A.; et al. *Gaussian 09, Revision A.1*; Gaussian, Inc.: Wallingford, CT, USA, 2009.
43. Piela, L. *Ideas of Quantum Chemistry*; Elsevier Science Publishers: Amsterdam, The Netherlands, 2007; pp. 684–691.
44. Boys, S.F.; Bernardi, F. The calculation of small molecular interactions by the differences of separate total energies. Some procedures with reduced errors. *Mol. Phys.* **1970**, *19*, 553–561.
45. Glendening, E.D.; Badenhoop, J.K.; Reed, A.E.; Carpenter, J.E.; Bohmann, J.A.; Morales, C.M.; Weinhold, F. *NBO 5.0.*; Theoretical Chemistry Institute, University of Wisconsin: Madison, WI, USA, 2001.
46. Schmidt, M.W.; Baldridge, K.K.; Boatz, J.A.; Elbert, S.T.; Gordon, M.S.; Jensen, J.H.; Koseki, S.; Matsunaga, N.; Nguyen, K.A.; Su, S.J.; et al. General Atomic and Molecular Electronic Structure System. *J. Comput. Chem.* **1993**, *14*, 1347–1363.

47. Keith, T.A. *AIMAll*; version 11.08.23; TK Gristmill Software: Overland Park, KS, USA, 2011.
48. Kitaigorodsky, A.I. *Molecular Crystals and Molecules*; Academic Press, Inc.: London, UK, 1973.
49. Dance, I. Distance criteria for crystal packing analysis of supramolecular motifs. *New J. Chem.* **2003**, *27*, 22–27.
50. Dunitz, J.D. *X-Ray Analysis and the Structure of Organic Molecules*; Cornell University Press: Ithaca, NY, USA, 1979.
51. Grabowski, S.J.; Sokalski, W.A. Different types of hydrogen bonds: Correlation analysis of interaction energy components. *J. Phys. Org. Chem.* **2005**, *18*, 779–784.
52. Grabowski, S.J. Triel Bonds, π-Hole-π-Electrons Interactions in Complexes of Boron and Aluminium Trihalides and Trihydrides with Acetylene and Ethylene. *Molecules* **2015**, *20*, 11297–11316.
53. Grabowski, S.J. What is the Covalency of Hydrogen Bonding? *Chem. Rev.* **2011**, *11*, 2597–2625.
54. Cremer, D.; Kraka, E. A Description of the Chemical Bond in Terms of Local Properties of Electron Density and Energy. *Croat. Chem. Acta* **1984**, *57*, 1259–1281.
55. Jenkins, S.; Morrison, I. The chemical character of the intermolecular bonds of seven phases of ice as revealed by ab initio calculation of electron densities. *Chem. Phys. Lett.* **2000**, *317*, 97–102.
56. Hunt, S.W.; Higgins, K.J.; Craddock, M.B.; Brauer, C.S.; Leopold, K.R. Influence of a Polar Near-Neighbor on Incipient Proton Transfer in a Strongly Hydrogen Bonded Complex. *J. Am. Chem. Soc.* **2003**, *125*, 13850–13860.
57. Reed, C.A. The Strongest Acid. *Chem. New Zealand* **2011**, *75*, 174–179.
58. Reed, C.A.; Kim, K.-C.; Stoyanov, E.S.; Stasko, D.; Tham, F.S.; Mueller, L.J.; Boyd, P.D.W. Isolating Benzenium Ion Salts. *J. Am. Chem. Soc.* **2003**, *125*, 1796–1804.

MDPI AG
St. Alban-Anlage 66
4052 Basel, Switzerland
Tel. +41 61 683 77 34
Fax +41 61 302 89 18
http://www.mdpi.com

Crystals Editorial Office
E-mail: crystals@mdpi.com
http://www.mdpi.com/journal/crystals

www.ingramcontent.com/pod-product-compliance
Lightning Source LLC
LaVergne TN
LVHW070051120526
838202LV00102B/2020